Springer Proceedings in Mathematics & Statistics

Volume 171

Springer Proceedings in Mathematics & Statistics

This book series features volumes composed of selected contributions from workshops and conferences in all areas of current research in mathematics and statistics, including operation research and optimization. In addition to an overall evaluation of the interest, scientific quality, and timeliness of each proposal at the hands of the publisher, individual contributions are all refereed to the high quality standards of leading journals in the field. Thus, this series provides the research community with well-edited, authoritative reports on developments in the most exciting areas of mathematical and statistical research today.

More information about this series at http://www.springer.com/series/10533

Vinai K. Singh · H.M. Srivastava
Ezio Venturino · Michael Resch
Vijay Gupta
Editors

Modern Mathematical Methods and High Performance Computing in Science and Technology

M3HPCST, Ghaziabad, India, December 2015

Springer

Editors
Vinai K. Singh
Department of Applied Mathematics
Raj Kumar Goel Institute of Technology
Ghaziabad, Uttar Pradesh
India

H.M. Srivastava
Department of Mathematics and Statistics
University of Victoria
Victoria, BC
Canada

Ezio Venturino
Dipartimento di Matematica
University of Turin
Torino
Italy

Michael Resch
High Performance Computing Center (HLRS)
University of Stuttgart
Stuttgart, Baden-Württemberg
Germany

Vijay Gupta
Department of Mathematics
Netaji Subhas Institute of Technology
New Delhi
India

ISSN 2194-1009 ISSN 2194-1017 (electronic)
Springer Proceedings in Mathematics & Statistics
ISBN 978-981-10-1453-6 ISBN 978-981-10-1454-3 (eBook)
DOI 10.1007/978-981-10-1454-3

Library of Congress Control Number: 2016941306

Mathematics Subject Classification (2010): 46-06, 47-06, 65-06, 68-06, 76-06

Printed on acid-free paper

This Springer imprint is published by Springer Nature
The registered company is Springer Nature Singapore Pte Ltd.
The registered company address is: 152 Beach Road, #22-06/08 Gateway East, Singapore 189721, Singapore

Preface

We are delighted to present the proceedings of the *1st International Conference on Modern Mathematical Methods and High Performance Computing in Science & Technology* (M3HPCST-2015) held at *Raj Kumar Goel Institute of Technology, Ghaziabad, India* from *December 27–29, 2015*. The three-day conference received an excellent response from number of national and international academicians. The pre-conference souvenir had 181 abstracts. Out of 130 papers discussed, 25 were selected for plenary talk and 106 for formal paper presentation. All papers were appropriately reviewed by well-known academicians and researchers. Finally, as many as 25 papers were selected for inclusion in the conference proceedings published by Springer-Verlag.

As we all are aware, mathematics has always been a discipline of interest not only to theoreticians but also to all professionals irrespective of their specific profession. Be it science, technology, economics, high-performance computing, or even sociology, new mathematical principles and models have been emerging and helping in new research and in drawing inferences from practical data as well as through scientific computing. The past few decades have seen enormous growth in applications of mathematics in different multidisciplinary areas.

M3HPCST-2015 covered a wide range of research interests: advances in the area of high-performance computing, which is applied to complex large-scale computational problems, numerical methods for partial differential equations, nonlinear problems, linear and nonlinear optimization, orthogonal polynomials and applications, functional analysis, fluid dynamics, vibration phenomena, and last but not least, biomathematics.

New problems with large-scale computing continually arise in many scientific and engineering applications. The development of new technologies is associated with the design of efficient algorithms in high-performance computing.

The theory of computation and its applications is one of the most important developments in modern science. Three technical sessions were devoted to scientific computing and computational methods for different engineering problems.

Many phenomena in science and engineering are modeled by partial or ordinary differential equations and nonlinear systems. They are usually treated numerically; therefore it is necessary to improve algorithm in terms of stability. Four technical sessions were devoted to represent trends in these areas of research.

Theoretical and practical applications pertaining to biological mathematics, functional analysis, operator theory, and orthogonal polynomials appeared in four technical sessions in which researchers presented the latest results in this area of investigation.

A conference of this kind would not have been possible without the support from different organizations and the people across different committees. We are indebted to the *Science and Engineering Research Board, Department of Science & Technology Govt. of India, Dr. A.P.J. Abdul Kalam Technical University Lucknow, U.P.* Cloud 9, Irish-Hindon, and HP India for sponsoring the event. Their support helped in significantly raising the profile of the conference.

All logistic and general organizational aspects were looked after locally by the organizing committee members from the institute who spent their time and energy in making the conference a grand success. The Technical Program Committee and external reviewers helped in selecting the paper for presentations and working out the technical program. We acknowledge the support and help from all of them.

Last but not least, our sincere thanks to all the authors, participants, and invited speakers, who submitted their papers and contributed to the in-depth discussions.

The organizers also express their hearty thanks to Springer for agreeing to publish the proceedings in its Mathematics and Computer Science series.

We sincerely hope that the reader will find the proceedings stimulating and inspiring.

Ghaziabad, India — Vinai K. Singh
Victoria, Canada — H.M. Srivastava
Torino, Italy — Ezio Venturino
Stuttgart, Germany — Michael Resch
New Delhi, India — Vijay Gupta
December 2015

Organizing Committee

Chief Patrons

Vinay Kumar Pathak
Vice Chancellor, Dr. APJ Abdul Kalam Technical University, Lucknow, India
Dinesh Kumar Goel, Chairman, RKG Group of Institutions, Ghaziabad, India

Patrons

B.K. Gupta, Advisor, RKG Group of Institutions, Ghaziabad, India
Laxman Prasad, Director R&D, RKG Group of Institutions, Ghaziabad, India
R.P. Maheshwari, Director, RKGIT, Ghaziabad, India

Conference Chairperson

Vinai K. Singh, Raj Kumar Goel Institute of Technology, Ghaziabad, India

Program Technical Committee

H.M. Srivastava, University of Victoria, Victoria, BC, Canada
Ezio Venturino, Universita' di Torino, Italy
Michael Resch, Center of HPC, University of Stuttgart, Germany
Heinrich Begehr, Freie Universitat Berlin, Germany
Wolfgang Sprobig, TU Bergakademie Freiberg, Germany
Vijay Gupta, NSIT, Dwarka, New Delhi, India

Sever S. Dragomir, Victroia University Melbourne, Australia
Andreas Fischer, TU Dursden, Germany
Qin Sheng, Baylor University, USA
R.P. Agrawal, A&M University, Kinggsvile, USA
Taekyun Kim, Kwangwoon Univeristy Seoul, South Korea
Vasil Berinde, North University of Baia Mare Victoriei, Romania
T.M. Rassias, National Technical University of Athens
Ali Aral, Kirikkale University, Kirikkale, Turkey
Arindam Singh, IIT Madras, India
Rajinder Jeet Hans Gill, Punjab University Chandigarh, India
Sudesh Kaur Khanduj, IISER Mohali, India
S.D. Adhikari, HRI Jhunshi Allahabad, India
Detlef H. Mache, University of Applied Science TFH Bochum, Germany
Rajen K. Sinha, IIT Guwahati, India
K.K. Shukla, IIT BHU, Varanasi, India
R.K. Mohanty, South Asian University, New Delhi, India
J.R. Torregrosa, Universided Politecnica, de Valancia, Spain
Kum Kum Dewan, Jamila Milia Islamia Central University, New Delhi, India
Phalguni Gupta, IIT Kanpur, India
Somaesh Kumar, IIT Kharagpur, India
R.G. Vyas, M.S. University, Baroda, India
C. Adiga, University of Mysore, India
F.D. Zaman, King Fahd University of Petroleum & Minerals, Saudi Arabia
Sanjeev Kumar, B.R. Ambedkar University, Agra, India
Shailesh K. Shivakar, Infosys R&D Center, New Delhi, India
V.K. Singh, IET Lucknow, India
S.B. Singh, G.B. Pant Agriculture University, Pant Nagar, India
J. Martinez-Moreno, University of Jaen, Spain
R.S. Gupta, KNIT Sultanpur, India
Bani Singh, JIIT, Noida, India
U.S. Gupta, IIT Roorkee, India
Adisak Pongpullponsank, King Mongnut's University of Technology, Thonburi, Thailand
S.K. Mishra, Banaras Hindu University, Varanasi, India
M. Khandaqji, Hashemite University, Zarqa-Jordan
Ibrahim Salim, University of Victoria, Victoria, BC, Canada
B.K. Das, Delhi University, New Delhi, India
Ashok Kumar Singh, Principal Scientific Officer, DST, Government of India
Frederic Magoules, Ecole Centrale Paris, France
Aravind Kumar Singh, Banaras Hindu University, Varanasi, India
T.D. Mahabaleswar, Indian Science Academy, Banglore, India

Invited Speakers

Qin Sheng, S.K. Mishra, Ali Aral, Michael Resch, Atma Sahu, Yulia Pronina, Vijay Gupta, Fairouz Kamareddine, Y.K. Gupta, Georgii Alexander Omelyanov, Jichun Li, Samar Aseeri, A.K. Singh, Tony W.H. Sheu, Heinrich Begehr, Sanjeev Kumar, Galina Filipuk, V.K. Singh, Bani Singh, R.G. Vyas, S.B. Singh, H.U. Siddiqui, Y.P. Singh, Andreas Fischer.

Local Organization Committee

Prag Singhal, Pankaj Singh Yadav, Navneet Kr. Yadav, Rajeev Kumar, Rahul Kumar, Pratima Sharma, Ms. Sarita Singh, Pavan Shukla, Manoj Mangal, P.C. Mishra, Jagdeep Singh, Ravi Ranjan, Deepak Kumar, Dharambeer Singh, Nitin Narula, Vineeta Singh, Neena Sharma, Anupam Sharma, Anand Tyagi, Garima Garg, Arvind Tiwari, Nikunj Kumar, Sujeet Kumar Singh, Pawan Pandey, Zatin Singhal, Satish Chhokar, Lalit Saraswat, Rohit Kumar, V.K. Tripathi, U.K. Jha, Vinod Chaudhary, Anuj Kumar, Sanjay Singh, Praveen Kumar, Nand Kishor Yadav, Alok Tyagi, Ajay Chauhan, Soniya Verma, Nishi Pathak, A.K.S. Yadav, Minakshi Kaushik, Ajeet Pal Singh, Shalini Gupta, Vishal Srivastava Upesh Bhatnagar, Santosh Mishra, Ashish Singh.

Contribution as Reviewer

A.S.M. Riyad
A. Ramachandran
Abdul Wafi
Abhishek Chakraborty
Al-Mutairi Dhaifalla
Alan D. Rendall
Ali Aral
Alkhalfan Laila
Andrea Erafini
Andreas Fischer
Aravind Kumar Mishra
Armando Fox
Armend Sh Shabani
Arzad Alam Kherani
Ashis K. Chatterjee
Asmatullah Chaudhry
Aydin Izgi

B.S.N. Raju
B.S. Sanjeev
Bani Singh
Bashir Ahmad
Bharat Adsul
C. Bardaro
C.K. Tripathy
C. Adiga
D.B. Ojha
D. Easwaramoorthy
Daniel
Dave Beulke
Dhananjay Madhav
Didem Aydin
Dimitrie D. Stancu
Dmitri Kuzmin
Dukka B. Kc
Dumitru Baleanu
Emmanuel M. Gutman
Esmaeil Najafi
Ezio Venturino
F. Saleki
Fatma Ayaz
Felipe A. Asenjo
G.C. Bohmer
G. Icoz
G.S. Srivastava
Gabriel Barrenechea
Gabriele Jost
George Babis
Ghulam Mohammad
Gilson G. Da Silva
Gleb Belov
H. Ligteringen
H.S. Guruprasad
Harish Chandra
Harish Chaudhry
Heinrich Begehr
Hilde De Ridder
Himani Dem
I Finocchi
Inderveer Channa
Istvan Farago
Ivan Zelinka
J. Duan

J.E. Tumblin
J.I. Romas
J.K. Srivastava
J. Rajsankar
J. Sandor
J.T. Teng
J.R. Torregrosa
Jagdish Chandra Bansal
Jean-Marie Ekoé
Jie Shen
Joao A. Ferreira
Jose A. Oliveira
Joseignacio Martinez
Jovana Dzunic
Jui-Jung Liao
Jun Bae Seo
Jun-An Lu
Jung Hee Cheon
Kejal Khatri
Koichi Saito
Krishna S.
Kunal Chakraborty
L.D.S. Colleo
L. Fanucci
L. Franco
L.V. Reddy
Lakdere Benkherouf
Lars M. Kristensen
Laszlo Stacho
Leo P. Franca
M.A. Noor
M. Ghalambaz
M. Jamil Amir
M. Kubo
M. Maiti
M.A. Hafiz
Maciej Besta
Mamood Otadi
Manuel C. Figueiredo
Mari Min
Maryam Mosleh
Matteo Filippi
Mehdi Mahmoodi
Mehmet Sezgin
Meraj Mustafa

Michael Resch
Michael Westergaard
Mioara Boncut
Mohammad Tamsir
Mohsen Alipour
Mousa Shamsi
Muhammad Alia Babar
Muhammed I. Syam
Murilo A. Vaz
N. Ispir
Nelson Faustini
Nipun Agarwal
Nita Shah
Ogün Doğru
Ozlem Ersoy
P. Chu
P.N. Agrawal
Pashanta Mahato
Paulo Santos
Petr Knobloch
Po-Cheng Chou
Pronina Yulia
Qin Tim Sheng
R. Balasubramanian
R. Ellahi
R.G. Vyas
Ram N. Mohapatra
Ram Naresh
Reza Ezzati
Robert Gerstenberger
Roger Miller
Rolf Rabenseifner
Roshan Lal
Ryusuke Nosaka
S.D. Maharaj
S.K. Goyal
S.M. Chow
S. Moradi
S.R. Fulton
S. Romaguera
S. Sakakibara
S.B. Singh
S.N. Singh
Saeid Najafalizadeh
Sahadeo Padhey

Samuel Chapman
Sandip Banerjee
Sankar Dutta
Santosh Senger
Sezgin Sucu
Shalini Chandra
Shigeyoshi Owa
Somnath Bhatacharya
Sridhar Iyer
Stephen J. Crothers
Sudeshna Banerjee
Sukumar Das Adhikari
Supriyo Roy
T. Acar
T. Vaitekhovich
Tayekum King
Tibor Krisztin
U.S. Gupta
Ulrich Abel
V.K. Singh
Valmir B. Krasniqi
Vasanth Krishnaswami
Vasile Berinde
Vijay G. Ukadgaonkar
Vijay Gupta
Vijay Krishan Singh
Vinay Joseph Ribeiro
Volker John
X. Chen
Y. Ordokhani
Zead Mustafa
Zohreh Molamohmadi
Zoltan Finta

Contents

About the Editors

Vinai K. Singh is a mathematician and holds a Ph.D. in Approximation Theory from Department of Applied Mathematics, Institute of Technology, Banaras Hindu University, Varanasi, India. He has been actively engaged in research activity since 1997, and is now Professor in the Department of Applied Mathematics at Raj Kumar Goel Institute of Technology, Ghaziabad, India. His areas of research interest include functional analysis, approximation theory, and different kinds of positive operators. He is author of three book chapters and over 24 research papers in the international journals. He referees articles for professional journals and serves as editorial member of many national and international journals.

H.M. Srivastava has held the position as Professor Emeritus in the Department of Mathematics and Statistics at the University of Victoria in Canada since 2006, having joined there as faculty in 1969, first as Associate Professor (1969–1974) and then as Full Professor (1974–2006). He began his university-level teaching career right after having received his M.Sc. degree in 1959 at the age of 19 years from the University of Allahabad in India. He earned his Ph.D. degree in 1965 while he was a full-time member of the teaching faculty at the J.N.V. University of Jodhpur in India. He has held numerous visiting research and honorary chair positions at many universities and research institutes in different parts of the world. Having received several D.Sc. (honoris causa) degrees as well as honorary memberships and fellowships of many scientific academies and learned societies around the world, he is also actively associated editorially with numerous international scientific research journals. His current research interests include several areas of Pure and Applied Mathematical Sciences such as (for example) real and complex analysis, fractional calculus and its applications, integral equations and transforms, higher transcendental functions and their applications, q-series and q-polynomials, analytic number theory, analytic and geometric inequalities, probability and statistics, and inventory modeling and optimization. He has published 24 books, monographs, and edited volumes, 30 book (and encyclopedia) chapters, 45 papers in international conference proceedings, and more than 1,000 scientific research journal articles, as well as he has written forewords and prefaces to many books and journals, and so on. Further details about his other professional achievements and scholarly accomplishments, as well as honors, awards, and distinctions can be found at the following website: http://www.math.uvic.ca/faculty/harimsri/.

Ezio Venturino has obtained the Ph.D. in Applied Mathematics, SUNY at Stony Brook, on December 21, 1984, with the dissertation entitled An analysis of some direct methods for the numerical solution of singular integral equations under the supervision of Prof. R.P. Srivastav. His research interests lie in numerical analysis and mathematical modeling. In numerical analysis he has studied quadrature formulae for singular integrals, methods for singular integral equations, lacunary interpolation problems. In mathematical modeling, he investigates biological and socioeconomic applications. Among his professional activities, he has coauthored a book for CRC and coedited another for Birkhaeuser. He has authored about 40 research papers in numerical analysis and more than 160 in mathematical modeling. He has taken part in more than 150 international conferences, with several plenary talks given upon invitation, contributing a paper to most, regularly organizing special sessions and chairing sessions and doing organizational work. He acts regularly as referee for international journals. His editorial activity involves the following journals: Electronic Proceedings of WSEAS, Conference on Mathematics and Computers in Biology and Chemistry 2004 (MCBC 2004), Venice, November 15–17, 2004, Guest Editor for the WSEAS Transactions on Biology and Biomedicine volume 1, no. 4, October 2004, Area Editor, Simulation Theory and Practice, Elsevier, 2008–2013, Advisory Editor, Mathematical Methods in the Applied Sciences, Wiley, since 2009, Editorial Board of Network Biology (ISSN 2220-8879), Computational Ecology and Software (ISSN 2220-721X) Shekhar (New Series) International Journal of Mathematics, Proceedings CMMSE 2008, 2009, 2010, 2011, 2012, 2013, 2014, Proceedings DWCAA 2009, Canazei, 4-9/9/2009, Associate Editor of Contemporary Mathematics and Statistics, Columbia International Publishing,

ISSN: 2163-1204 (Online) since 2012, Rendiconti del Seminario Matematico dell'Universita' e del Politecnico di Torino, Heliyon, Elsevier's open access journal, since 2015, Member of BIOMAT Consortium Director Board, since 2015, Editorial board of Applied Mathematics and Nonlinear Sciences since 2016.

Michael Resch is the Director of the High Performance Computing Center Stuttgart (HLRS) and the Department for HPC (IHR) at the University of Stuttgart, holding a full professorship for HPC. Michael Resch was an invited plenary speaker at SC'07 in Reno, USA. He won HPC Wire awards for industrial applications in 2015 and 2013 and won with his team the HPC Challenge Award in 2003. In 1999 his team received the NSF Award for High Performance Distributed Computing. He holds honorable doctoral degrees from the Technical University of Donezk/Ukraine (2009) and the Russian Academy of Science (2011) as well as an honorary professorship from the Russian Academy of Sciences (2014). Michael Resch is a PI in the cluster of excellence for Simulation Technology as part of the German national Initiative for Excellence in Research. Michael Resch holds an M.Sc. in Technical Mathematics (Technical University of Graz/Austria) and a Ph.D. in Engineering from the University of Stuttgart. In 2002 he was Assistant Professor at the Department of Computer Science of the University of Houston, TX.

Vijay Gupta is working presently as Professor in Department of Mathematics at Netaji Subhas Institute of Technology, New Delhi. He obtained M.Sc. from Christ Church College, Kanpur, M.Phil. from Meerut University, Meerut and Ph.D. from University of Roorkee (now IIT Roorkee, India). His area of research interest includes approximation theory, especially on convergence of linear positive operators. He is the author of two books, 11 book chapters, and over 260 research papers in journals of international repute. He visited several universities globally for academic activities and delivered many lectures. He is actively associated editorially with over 25 international scientific research journals.

On Approximation Properties of Generalized Durrmeyer Operators

Ali Aral and Tuncer Acar

Abstract The concern of this paper is to introduce new generalized Durrmeyer-type operators from which classical operators can be obtained as a particular case, inspiring from the Ibragimov–Gadjiev operators (Gadjiev and Ibragimov, Soviet Math. Dokl. 11, 1092–1095, (1970) [8]). After the construction of new Durrmeyer operators is given, we obtain some pointwise convergence theorems and Voronovskaya-type asymptotic formula for new Durrmeyer-type operators. We establish a quantitative version of the Voronovskaya-type formula with the aid of the weighted modulus of continuity. Some special cases of new operators are presented as examples.

Keywords Ibragimov–Gadjiev operators · Durrmeyer operators · Voronovskaya theorem · Weighted modulus of continuity

1 Introduction

In the field of approximation theory, several researchers have defined general sequences of linear positive operators with the purpose of obtaining results which are valid for the wide class of such sequences, one of the most important type of generalized sequences introduced by Ibragimov and Gadjiev in 1970 [8]. From the Gadjiev–Ibragimov operators, we can derive many of the classical sequences of linear positive operators by means of a suitable transformation. Now we recall these operators called the Gadjiev–Ibragimov operators.

Let $(\varphi_n(t))_{n\in\mathbb{N}}$ and $(\psi_n(t))_{n\in\mathbb{N}}$ be sequences of functions in $C\left(\mathbb{R}^+\right)$ which is the space of continuous function on $\mathbb{R}^+ := [0, \infty)$, such that $\varphi_n(0) = 0$, $\psi_n(t) > 0$, for all t and $\lim_{n\to\infty} 1/n^2\psi_n(0) = 0$. Also let $(\alpha_n)_{n\in\mathbb{N}}$ denote a sequence of positive numbers which satisfy the following conditions:

A. Aral · T. Acar (✉)
Department of Mathematics, Kirikkale University, 71450 Yahsihan, Kirikkale, Turkey
e-mail: tunceracar@ymail.com

A. Aral
e-mail: aliaral73@yahoo.com

V.K. Singh et al. (eds.), *Modern Mathematical Methods and High Performance Computing in Science and Technology*, Springer Proceedings in Mathematics & Statistics 171, DOI 10.1007/978-981-10-1454-3_1

$$\lim_{n\to\infty}\frac{\alpha_n}{n}=1 \text{ and } \lim_{n\to\infty}\alpha_n\psi_n(0)=l_1,\ l_1\geq 0. \tag{1}$$

The Gadjiev–Ibragimov operators are defined by

$$G_n(f;x)=\sum_{\nu=0}^{\infty}f\left(\frac{\nu}{n^2\psi_n(0)}\right)\frac{\partial^\nu}{\partial u^\nu}K_n(x,t,u)\bigg|_{u=\alpha_n\psi_n(t),t=0}\frac{(-\alpha_n\psi_n(0))^\nu}{\nu!}, \tag{2}$$

where $K_n(x,t,u)$ $\left(x,t\in\mathbb{R}^+ \text{ and } -\infty<u<\infty\right)$ is a sequence of functions of three variable and have to meet the following conditions:

1. Every function of this sequence is an entire function with respect to u for fixed $x,t\in\mathbb{R}^+$ and $K_n(x,0,0)=1$ for $x\in\mathbb{R}^+$ and $n\in\mathbb{N}$,
2. $\left[(-1)^\nu\frac{\partial^\nu}{\partial u^\nu}K_n(x,t,u)\big|_{u=u_1,t=0}\right]\geq 0$ for $\nu=0,1,\ldots$, any fixed $u=u_1$ and $x\in\mathbb{R}^+$,
   (This notation means that the derivative with respect to u is taken ν times, then one set $u=u_1$ and $t=0$)
3. $\frac{\partial^\nu}{\partial u^\nu}K_n(x,t,u)\big|_{u=u_1,t=0}=-nx\left[\frac{\partial^{\nu-1}}{\partial u^{\nu-1}}K_{m+n}(x,t,u)\big|_{u=u_1,t=0}\right]$ for all $x\in\mathbb{R}^+$ and $n\in\mathbb{N}$, $\nu=0,1,\ldots$, where m is a number such that $m+n=0$ or a natural number.

Evidently, by ν-times application of property (3) with $u_1=\alpha_n\psi_n(t)$ and $t=0$, operator (2) can be reduced to the form

$$G_n(f;x)=\sum_{\nu=0}^{\infty}f\left(\frac{\nu}{n^2\psi_n(0)}\right)\frac{n(n+m)\ldots(n+(\nu-1)m)(\alpha_n\psi_n(0))^\nu}{\nu!}K_{n+\nu m}(x,0,\alpha_n\psi_n(0))\,x^\nu,$$

One can obtain the well-known operators in particular cases (see [8]).

The operators G_n are linear and positive. It is well known that the operator G_n preserves the degree of the polynomials and they approximate not only the function but also its derivatives. They also have some shape preserving and weighted approximation properties. We can find a large number of papers devoted to study of the properties of convergence of these operators such as [1, 3, 5, 6, 9, 16].

On the other hand, Durrmeyer-type generalizations of approximation operators is an important subject in approximation theory and they are the method to approximate Lebesgue integrable functions see [4, 7]. Now we recall some of them.

Mazhar and Totik [15] modified Szasz–Mirakyan operators and introduced Szasz–Durrmeyer operators as

$$L_n(f,x)=n\sum_{k=0}^{\infty}b_{n,k}(x)\int_0^\infty b_{n,k}(t)\,f(t)\,dt, \tag{3}$$

where $b_{n,k}(x) = e^{-nx}(nx)^k/k!$, and Durrmeyer-type modification of Baskakov operators [17] was introduced as

$$V_n(f,x) = (n-1)\sum_{k=0}^{\infty} v_{n,k}(x)\int_0^{\infty} v_{n,k}(t) f(t)\,dt, \tag{4}$$

where $v_{n,k}(x) = \binom{n+k-1}{k}x^k(1+x)^{-n-k}$. Furthermore, as direct generalizations of these Durrmeyer-type operators, producing summation-integral type operators having different basis functions has been studied intensively. For details, we refer the readers to [12, 18].

In the present paper, we introduce Durrmeyer modification of the operator (2). It is defined by replacing the discrete values $f\left(\nu/n^2\psi_n(0)\right)$ in (2) by an integral over the weighted function in the mean of (7). In the next section, we will show that many classical sequences of Durrmeyer type can be obtained by making an special selection of K_n. Then we give some auxiliary results to construct the new operators and calculate moments for these operators. We obtain pointwise convergence theorem at continuity point of f and quantitative version of the approximation by using suitable weighted modulus of continuity for new Durrmeyer operators. We also study Voronovskaya-type asymptotic formula and its quantitative version with the aid of same weighted modulus of continuity. For more details Voronovskaya type theorems, we refer the readers to [2, 10, 11, 19] We note that, since our new operator is defined on unbounded interval, use of such weighted modulus of continuity for estimation of convergence is required.

2 Construction of Ibragimov–Gadjiev–Durrmeyer Operators

By similar consideration constructed by Ibragimov and Gadjiev, we purpose to define general Durrmeyer-type operators including well-known Durrmeyer operators. In order to achieve this in addition to the mentioned three conditions in introduction, we also assume the following two conditions:

4. $K_n(0,0,u) = 1$ for any $u \in \mathbb{R}$, and

$$\lim_{x\to\infty} x^p \left.\frac{\partial^\nu}{\partial u^\nu} K_n(x,t,u)\right|_{u=u_1,t=0} = 0,$$

   for any $p \in \mathbb{N}$ and fixed $u = u_1$.
5. For any fixed t and u the function $K_n(x,t,u)$ is continuously differentiable with respect to variable $x \in \mathbb{R}^+$ and satisfying the equality

$$\frac{d}{dx}K_n(x,0,u_1) = -nu_1 K_{m+n}(x,0,u_1)$$

for fixed $u = u_1$.

For simplicity, we use the notation $K_n^{(\nu)}(x,0,\alpha_n\psi_n(0))$ in place of $\frac{\partial^\nu}{\partial u^\nu}K_n(x,t,u)|_{u=\alpha_n\psi_n(0),t=0}$. By the Taylor expansion of entire function $K_n(x,t,u)$ in the powers of $(\varphi_n(t) - \alpha_n\psi_n(t))$ and taking $t = 0$, since $\varphi_n(0) = 0$, we have the following representation:

$$K_n(x,0,0) = \sum_{\nu=0}^{\infty} K_n^{(\nu)}(x,0,\alpha_n\psi_n(0)) \frac{(-\alpha_n\psi_n(0))^\nu}{\nu!}$$

Taking into account that $K_n(x,0,0) = 1$ by the condition (1), we have

$$\sum_{\nu=0}^{\infty} K_n^{(\nu)}(x,0,\alpha_n\psi_n(0)) \frac{(-\alpha_n\psi_n(0))^\nu}{\nu!} = 1.$$

Lemma 1 *Let ν be a nonnegative integer, $x \in \mathbb{R}^+$ and $m, n \in \mathbb{N}$. Then the condition (5) is equivalent to the equality*

$$\frac{d}{dx}K_n^{(\nu)}(x,0,u_1) = \frac{\nu}{x}K_n^{(\nu)}(x,0,u_1) - nu_1 K_{n+m}^{(\nu)}(x,0,u_1).$$

Proof By ν-multiple application of condition (3), we obtain

$$K_n^{(\nu)}(x,0,u_1) = (-1)^\nu n(n+m)\dots(n+(\nu-1)m)x^\nu K_{n+\nu m}(x,0,u_1). \quad (5)$$

Applying condition (5) we get

$$(-1)^\nu \frac{d}{dx}K_n^{(\nu)}(x,0,u_1) = n(n+m)\dots(n+(\nu-1)m)$$
$$\times \left\{\nu x^{\nu-1}K_{n+\nu m}(x,0,u_1) - x^\nu(n+\nu m)u_1 K_{n+(\nu+1)m}(x,0,u_1)\right\}.$$

Using (5) we get desired result.

Lemma 2 *Let ν be a nonnegative integer, $x \in \mathbb{R}^+$ and $m, n \in \mathbb{N}$. Then we have*

$$\int_0^\infty K_n^{(\nu)}(x,0,u_1)\,dx = (-1)^\nu \frac{\nu!}{(n-m)u_1^{\nu+1}}.$$

Proof Using integration by parts and conditions (1) and (4) we have

$$\int_0^\infty K_n^{(\nu)}(x,0,u_1)\,dx = -\int_0^\infty x\frac{d}{dx}K_n^{(\nu)}(x,0,u_1)\,dx.$$

Using Lemma 1, we get

$$\int_0^\infty K_n^{(\nu)}(x,0,u_1)\,dx$$
$$= -\nu \int_0^\infty K_n^{(\nu)}(x,0,u_1)\,dx + nu_1 \int_0^\infty x K_{n+m}^{(\nu)}(x,0,u_1)\,dx.$$

Also by condition (3) , we have

$$\int_0^\infty K_n^{(\nu)}(x,0,u_1)\,dx$$
$$= -\nu \int_0^\infty K_n^{(\nu)}(x,0,u_1)\,dx - u_1 \int_0^\infty K_n^{(\nu+1)}(x,0,u_1)\,dx.$$

Hence we can write

$$\int_0^\infty K_n^{(\nu)}(x,0,u_1)\,dx = \frac{-u_1}{\nu+1}\int_0^\infty K_n^{(\nu+1)}(x,0,u_1)\,dx.$$

By ν-times application of above equality and using condition (1) and (5), we get

$$\begin{aligned}\int_0^\infty K_n^{(\nu)}(x,0,u_1)\,dx &= -\frac{\nu}{u_1}\int_0^\infty K_n^{(\nu-1)}(x,0,u_1)\,dx \\ &\vdots \\ &= (-1)^\nu \frac{\nu!}{u_1^\nu}\int_0^\infty K_n(x,0,u_1)\,dx \\ &= \frac{(-1)^{\nu+1}\,\nu!}{(n-m)\,u_1^{\nu+1}}\int_0^\infty \frac{d}{dx}K_{n-m}(x,0,u_1)\,dx \\ &= (-1)^\nu \frac{\nu!}{(n-m)\,u_1^{\nu+1}}. \end{aligned} \tag{6}$$

Definition 1 Let ν be a nonnegative integer, $x \in \mathbb{R}^+$ and $m, n \in \mathbb{N}$. Choosing $u_1 = \alpha_n \psi_n(0)$ and $t = 0$,we can define Ibragimov–Gadjiev–Durrmeyer operators as follows:

$$\begin{aligned} M_n(f;x) = (n-m)\,\alpha_n\psi_n(0)\sum_{\nu=0}^{\infty} K_n^{(\nu)}(x,0,\alpha_n\psi_n(0))\,\frac{[-\alpha_n\psi_n(0)]^\nu}{(\nu)!} \\ \times \int_0^\infty f(y)\,K_n^{(\nu)}(y,0,\alpha_n\psi_n(0))\,\frac{[-\alpha_n\psi_n(0)]^\nu}{(\nu)!}dy \end{aligned} \tag{7}$$

We call these new operators as Ibragimov–Gadjiev–Durrmeyer operators. The family of operators $M_n(f;x)$ is linear and positive. Also, we have the following operators in special cases.

$M_n(f;x)$	$K_n(x,t,u)$	α_n	$\psi_n(0)$	m
Baskakov–Durrmeyer	$[1+t+ux]^{-n}$	n	1/n	1
Szasz–Durrmeyer	$e^{-n(t+ux)}$	n	1/n	0
Generalized Baskakov–Durrmeyer	$K_n(t+ux)$	n	1/n	1

For the above special cases, see the papers [13, 15, 17], respectively.

Lemma 3 *Let* ν *be a nonnegative integer,* $n, m \in \mathbb{N}$. *For any natural number* r *we have*

$$\int_0^\infty x^r K_n^{(\nu)}(x,0,u_1)\,dx = \frac{(-1)^\nu(\nu+r)!}{(n-m)(n-2m)\dots(n-pm)(n-(r+1)m)\,u_1^{\nu+r+1}}. \tag{8}$$

Proof Using the condition (3) recursively ν-times we get

$$\begin{aligned}\int_0^\infty x^r K_n^{(\nu)}(x,0,u_1)\,dx &= -\frac{1}{n-m}\int_0^\infty x^{r-1}K_{n-m}^{(\nu+1)}(x,0,u_1)\,dx.\\ &= \frac{1}{(n-m)(n-2m)}\int_0^\infty x^{r-2}K_{n-2m}^{(\nu+2)}(x,0,u_1)\,dx\\ &\vdots\\ &= \frac{(-1)^r}{(n-m)(n-2m)\dots(n-rm)}\int_0^\infty K_{n-rm}^{(\nu+r)}(x,0,u_1)\,dx.\end{aligned}$$

Using (6) it follows

$$\begin{aligned}&\int_0^\infty x^r K_n^{(\nu)}(x,0,u_1)\,dx\\ &= \frac{(-1)^\nu(\nu+r)!}{(n-m)(n-2m)\dots(n-rm)(n-(r+1)m)\,u_1^{\nu+r+1}}.\end{aligned}$$

Lemma 4 *Let* ν *be a nonnegative integer,* $m, n \in \mathbb{N}$. *For any natural number* r *we have*

$$\begin{aligned}M_n(y^r;x) &= \frac{n^{2r}}{(n-2m)\dots(n-pm)(n-(r+1)m)(\alpha_n)^r\left(n^2\psi_n(0)\right)^r}\\ &\times\sum_{j=0}^r n(n+m)\dots(n+(j-1)m)\,C_{j,r}\,[\alpha_n\psi_n(0)]^j\,x^j,\end{aligned}$$

where $C_{j,r} = \frac{r!}{j!}\binom{r}{j}$. *Also,*

$$M_n(1;x) = 1,\ M_n(y;x) = \frac{n^2}{(n-2m)\alpha_n}\left(\frac{\alpha_n}{n}x + \frac{1}{n^2\psi_n(0)}\right),$$

$$M_n\left(y^2;x\right) = \frac{n^4}{(n-2m)(n-3m)\alpha_n^2}\left(\left(\frac{\alpha_n}{n}x\right)^2\frac{(m+n)}{n} + \frac{\alpha_n}{n}\frac{4}{n^2\psi_n(0)}x + \frac{2}{\left(n^2\psi_n(0)\right)^2}\right). \tag{9}$$

Proof Directly from the definition of operator (7) we can write

$$M_n\left(y^r;x\right) = (n-m)\alpha_n\psi_n(0)\sum_{\nu=0}^{\infty}K_n^{(\nu)}(x,0,\alpha_n\psi_n(0))\frac{[-\alpha_n\psi_n(0)]^\nu}{(\nu)!}$$
$$\times\int_0^\infty y^r K_n^{(\nu)}(y,0,\alpha_n\psi_n(0))\frac{[-\alpha_n\psi_n(0)]^\nu}{(\nu)!}dy.$$

Using (8) with $u_1 = \alpha_n\psi_n(0)$ we conclude that

$$M_n\left(y^r;x\right) = (n-m)\alpha_n\psi_n(0)\sum_{\nu=0}^{\infty}K_n^{(\nu)}(x,0,\alpha_n\psi_n(0))\frac{[-\alpha_n\psi_n(0)]^\nu}{(\nu)!}$$
$$\times\frac{(-1)^\nu(\nu+r)!}{(n-m)(n-2m)\ldots(n-rm)(n-(r+1)m)(\alpha_n\psi_n(0))^{\nu+r+1}}\frac{[-\alpha_n\psi_n(0)]^\nu}{(\nu)!}.$$
$$=\sum_{\nu=0}^{\infty}K_n^{(\nu)}(x,0,\alpha_n\psi_n(0))\frac{[-\alpha_n\psi_n(0)]^\nu}{(\nu)!}$$
$$\times\frac{1}{(n-2m)\ldots(n-rm)(n-(r+1)m)(\alpha_n\psi_n(0))^r}(\nu+r)\ldots(\nu+1).$$

Using the equality

$$(\nu+r)\ldots(\nu+1) = \sum_{j=1}^{r}C_{j,r}\prod_{l=0}^{j-1}(\nu-l),$$

where $C_{j,r} = \frac{r!}{j!}\binom{r}{j}$ and (6) we have

$$M_n\left(y^r;x\right) = \sum_{\nu=0}^{\infty}K_n^{(\nu)}(x,0,\alpha_n\psi_n(0))\frac{[-\alpha_n\psi_n(0)]^\nu}{(\nu)!}$$
$$\times\frac{1}{(n-2m)\ldots(n-rm)(n-(r+1)m)(\alpha_n\psi_n(0))^r}\sum_{j=1}^{r}C_{j,r}\prod_{l=0}^{j-1}(\nu-l)$$

$$= \frac{1}{(n-2m)\dots(n-rm)(n-(r+1)m)(\alpha_n\psi_n(0))^r}$$
$$\times \sum_{j=0}^{r} C_{j,r} \sum_{\nu=0}^{\infty} \prod_{l=0}^{j-1} (\nu - l)\, K_n^{(\nu)}(x, 0, \alpha_n\psi_n(0)) \frac{[-\alpha_n\psi_n(0)]^\nu}{(\nu)!}$$
$$= \frac{1}{(n-2m)\dots(n-rm)(n-(r+1)m)(\alpha_n\psi_n(0))^r}$$
$$\times \sum_{j=0}^{r} C_{j,r} \sum_{\nu=j}^{\infty} K_n^{(\nu)}(x, 0, \alpha_n\psi_n(0)) \frac{[-\alpha_n\psi_n(0)]^\nu}{(\nu - j)!}$$
$$= \frac{1}{(n-2m)\dots(n-rm)(n-(r+1)m)\, u_1^{r+1}}$$
$$\times \sum_{j=1}^{r} C_{j,r} x^j \sum_{\nu=0}^{\infty} K_n^{(\nu)}(x, 0, \alpha_n\psi_n(0)) \frac{(-1)^j [-\alpha_n\psi_n(0)]^{\nu+j}}{(\nu)!}$$
$$= \frac{n^{2r}}{(n-2m)\dots(n-rm)(n-(r+1)m)(\alpha_n)^r \left(n^2\psi_n(0)\right)^r}$$
$$\times \sum_{j=1}^{r} n(n+m)\dots(n+(j-1)m)\, C_{j,r} [\alpha_n\psi_n(0)]^j x^j.$$

Lemma 5 *For each* $x \geq 0$ *and* $n > 3m$ *we have*

(i) $M_n(y-x;x) = \frac{2mx}{(n-2m)} + \frac{1}{(n-2m)\alpha_n\psi_n(0)}$,

(ii) $M_n\left((y-x)^2;x\right) = x^2\left[\frac{m(2n+6m)}{(n-2m)(n-3m)}\right] + \frac{(2n+6m)\alpha_n\psi_n(0)x+2}{(n-2m)(n-3m)\alpha_n^2\psi_n^2(0)}$,

(iii) $M_n\left((y-x)^r;x\right) = \mathcal{O}\left(\left(\frac{1}{n\alpha_n\psi_n(0)}\right)^{\left[\frac{r+1}{2}\right]}\right)(x^r + \dots + x + 1)$, where $[.]$ is integral part of $(r+1)/2$.

Proof Proof is clear from the Lemma 4.

Remark 1 In the paper [18], Srivastava and Gupta considered a general sequence of Durrmeyer operators. But the operators introduced there reduce to restricted number of Durrmeyer type operators, but one can obtain any Durrmeyer operators from the operators M_n as special cases.

3 Pointwise Convergence Results

For our main results, we consider the following function spaces. We will denote by $L^\infty[0,\infty)$, the Lebesgue space of all essentially bounded functions and by $\|f\|_\infty$ the corresponding norm.

Let $B_{x^2}[0,\infty)$ be the set of all functions f defined on $[0,\infty)$ satisfying the condition $|f(x)| \leq M_f\left(1+x^2\right)$ with some constant M_f, depending only on f. $C_{x^2}[0,\infty)$

denotes the subspace of all continuous function in $B_{x^2}[0,\infty)$. By $C^k_{x^2}[0,\infty)$, we denote subspace of all functions $f\in C_{x^2}[0,\infty)$ for which $\lim_{x\to\infty}\frac{f(x)}{1+x^2}=k$, where k is a constant depending on f.

Weighted modulus of smoothness is denoted by $\Omega(f;\delta)$ and given by

$$\Omega(f;\delta)=\sup_{0\le h<\delta,\ x\in[0,\infty)}\frac{|f(x+h)-f(x)|}{\left(1+h^2\right)\left(1+x^2\right)} \tag{10}$$

for $f\in C^k_{x^2}[0,\infty)$ (see [14]).

We know that for every $f\in C^k_{x^2}[0,\infty)$, $\Omega(.;\delta)$ has the properties

$$\lim_{\delta\to 0}\Omega(f;\delta)=0$$

and

$$\Omega(f;\lambda\delta)\le 2(1+\lambda)\left(1+\delta^2\right)\Omega(f;\delta),\quad \lambda>0. \tag{11}$$

For $f\in C^k_{x^2}[0,\infty)$, from (10) and (11) we can write

$$\begin{aligned}|f(y)-f(x)| &\le \left(1+(y-x)^2\right)\left(1+x^2\right)\Omega(f;|y-x|)\\ &\le 2\left(1+\frac{|y-x|}{\delta}\right)\left(1+\delta^2\right)\Omega(f;\delta)\left(1+(y-x)^2\right)\left(1+x^2\right)\end{aligned} \tag{12}$$

Theorem 1 *If $f\in L^\infty[0,\infty)$ then at every point x of continuity of f we have*

$$\lim_{n\to\infty}M_n(f;x)=f(x).$$

Moreover if the function f is uniformly continuous and bounded in $[0,\infty)$ then on every compact interval $J\subset[0,\infty)$ we have

$$\lim_{n\to\infty}\|(M_n(f;x)-f(x))\chi_J\|_\infty=0$$

where χ_J is the characteristic function of J.

Proof Since $M_n(1;x)=1$ we can write

$$\begin{aligned}M_n(f;x)-f(x) &= (n-m)\alpha_n\psi_n(0)\sum_{\nu=0}^{\infty}K_n^{(\nu)}(x,0,\alpha_n\psi_n(0))\frac{[-\alpha_n\psi_n(0)]^\nu}{(\nu)!}\\ &\times\int_0^\infty[f(y)-f(x)]K_n^{(\nu)}(y,0,\alpha_n\psi_n(0))\frac{[-\alpha_n\psi_n(0)]^\nu}{(\nu)!}dy.\end{aligned}$$

Let $\varepsilon>0$ be given. By the continuity of f at the point x there exists $\delta>0$ such that $|f(y)-f(x)|<\varepsilon$ whenever $|y-x|<\delta$. For this $\delta>0$ we can write

$$M_n(f;x)-f(x) = (n-m)\alpha_n\psi_n(0)\sum_{\nu=0}^{\infty} K_n^{(\nu)}(x,0,\alpha_n\psi_n(0))\frac{[-\alpha_n\psi_n(0)]^{\nu}}{(\nu)!}$$
$$\times\left(\int_{|y-x|<\delta}+\int_{|y-x|\geq\delta}\right)[f(y)-f(x)]K_n^{(\nu)}(y,0,\alpha_n\psi_n(0))\frac{[-\alpha_n\psi_n(0)]^{\nu}}{(\nu)!}dy$$
$$= I_1+I_2.$$

It is obvious that

$$|I_1| \leq \varepsilon M_n(1;x) = \varepsilon.$$

It remains to estimate I_2. We can write

$$|I_2| \leq 2\|f\|_{\infty}(n-m)\alpha_n\psi_n(0)\sum_{\nu=0}^{\infty} K_n^{(\nu)}(x,0,\alpha_n\psi_n(0))\frac{[-\alpha_n\psi_n(0)]^{\nu}}{(\nu)!}$$
$$\times\int_{|y-x|\geq\delta} K_n^{(\nu)}(y,0,\alpha_n\psi_n(0))\frac{[-\alpha_n\psi_n(0)]^{\nu}}{(\nu)!}dy$$
$$\leq 2\frac{\|f\|_{\infty}}{\delta^2}M_n\left((t-x)^2;x\right).$$

If we choose $\delta = \frac{1}{\sqrt[3]{n\alpha_n\psi_n(0)}}$ and use Lemma 5 we have

$$|I_2| \leq 2\|f\|_{\infty}\left\{x^2\frac{m\left(n\alpha_n\psi_n^2(0)\right)^{1/3}(2n+6m)}{(n-2m)(n-3m)}+\frac{n^{1/3}(2n+6m)\alpha_n\psi_n(0)x+2n^{1/3}}{(n-2m)(n-3m)\left(\alpha_n\psi_n^2(0)\right)^{5/3}}\right\},$$

which proves the theorem. The second part of the theorem is proved similarly.

Theorem 2 *If $f\in C_{x^2}^k[0,\infty)$ then we have*

$$|M_n(f;x)-f(x)| \leq 16C\left(1+x^2\right)^3\Omega\left(f;\frac{1}{\sqrt{n\alpha_n\psi_n(0)}}\right),$$

where C is a positive constant independent of f and n.

Proof Using the inequality (12), we can write that

$$|f(y)-f(x)| \leq \begin{cases} 2\left(1+\delta^2\right)^2\left(1+x^2\right)\Omega(f;\delta), & |y-x|<\delta \\ 2\left(1+\delta^2\right)^2\left(1+x^2\right)\frac{(y-x)^4}{\delta^4}\Omega(f;\delta), & |y-x|\geq\delta \end{cases}$$

and choosing $\delta<1$, we have

$$|f(y)-f(x)| \leq 2\left(1+\delta^2\right)^2\left(1+x^2\right)\Omega(f;\delta)\left(1+\frac{(y-x)^4}{\delta^4}\right)$$
$$\leq 8\left(1+x^2\right)\Omega(f;\delta)\left(1+\frac{(y-x)^4}{\delta^4}\right).$$

Using above inequality we deduce that

$$
\begin{aligned}
|M_n(f;x) - f(x)| &= (n-m)\,\alpha_n\psi_n(0)\sum_{\nu=0}^{\infty} K_n^{(\nu)}(x,0,\alpha_n\psi_n(0))\frac{[-\alpha_n\psi_n(0)]^{\nu}}{(\nu)!} \\
&\quad\times\int_0^{\infty}|f(y)-f(x)|\,K_n^{(\nu)}(y,0,\alpha_n\psi_n(0))\frac{[-\alpha_n\psi_n(0)]^{\nu}}{(\nu)!}dy. \\
&\le 8\left(1+x^2\right)\Omega(f;\delta)\left(1+\frac{1}{\delta^4}M_n\left((y-x)^4;x\right)\right).
\end{aligned}
$$

By Lemma 5, it follows that

$$
|M_n(f;x) - f(x)| \le 8\left(1+x^2\right)\Omega(f;\delta)\left(1+\frac{\left(x^4+\cdots+x+1\right)}{\delta^4}\mathcal{O}\left(\frac{1}{(n\alpha_n\psi_n(0))^2}\right)\right).
$$

Choosing $\delta = \frac{1}{\sqrt{n\alpha_n\psi_n(0)}}$ we have desired result.

4 Voronovskaya-Type Results

In this section, we obtain some asymptotic estimates of the pointwise convergence in case the function f is regular at the point $x \in [0,\infty)$.

Theorem 3 *Let $f \in C_{x^2}[0,\infty)$. Suppose that the second derivative f'' exists at a point $x \in [0,\infty)$, then we have*

$$
\lim_{n\to\infty} n\alpha_n\psi_n(0)\left[M_n(f;x) - f(x)\right] = (2xml_1+1)\,f'(x) + x\,(xml_1+1)\,f''(x).
$$

Proof Let $f, f', f'' \in C_{x^2}[0,\infty)$ and $x \in [0,\infty)$ be fixed. By Taylor expansion of f we can write

$$
f(t) = f(x) + (t-x)\,f'(x) + \frac{(t-x)^2}{2!}f''(x) + \frac{\left(f''(\eta_x) - f''(x)\right)}{2}(t-x)^2. \tag{13}
$$

After setting

$$
r(t,x) := \frac{\left(f''(\eta_x) - f''(x)\right)}{2}
$$

from the condition of the theorem we have $\lim_{t\to x} r(t,x) = 0$. Applying M_n to both sides of (13) and multiplying $n\alpha_n\psi_n(0)$ respectively, we deduce that

$$n\alpha_n\psi_n(0)[M_n(f;x)-f(x)] = n\alpha_n\psi_n(0)f'(x)M_n(t-x;x) + \frac{n\alpha_n\psi_n(0)f''(x)}{2!}M_n\left((t-x)^2;x\right) + n\alpha_n\psi_n(0)M_n\left(r(t,x)(t-x)^2;x\right).$$

Using Lemma 5 and passing to limit with $n\to\infty$, we have

$$\begin{aligned}\lim_{n\to\infty} n\alpha_n\psi_n(0)[M_n(f;x)-f(x)] &= f'(x)\lim_{n\to\infty} n\alpha_n\psi_n(0)M_n(t-x;x) \\ &\quad + \frac{f''(x)}{2!}\lim_{n\to\infty} n\alpha_n\psi_n(0)M_n\left((t-x)^2;x\right) \\ &\quad + \lim_{n\to\infty} n\alpha_n\psi_n(0)M_n\left(r(t,x)(t-x)^2;x\right) \\ &= (2xml_1+1)f'(x)+x(xml_1+1)f''(x)+\lim_{n\to\infty}F_n.\end{aligned}$$

Now, it suffices to show that $F_n\to 0$ as $n\to\infty$. Let $\varepsilon>0$ be given. Since $r(t,x)\to 0$ as $t\to x$, then there exists $\delta>0$ such that we have $|r(t,x)|<\varepsilon$ when $|t-x|<\delta$ and we can write

$$|r(t,x)|\le C\le C\frac{(t-x)^2}{\delta^2},$$

when $|t-x|\ge\delta$. Thus, for all $x,t\in[0,\infty)$, we have

$$|r(t,x)|\le\varepsilon+C\frac{(t-x)^2}{\delta^2},$$

and

$$\begin{aligned}F_n &\le n\alpha_n\psi_n(0)M_n\left((t-x)^2\left(\varepsilon+C\frac{(t-x)^2}{\delta^2}\right),x\right) \\ &\le n\alpha_n\psi_n(0)\,\varepsilon M_n\left((t-x)^2,x\right)+\frac{C}{\delta^2}n\alpha_n\psi_n(0)M_n\left((t-x)^4,x\right).\end{aligned}$$

By the property (iii) of Lemma 5, we can say that

$$M_n\left((t-x)^4,x\right)=\left(x^4+\cdots+1\right)\mathcal{O}\left(\frac{1}{(n\alpha_n\psi_n(0))^2}\right).$$

So, we get that $F_n\to 0$ as $n\to\infty$ which completes the proof.

Now we give quantitative version of Voronovskaya theorem for the operator (7).

Theorem 4 *Let $f''\in C_{x^2}^k[0,\infty)$ and $x>0$ be fixed, $m,n\in\mathbb{N}$. Then we have*

$$\left| n\alpha_n \psi_n(0)\left[M_n(f;x) - f(x)\right] - f'(x)(2xml_1+1) - f''(x)\,x(xml_1+1)\right|$$
$$\le f'(x)\left|n\alpha_n\psi_n(0)\,M_n(y-x;x) - (2xml_1+1)\right|$$
$$+\frac{f''(x)}{2}\left|n\alpha_n\psi_n(0)\,M_n\left((y-x)^2;x\right) - 2x(xml_1+1)\right|$$
$$+8C\left(1+x^2\right)^4 \Omega\left(f'';\frac{1}{(n\alpha_n\psi_n(0))^{1/2}}\right),$$

where C is a positive constant.

Proof By the local Taylor's formula there exists η lying between x and y such that

$$f(y) = f(x) + f'(x)(y-x) + \frac{f''(x)}{2}(y-x)^2 + h(y,x)(y-x)^2,$$

where

$$h(y,x) := \frac{\left(f''(\eta) - f''(x)\right)}{2}$$

and h is a continuous function which vanishes at 0. Applying the operator M_n to above equality, we obtain the equality

$$M_n(f;x) - f(x) = f'(x)\,M_n(y-x;x) + \frac{f''(x)}{2}M_n\left((y-x)^2;x\right) + M_n\left(\left(h(y,x)(y-x)^2\right);x\right)$$

also we can write that

$$\left|M_n(f;x) - f(x) - \frac{f'(x)}{n\alpha_n\psi_n(0)}(2xml_1+1) - \frac{f''(x)}{n\alpha_n\psi_n(0)}x(xml_1+1)\right|$$
$$\le f'(x)\left(M_n(y-x;x) - \frac{1}{n\alpha_n\psi_n(0)}(2xml_1+1)\right)$$
$$+\frac{f''(x)}{2}\left(M_n\left((y-x)^2;x\right) - \frac{2}{n\alpha_n\psi_n(0)}x(xml_1+1)\right)$$
$$+M_n\left(|h(y,x)|(y-x)^2;x\right)$$

To estimate last inequality, using the inequality (12) and the inequality $|\eta - x| \le |y-x|$, we can write that

$$|h(y,x)| = \frac{\left|f''(\eta) - f''(x)\right|}{2}$$
$$\le \frac{1}{2}\Omega\left(f;|\eta - x|\right)\left(1+(\eta-x)^2\right)\left(1+x^2\right)$$
$$\le \frac{1}{2}\Omega\left(f;|y-x|\right)\left(1+(y-x)^2\right)\left(1+x^2\right)$$

$$\leq \left(1+\frac{|y-x|}{\delta}\right)\left(1+\delta^2\right)\Omega\left(f'';\delta\right)\left(1+(y-x)^2\right)\left(1+x^2\right).$$

Since

$$|h(y,x)| \leq \begin{cases} 2\left(1+\delta^2\right)^2\left(1+x^2\right)\Omega\left(f'';\delta\right), & |y-x|<\delta \\ 2\left(1+\delta^2\right)^2\left(1+x^2\right)\frac{(y-x)^4}{\delta^4}\Omega\left(f'';\delta\right), & |y-x|\geq\delta \end{cases}$$

choosing $\delta < 1$, we have

$$\begin{aligned} |h(y,x)| &\leq 2\left(1+\delta^2\right)^2\left(1+x^2\right)\Omega\left(f'';\delta\right)\left(1+\frac{(y-x)^4}{\delta^4}\right) \\ &\leq 8\left(1+x^2\right)\Omega\left(f'';\delta\right)\left(1+\frac{(y-x)^4}{\delta^4}\right). \end{aligned}$$

We deduce that

$$\begin{aligned} & M_n\left((y-x)^2|h(y,x)|\right) \\ &= 8\left(1+x^2\right)\Omega\left(f'';\delta\right)\left(M_n\left((y-x)^2\right)+\frac{1}{\delta^4}M_n\left((y-x)^6\right)\right). \end{aligned}$$

From Lemma 5 we know that

$$M_n\left((y-x)^2\right) = \mathcal{O}\left(\frac{1}{n\alpha_n\psi_n(0)}\right)\left(x^2+x+1\right),$$

and

$$M_n\left((y-x)^6\right) = \mathcal{O}\left(\frac{1}{(n\alpha_n\psi_n(0))^3}\right)\left(x^6+\cdots+x+1\right).$$

Choosing $\delta = \frac{1}{(n\alpha_n\psi_n(0))^{1/2}}$ we have

$$\begin{aligned} & M_n\left((y-x)^2|h(y,x)|\right) \\ &\leq 8\left(1+x^2\right)\Omega\left(f'';\frac{1}{(n\alpha_n\psi_n(0))^{1/2}}\right)\frac{\left(x^6+\cdots x^3+2x^2+2x+2\right)}{n\alpha_n\psi_n(0)} \end{aligned}$$

Corollary 1 *Let $f'' \in C^k_{x^2}[0,\infty)$ and $x>0$ be fixed, $m,n\in\mathbb{N}$. Then there holds*

$$\lim_{n\to\infty} n\alpha_n\psi_n(0)\left[M_n(f;x)-f(x)\right] = (2m\ell_1 x+1)f'(x)+\left(m\ell_1x^2+x\right)f''(x)$$

As a corollary, we reach that the convergence in the Voronovskaya formula is uniform on every compact interval in $(0,\infty)$.

References

1. Aral, A.: Approximation by Ibragimov-Gadjiyev operators in polynomial weighted space. Proc. IMM NAS Azerbaijan **XIX**, 35–44 (2003)
2. Bardaro, C., Mantellini, I.: A quantitative Voronovskaya formula for Mellin convolution operators. Mediterr. J. Math. **7**(4), 483–501 (2010)
3. Coskun, T.: On a construction of positive linear operators for approximation of continuous functions in the weighted spaces. J. Comp. Anal. Appl. **13**(4), 756–770 (2011)
4. Derriennic, M.M.: Sur l'approximation de functions integrable sur [0; 1] par des polynomes de Bernstein modifies. J. Approx. Theory **31**, 323–343 (1981)
5. Dogru, O.: On a certain family linear positive operators. Turk. J. Math. **21**, 387–399 (1997)
6. Dogru, O., On the order of approximation of unbounded functions by the family of generalized linear positive operators. Commun. Fac. Sci. Univ. Ank., Ser. A1, **46**, 173–181 (1997)
7. Durrmeyer, J.L.: Une formule d' inversion de la Transformee de Laplace, Applications a la Theorie des Moments, These de 3e Cycle, Faculte des Sciences de l' Universite deParis (1967)
8. Gadjiev, A.D., Ibragimov, I.I.: On a sequence of linear positive operators. Soviet Math. Dokl. **11**, 1092–1095 (1970)
9. Gadjiev, A.D., İspir, N.: On a sequence of linear positive operators in weighted spaces. Proc. IMM Azerbaijan AS **XI**(XIX), 45–56 (1999)
10. Gonska, H., Pitul, P., Rasa, I.: On Peano's form of the Taylor remainder,Voronovskaja's theorem and the commutator of positive linear operators. In: Proceedings of the International Conference on Numerical Analysis and Approximation Theory, pp. 55–80. Cluj-Napoca, Romania, 5–8 July 2006
11. Gonska, H., Pitul, P., Rasa, I.: On differences of positive linear operators. Carpathian J. Math. **22**(1–2), 65–78 (2006)
12. Gupta, V., Mohapatra, R.N., Finta, Z.: A certain family of mixed summation-integral type operators. Math. Comput. Model. **42**(1–2), 181–191 (2005)
13. Heilmann, M.: Direct and converse results for operators of Baskakov-Durrmeyer type. Approx. Theory Appl. **5**(1), 105–127 (1988)
14. Isir, N.: On modifed Baskakov operators on weighted spaces. Turk. J. Math. **25**, 355–365 (2001)
15. Mazhar, S.M., Totik, V.: Approximation by modified Szasz operators. Acta Sci. Math. **49**, 257–269 (1985)
16. Radatz, P., Wood, B., Approximating derivatives of functions unbounded on the positive axis with lineare operators, Rev. Roum. Math. Pures et Appl., Bucarest, Tome **XXIII**(5), 771–781 (1978)
17. Sahai, A., Prasad, G.: On simultaneous approximation bymodified Lupas operators. J. Approx. Theory **45**(12), 122–128 (1985)
18. Srivastava, H.M., Gupta, V.: A certain family of summation integral type operators. Math. Comput. Model. **37**(12–13), 1307–1315 (2003)
19. Voronovskaya, E.V.: Determination of the asymptotic form of approximation of functions by the polynomials of S.N. Bernstein, Dokl. Akad. Nauk SSSR, A, 79–85 (1932)

Regression-Based Neural Network Simulation for Vibration Frequencies of the Rotating Blade

Atma Sahu and S. Chakravarty

Abstract The aim of this paper is to demonstrate the use of regression-based neural network (RBNN) method to study the problem of the natural frequencies of the rotor blade for micro-unmanned helicopter [3]. The training of the traditional artificial neural network (ANN) model and proposed RBNN model has been implemented in the MATLAB environment using neural network tools (NNT) built-in functions. The graphs for angular velocity (Omega) of the micro-unmanned helicopter are plotted for estimation of the natural frequencies (f1, f2, f3) of transverse vibrations. The results obtained in this research show that the RBNN model, when trained, can give the vibration frequency parameters directly without going through traditional and lengthy numerical solutions procedures. Succeeding this, the numerical results, when plotted, show that with the increase in Omega, there is increase in lagging motion frequencies. Additionally, it is found that the increase in the lower mode natural frequencies is smaller than that of the higher modes. This finding is in agreement with the results reported in earlier research [3–5] carried out by employing Rayleigh–Ritz and FEM, respectively.

Keywords Transverse vibrations · Artificial neural network · Harmonic motion · Mean square error · Micro-unmanned helicopters · Rotor blade vibrations

A. Sahu (✉)
Professor of Mathematics, Coppin State University, Baltimore, MD 21216, USA
e-mail: asahu@coppin.edu

S. Chakravarty
Professor & Head Mathematics Department, National Institute of Technology, Rourkela 769 008, Orissa, India
e-mail: sne_chak@yahoo.com

V.K. Singh et al. (eds.), *Modern Mathematical Methods and High Performance Computing in Science and Technology*, Springer Proceedings in Mathematics & Statistics 171, DOI 10.1007/978-981-10-1454-3_2

1 Introduction

The micro-unmanned helicopters are quite different from the conventional manned helicopters in their design scheme. Therefore, in the case of micro-unmanned helicopter, the rotor mechanism is altered in order to optimize the manufacturing costs [3] without compromising on its needed functionality. In this paper, however, for the prototype engineering design requirements, the vibrations of helicopter rotor blades, whether manned or unmanned, are of a major concern. The purpose of this paper is to use a regression-based neural network (RBNN) method [1] to solve the problem of studying the natural frequencies of the rotor blade for micro-unmanned helicopter [3]. With this R&D effort resulting in appropriate mathematical calculations, the design engineers are able to overcome blade resonance problems (maybe by putting damper on the blade or any other vibrations correction method). The authors choose not to go into the fluid (air) resistance motion problem of blades' airfoil system (Appendix Fig. 5).

2 Transverse Vibrations Analysis

In this paper, the rotor manipulation mechanism is based on the use of the inertia characteristic of the rotor and its elastic features as considered by J Lu [3]. Also, an equally important characteristic in rotor parametric manipulation is the blade shape change that can be affected by the leading and trailing edges of the entire airfoil system (Appendix Fig. 8). However, the authors in this research paper will limit the scope to RBNN-based analysis of the transverse vibrations of the rotor blade. Also, it is reasonable to assume that the blade length is very large compared to its width. For this reason, Euler–Bernoulli beam theory is adequate for our purposes. The scheme of the blade and notations (see Appendix Fig. 8) is adopted in this paper from Lü [3] to make comparisons of this work easier and comprehensible. In this paper, ANN model for blade vibrations has been undertaken. The training of network is performed using the pattern calculated with the help of Boundary Characteristic Orthogonal Polynomials (BCOPs) in the Rayleigh–Ritz method.

We adopt below the kinetic energy (KE) and potential energy (PE) equations as derived by Lü et al. [3]. Considering the conditions of small deflections, the KE of the rotor blade of length L is given by T as follows:

$$T = \frac{1}{2}\int_0^L \rho\left\langle \dot{u}^2 + \left[u^2 + (a+x)^2\right]\dot{\theta} + 2\dot{\theta}\dot{u}(a+x)\right\rangle dx;$$

and PE is given by U as follows:

$$U = \frac{1}{2}\int_0^L EIu_x^2 dx + \frac{1}{2}\int_0^L \left\langle \frac{1}{2}\rho\dot{\theta}^2(L^2 - x^2) + \rho\dot{\theta}^2 a(L-x)\right\rangle (u_x)^2 dx$$

For harmonic motion, the blade deflection is given by $u(x,t) = Y(x)\sin(\omega t + \phi)$; using u(x) in T and U above, Lagrangian is obtained. Following notations are used:

L = Length of the blade (m), ρ = mass in unit length of the blade (kg/m), $\omega = \dot{\theta}$ = angular velocity of rotor, EI = flexural rigidity of the blade (N^{m^2}), XO'Y = Inertia reference frame, xOu = Flying reference frame, V_{ji} = Hidden layer weights, and W_{kj} = Output layer weights.

Substituting the linear combination of BCOPs in the Rayleigh–Ritz method for T and U, we may turn it to a standard eigenvalue problem. The solution of the standard eigenvalue problem then gives the natural frequencies at various rotational speeds [2]. The computations have been carried out by taking EI = 1.392 Nm, L = 0.15 m, ρ = 0.1260 Kg/m. [2]. As such natural frequencies have been computed for the blade at various rotational speeds for the simulation in RBNN model. In the following paragraphs, ANN architecture is described for the estimation of natural frequencies for given values of Omega which is the angular velocity parameter.

3 Identification of the RBNN Model: Solution Technique

Three-layer architecture for regression-based artificial neural network approach is considered here to understand the proposed model for solving the present problem. Figure 1 show the neural network used in the process. The input layer consists of single input as Omega and the output layer consists of three outputs in the form of the corresponding frequency parameters f1, f2, and f3. Three cases of the number of nodes depending upon the proposed parameter of the methodology have been considered in the hidden layer to facilitate a comparative study on the architecture of the network. The output of the network is computed by regression analysis combined

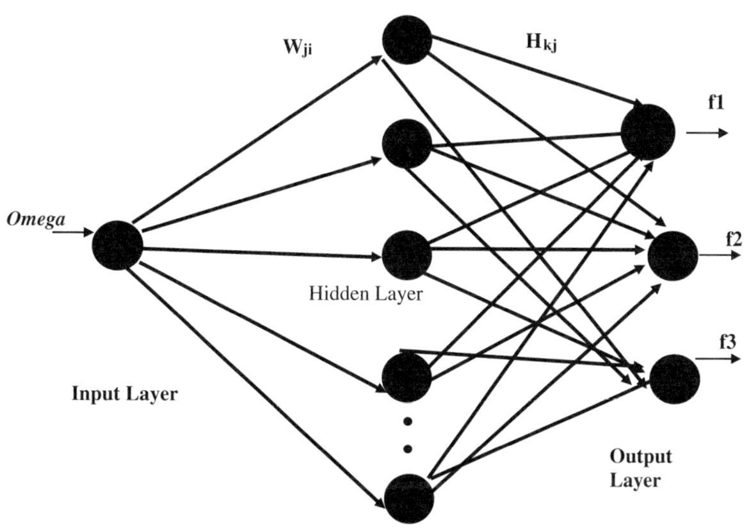

Fig. 1 ANN architecture used for estimation of frequencies for given values of Omega

with neural activation function performed at two stages, i.e., the stage of hidden layer and the stage of output layer. Number of neurons in the hidden layer depends upon the degree of the regression polynomial that is used to fit the data between input and output. If we consider a polynomial of degree n, then number of nodes in hidden layer will be (n + 1) and the corresponding (n + 1) coefficients of this polynomial (say, $a_i, i = 0, 1, \ldots, n$) are taken as the initial weights from input layer to the hidden layer (H_{kj}). Architecture of the network for a polynomial of nth degree is shown in Fig. 1.

4 Numerical Results

The training of the traditional artificial neural network (ANN) model and proposed RBNN model has been implemented in the MATLAB environment using neural network tools (NNT) built-in functions. Also, in the following paragraphs, the graphs for angular velocities (Omega) of the micro-unmanned helicopter are plotted for estimation of the natural frequencies (f1, f2, f3).

4.1 The Experiment 1

The training of the traditional ANN model and proposed RBNN model has been implemented for estimation of the frequencies with respect to omega values. In the traditional model, the output of the network is computed by built-in transfer functions, namely, tan-sigmoid (tansig) and linear (purelin) of the neural network tool (NNT) performed at two stages, i.e., the stage of hidden layer and the stage of output layer. The connection weights interconnecting the neurons between different layers are taken through a random number generator built-in function in the NNT. The neural network based on this feedforward back propagation algorithm has been trained with Levenberg–Marquardt training function of the NNT.

4.2 The Experiment 2

In proposed RBNN model, regression polynomials of degree three are fitted to the training patterns. The coefficients of this polynomial are taken as the connecting weights for the hidden layer, as described earlier. The output of the neurons in the hidden layer is calculated using activation function. At this stage, the error of the RBNN model is calculated and a decision is taken as to whether the network has been trained or not. If the tolerance level of the error is not achieved, the procedure is repeated; otherwise, we say that the network has got trained. In this case, the network has been converged with the desired accuracy as shown in the Fig. 1 for

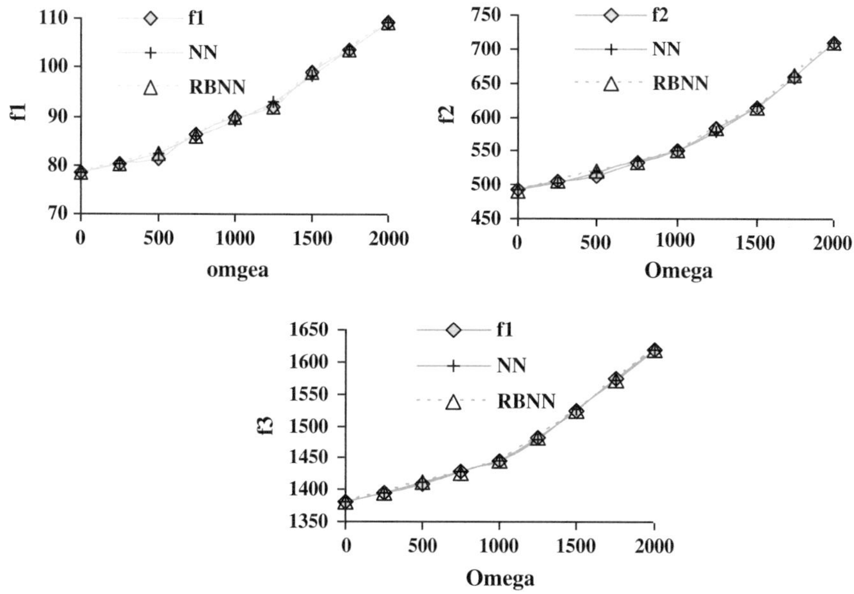

Fig. 2 Results of training of *RBNN* and *ANN* models for Omega verus Frequency F1, F2, and F3 (D-3)

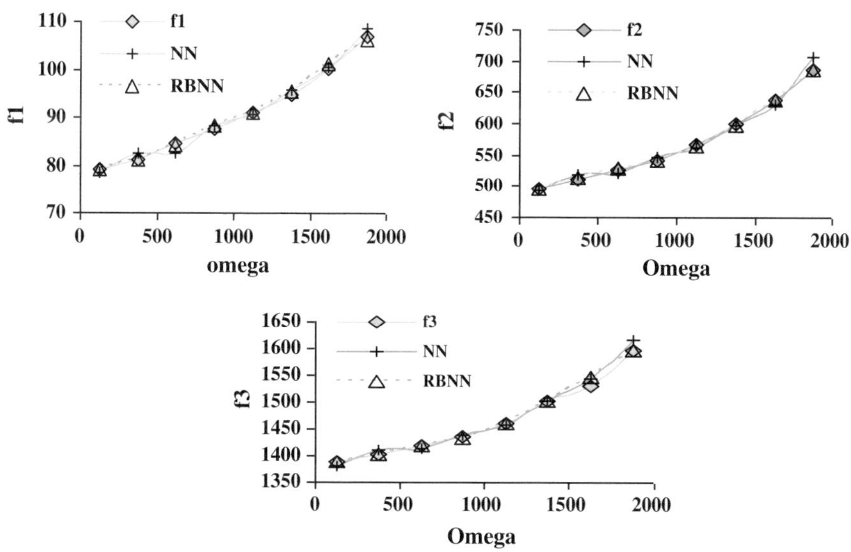

Fig. 3 Performance of the proposed regression-based neural network for Omega versus Frequency F1, F2, and F3 (D-3)

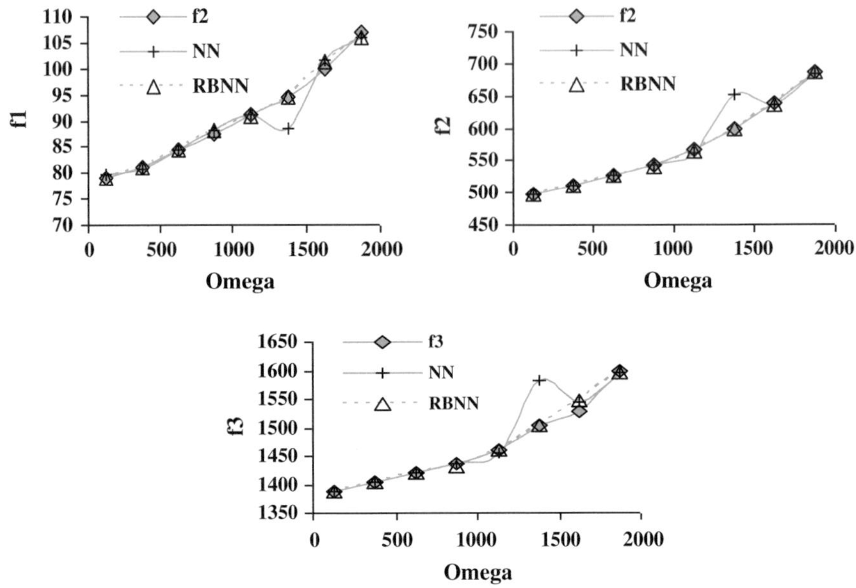

Fig. 4 Performance of the proposed regression-based neural network for Omega versus Frequency F1, F2, and F3 (D-4)

the problems under consideration. The output of the network f1, f2, and f3 and the mean square error (MSE) between neural and desired output are calculated. In this figure, f1, f2, and f3 represent the desired output values, NN represents these values obtained by the traditional ANN models, and RBNN represents the values of these parameters obtained from the proposed model with four nodes in the hidden layer. The performance of the proposed model is given in the Fig. 2. The pattern characteristics of the traditional ANN model and RBNN model for degree four are incorporated in Fig. 3. The performance of the proposed model for degree four is given in the Fig. 4.

5 Conclusion

The RBNN method employed to solve fourth order partial differential equation for rotor blade in this paper gives a direct estimation of frequencies without going through traditional and lengthy numerical solutions procedures. The numerical results, when plotted, show that with the increase in Omega (angular velocity), there is increase in lagging motion frequencies. The increase in the lower mode natural frequencies is smaller than that of the higher modes. This finding is in agreement with the results reported in earlier researches [3–5] that have been carried out by employing Rayleigh–Ritz and FEM, respectively. Furthermore, RBNN soft computing method used in this research is useful to solve other beam, plates, and shell vibration problems and guide engineers immensely in their structures design needs. Last of all, NN methods in general [2, 6] have attracted extensive attention in recent past as NN approaches have led many efficient algorithms help in exploring the intrinsic structure of data set.

Appendix

See Appendix Fig. 5.

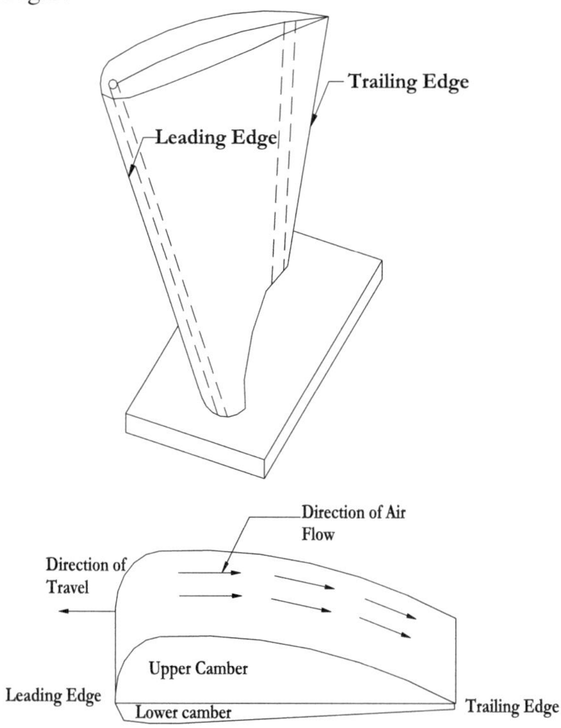

Leading and trailing edges of a blade

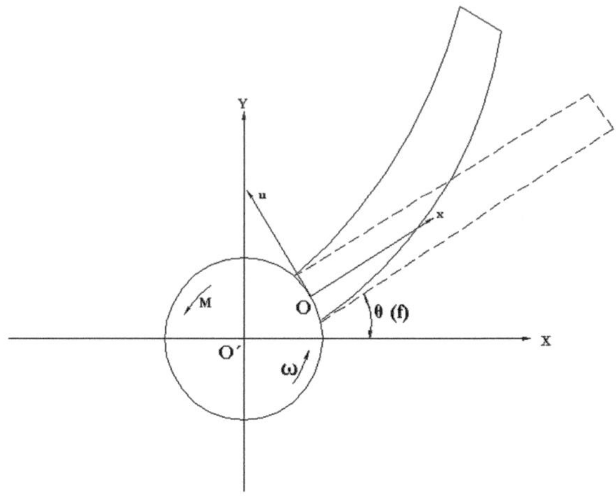

Fig. 5 Helicopter blade scheme

References

1. Chakravarty, S., Singh, V.P., Sharma, R.K.: Regression-based weight generation algorithm in network for estimation of frequencies of vibrating plates. Comput. Methods Appl. Mech. Eng. J. **195**, 4194–4202 (2006)
2. Kumar, S., MittaL, G.: Rapid detection of microorganism using image processing parameters and neuro network. Food Bio Process Technol. **3**, 741–751 (2010)
3. Lü, J., Jiadao, W., Chan, D.: Transverse vibration of blade for unmanned micro helicopter using Rayleigh–Ritz method. Univ. Sci. Technol. Beijing J. **10**(6), 40–43 (2003)
4. Rao, R.S., Gupta, S.S.: Finite element vibration analysis of rotating Timoshenko beams. Sound Vib. J. **242**(1), 103–124 (2001)
5. Sahu, A.: Theoretical frequency equation of bending vibrations of an exponentially tapered beam under rotation. J. Vib. Control **7**(6), 775–780 (2001)
6. Wang, J., Liao, X., Yi, Z. (eds.): Advances in Neural Networks, International Symposium on Neural Network Proceeding. Springer, Heidelberg (2005). NY 2005

Approximation by a New Sequence of Operators Involving Charlier Polynomials with a Certain Parameter

D.K. Verma and Vijay Gupta

Abstract In the present paper, we propose a certain integral modification of the operator, which involve Charlier polynomials with the weight function of generalized Baskakov and Szász basis functions. We estimate some approximation properties and asymptotic formula for these operators. Also, the weighted approximation for these is given.

Keywords Charlier polynomials · Modulus of continuity · Voronovskaja-type asymptotic formula · Weighted approximation

AMS MSC 2010 41A25 · 41A30

1 Introduction

Very recently, Varma and Taşdelen [12] introduced the Szász-type operators involving Charlier polynomials (1.1). Also, they estimated some results for the Kantorovich-type generalization of these operators and established the convergence properties for their operators with the help of Korovkin's theorem and the order of approximation by using the classical modulus of continuity. The operators discussed in [12] are defined as

$$L_n(f;x,a) = e^{-1}\left(1-\frac{1}{a}\right)^{(a-1)nx}\sum_{k=0}^{\infty}\frac{C_k^{(a)}(-(a-1)nx)}{k!}f\left(\frac{k}{n}\right) \tag{1.1}$$

D.K. Verma (✉)
Department of Mathematics, Ramjas College, University of Delhi,
New Delhi 110007, India
e-mail: durvesh.kv.du@gmail.com

V. Gupta
Department of Mathematics, Netaji Subhash Institute of Technology,
Sector 3 Dwarka, New Delhi 110078, India
e-mail: vijaygupta2001@hotmail.com

V.K. Singh et al. (eds.), *Modern Mathematical Methods and High Performance Computing in Science and Technology*, Springer Proceedings in Mathematics & Statistics 171, DOI 10.1007/978-981-10-1454-3_3

where $a > 0$, $x \in [0, \infty)$ and $C_k^{(a)}$ be the Charlier polynomials, which have the generating functions of the type

$$e^t\left(1-\frac{t}{a}\right)^x = \sum_{k=0}^{\infty} \frac{C_k^{(a)}(x)}{k!} t^k, \quad |t| < a,$$

and the explicit representation

$$C_k^{(a)}(u) = \sum_{r=0}^{k} \binom{n}{r} (-u)_r \left(\frac{1}{a}\right)_r,$$

where $(\alpha)_k$ is the Pochhammer's symbol given by

$$(\alpha)_0 = 1, \quad (\alpha)_r = \alpha(\alpha+1)\ldots(\alpha+r-1) \qquad r = 1, 2, \ldots$$

Note that Charlier polynomials are positive if $a > 0$, $u \leq 0$.

In order to approximate Lebesgue integrable functions, several new modifications of the discrete operators were discovered by the researchers in the last five decades. We mention the recent book [8] for some of the work on the integral operators in this direction and the references therein. Some other integral operators we mention in the papers [5, 6, 9], etc.

Also, recently with an idea of generalization of the Phillips operators [11] (see also [2, 3, 7]), Păltănea in [10] proposed the modified form of the Phillips operators based on certain parameter $\rho > 0$, which provide the link with the well-known Szász–Mirakyan operators as $\rho \to \infty$ for some class of functions. Motivated by such modifications we propose here for $a > 0$, $\rho \geq 0$ the integral-type generalization of the operator (1.1) as follows:

$$T_{n,\rho,c}(f; x, a) = e^{-1}\left(1-\frac{1}{a}\right)^{(a-1)nx}\Bigg[C_0^{(a)} f(0) + \sum_{k=1}^{\infty} \frac{C_k^{(a)}(-(a-1)nx)}{k!} \int_0^{\infty} \Theta_{n,k}^{\rho}(t, c) f(t)dt\Bigg] \tag{1.2}$$

where $C_k^{(a)}(u)$ is the Charlier polynomial and

$$\Theta_{n,k}^{\rho}(t, c) = \begin{cases} \frac{n\rho}{\Gamma(k\rho)} e^{-n\rho t} (n\rho t)^{k\rho-1}, & c = 0 \\ \frac{\Gamma(\frac{n\rho}{c}+k\rho)}{\Gamma(k\rho)\Gamma(\frac{n\rho}{c})} \frac{c^{k\rho} t^{k\rho-1}}{(1+ct)^{\frac{n\rho}{c}+k\rho}}, & c = 1, 2, 3, \ldots, \end{cases}$$

Remark 1 For $f \in \overline{\Pi}$, where $\overline{\Pi}$ be the closure of the space of polynomials, we have

$$\lim_{\rho\to\infty} T_{n,\rho,c}(f; x, a) = L_n(f; x, a), \quad \text{for all } x \in [0, \infty).$$

Since,

$$\int_0^\infty \Theta_{n,k}^{\rho}(t, c)t^r dt = \frac{\Gamma(k\rho+r)}{\Gamma(k\rho)} \frac{1}{\prod_{i=1}^{r}(n\rho - ic)}, \quad n\rho > rc$$

and

$$\lim_{\rho\to\infty} \frac{\Gamma(k\rho+r)}{\Gamma(k\rho)} \frac{1}{\prod_{i=1}^{r}(n\rho - ic)} = \left(\frac{k}{n}\right)^r, \quad n\rho > rc.$$

From this the result follows immediately.

Remark 2 We obtain Szász–Mirakyan operators by applying, respectively, the following operations to the both sides of (1.2)

(i) $\rho \to \infty$,
(ii) $a \to \infty$ and write $x - \frac{1}{n}$ instead of x.

In the present article, we first obtain the moments of the operators $T_{n,\rho,c}(f; x, a)$. Then we establish some direct results in ordinary approximation, which include the asymptotic formula, direct estimate in terms of modulus of continuity and the weighted approximation.

2 Auxiliary Results

In this section we provide the following set of lemmas.

Lemma 1 ([12]) *For $L_n(t^m; x, a)$, $m = 0, 1, 2$, we have*

$$\begin{aligned} L_n(1; x, a) &= 1, \\ L_n(t; x, a) &= x + \frac{1}{n} \\ L_n(t^2; x, a) &= x^2 + \frac{x}{n}\left(3 + \frac{1}{a-1}\right) + \frac{2}{n^2}. \end{aligned}$$

Lemma 2 *For* $T_{n,\rho,c}(t^m;x,a)$, $m=0,1,2$, *we have*

$$\begin{aligned}
T_{n,\rho,c}(1;x,a) &= 1,\\
T_{n,\rho,c}(t;x,a) &= \frac{\rho(nx+1)}{n\rho-c}\\
T_{n,\rho,c}(t^2;x,a) &= \frac{n^2}{(n\rho-c)(n\rho-2c)}\left[\rho^2x^2+\frac{\rho x}{n}\left(3\rho+1+\frac{\rho}{a-1}\right)+\frac{\rho(2\rho+1)}{n^2}\right]
\end{aligned}$$

Proof It is easy to see

$$\int_0^\infty \Theta_{n,k}^{\rho}(t,c)t^r\,dt=\frac{\Gamma(k\rho+r)}{\Gamma(k\rho)}\frac{1}{\prod_{i=1}^{r}(n\rho-ic)}.$$

In view of Lemma 1, the zeroth order moment is

$$\begin{aligned}
T_{n,\rho,c}(1;x,a) &= e^{-1}\left(1-\frac{1}{a}\right)^{(a-1)nx}\left(C_0^{(a)}f(0)+\sum_{k=1}^{\infty}\frac{C_k^{(a)}(-(a-1)nx)}{k!}\int_0^\infty \Theta_{n,k}^{\rho}(t)dt\right)\\
&= L_n(1;x,a)=1.
\end{aligned}$$

First-order moment is

$$\begin{aligned}
T_{n,\rho,c}(t;x,a) &= e^{-1}\left(1-\frac{1}{a}\right)^{(a-1)nx}\sum_{k=1}^{\infty}\frac{C_k^{(a)}(-(a-1)nx)}{k!}\int_0^\infty \Theta_{n,k}^{\rho}(t)t\,dt\\
&= e^{-1}\left(1-\frac{1}{a}\right)^{(a-1)nx}\sum_{k=1}^{\infty}\frac{C_k^{(a)}(-(a-1)nx)}{k!}\frac{\Gamma(k\rho+1)}{\Gamma(k\rho)}\frac{1}{(n\rho-c)}\\
&= \frac{n\rho}{(n\rho-c)}e^{-1}\left(1-\frac{1}{a}\right)^{(a-1)nx}\sum_{k=0}^{\infty}\frac{C_k^{(a)}(-(a-1)nx)}{k!}\frac{k}{n}\\
&= \frac{n\rho}{(n\rho-c)}L_n(t;x,a)\\
&= \frac{\rho(nx+1)}{n\rho-c}.
\end{aligned}$$

Second-order moment is

$$\begin{aligned}
T_{n,\rho,c}(t^2;x,a) &= e^{-1}\left(1-\frac{1}{a}\right)^{(a-1)nx}\sum_{k=1}^{\infty}\frac{C_k^{(a)}(-(a-1)nx)}{k!}\int_0^\infty \Theta_{n,k}^{\rho}(t)t^2dt\\
&= e^{-1}\left(1-\frac{1}{a}\right)^{(a-1)nx}\sum_{k=1}^{\infty}\frac{C_k^{(a)}(-(a-1)nx)}{k!}\frac{\Gamma(k\rho+2)}{\Gamma(k\rho)}\frac{1}{(n\rho-c)(n\rho-2c)}
\end{aligned}$$

$$= \frac{n^2\rho^2}{(n\rho-c)(n\rho-2c)}e^{-1}\left(1-\frac{1}{a}\right)^{(a-1)nx}\sum_{k=0}^{\infty}\frac{C_k^{(a)}(-(a-1)nx)}{k!}\frac{k^2}{n^2}$$
$$+\frac{n\rho}{(n\rho-c)(n\rho-2c)}e^{-1}\left(1-\frac{1}{a}\right)^{(a-1)nx}\sum_{k=0}^{\infty}\frac{C_k^{(a)}(-(a-1)nx)}{k!}\frac{k}{n}$$
$$= \frac{n^2\rho^2}{(n\rho-c)(n\rho-2c)}L_n(t^2;x,a)+\frac{n\rho}{(n\rho-c)(n\rho-2c)}L_n(t;x,a)$$
$$= \frac{n^2}{(n\rho-c)(n\rho-2c)}\left[\rho^2x^2+\frac{\rho x}{n}\left(3\rho+1+\frac{\rho}{a-1}\right)+\frac{\rho(2\rho+1)}{n^2}\right].$$

Remark 3 By simple computation, we have

$$T_{n,\rho,c}(t-x;x,a) = \frac{cx+\rho}{n\rho-c}$$
$$T_{n,\rho,c}((t-x)^2;x,a) = \frac{1}{(n\rho-c)(n\rho-2c)}\Bigg[(n\rho+2c)cx^2+n\rho x\left(\rho+1+\frac{\rho}{a-1}\right)$$
$$+4\rho cx+\rho(2\rho+1)\Bigg]$$

3 Direct Result and Asymptotic Formula

In this section we discuss the direct result and Voronovskaja-type asymptotic formula.

Let the space $C_B[0,\infty)$ of all continuous and bounded functions be endowed with the norm $\|f\| = \sup\{|f(x)| : x \in [0,\infty)\}$. Further let us consider the following K-functional:

$$K_2(f,\delta) = \inf_{g\in W^2}\{\|f-g\|+\delta\|g''\|\},$$

where $\delta > 0$ and $W^2 = \{g \in C_B[0,\infty) : g', g'' \in C_B[0,\infty)\}$. By ([1] p. 177, Th. 2.4), there exists an absolute constant $C > 0$ such that

$$K_2(f,\delta) \le C\,\omega_2(f,\sqrt{\delta}), \tag{3.1}$$

where

$$\omega_2(f,\sqrt{\delta}) = \sup_{0<h\le\sqrt{\delta}}\ \sup_{x\in[0,\infty)}|f(x+2h)-2f(x+h)+f(x)|$$

is the second-order modulus of smoothness of $f \in C_B[0,\infty)$.

Theorem 1 *For $f \in C_B[0,\infty)$ and $a > 1$, we have*

$$|T_{n,\rho,c}(f;x,a) - f(x)| \le C\omega_2(f,\sqrt{\delta}) + \omega\left(f, \left|\frac{cx+\rho}{n\rho - c}\right|\right)$$

where C is a positive constant and $\delta = \left|T_{n,\rho,c}((t-x)^2;x,a)\right| + \frac{1}{2}\left(\frac{cx+\rho}{n\rho-c}\right)^2$. Also, the both $\omega(f,\delta)$ and $\omega_2(f,\sqrt{\delta})$ tends to zero as $\delta \to 0$.

Proof We introduce auxiliary operators $\overline{T}_{n,\rho,c}$ as follows:

$$\overline{T}_{n,\rho,c}(f;x,a) = T_{n,\rho,c}(f;x,a) - f\left(x + \frac{cx+\rho}{n\rho - c}\right) + f(x).$$

These operators are linear and preserve the linear functions in view of Lemma 2. Let $g \in W^2$. From the Taylor's expansion of g we have

$$g(t) = g(x) + (t-x)g'(x) + \int_x^t (t-u)g''(u)du.$$

Applying the operator $\overline{T}_{n,\rho,c}$ on above

$$\overline{T}_{n,\rho,c}(g;x,a) = g(x) + g'(x)\overline{T}_{n,\rho,c}((t-x);x,a) + \overline{T}_{n,\rho,c}\left(\int_x^t (t-u)g''(u)du;x,a\right)$$

$$\begin{aligned}
|\overline{T}_{n,\rho,c}(g;x,a) - g(x)| &= \left|\overline{T}_{n,\rho,c}\left(\int_x^t (t-u)g''(u)du;x,a\right)\right| \\
&\le \left|T_{n,\rho,c}\left(\int_x^t (t-u)g''(u)du;x,a\right)\right| \\
&\quad + \left|\int_x^{x+\frac{cx+\rho}{n\rho-c}} \left(x + \frac{cx+\rho}{n\rho-c} - u)g''(u)\right) du\right| \\
&\le \left[\left|T_{n,\rho,c}\left(\left|\int_x^t |t-u|\,du\right|;x,a\right)\right|\right. \\
&\quad \left. + \left|\int_x^{x+\frac{cx+\rho}{n\rho-c}} \left|x + \frac{cx+\rho}{n\rho-c} - u\right| du\right|\right] \|g''\| \\
&\le \left[\left|T_{n,\rho,c}((t-x)^2;x,a)\right| + \frac{1}{2}\left(\frac{cx+\rho}{n\rho-c}\right)^2\right] \|g''\| && (3.2) \\
&= \delta\|g''\|, && (3.3)
\end{aligned}$$

where $\delta = \left|T_{n,\rho,c}((t-x)^2;x,a)\right| + \frac{1}{2}\left(\frac{cx+\rho}{n\rho-c}\right)^2$.

$$
\begin{aligned}
|T_{n,\rho,c}(f;x,a)-f(x)| &\le |\overline{T}_{n,\rho,c}(f-g;x,a)-(f-g)(x)|+|\overline{T}_{n,\rho,c}(g;x,a)-g(x)| \\
&\quad +\left|f\left(x+\frac{cx+\rho}{n\rho-c}\right)-f(x)\right| \\
&\le 2\|f-g\|+\delta\|g''\|+\omega\left(f,\left|\frac{cx+\rho}{n\rho-c}\right|\right).
\end{aligned}
$$

Taking infimum over all $g\in W^2$, we get

$$
|T_{n,\rho,c}(f;x,a)-f(x)|\le 2K_2(f,\delta)+\omega\left(f,\left|\frac{cx+\rho}{n\rho-c}\right|\right).
$$

In view of (3.1), we obtain

$$
|T_{n,\rho,c}(f;x,a)-f(x)|\le C\omega_2(f,\sqrt{\delta})+\omega\left(f,\left|\frac{cx+\rho}{n\rho-c}\right|\right),
$$

which proves the theorem.

Our next result in this section is the Voronovskaja-type asymptotic formula:

Theorem 2 *For any function $f\in C_B[0,\infty)$ and $a>1$ such that $f',\ f''\in C_B[0,\infty)$, we have*

$$
\lim_{n\to\infty} n[T_{n,\rho,c}(f;x,a)-f(x)]=\frac{cx+\rho}{\rho}f'(x)+\frac{x}{2\rho}\left(cx+\rho+1+\frac{\rho}{a-1}\right)f''(x)
$$

for every $x\ge 0$.

Proof Let $f,\ f',\ f''\in C_B[0,\infty)$ and $x\in[0,\infty)$ be fixed. By Taylor expansion we can write

$$
f(t)=f(x)+(t-x)f'(x)+\frac{(t-x)^2}{2!}f''(x)+r(t,x)(t-x)^2,
$$

where $r(t,x)$ is the Peano form of the remainder, $r(t,x)\in C_B[0,\infty)$ and $\lim_{t\to x} r(t,x)=0$. Applying $T_{n,\rho,c}$, we get

$$
\begin{aligned}
n[T_{n,\rho,c}(f;x,a)-f(x)] &= f'(x)nT_{n,\rho,c}(t-x;x,a)+\frac{f''(x)}{2!}nT_{n,\rho,c}((t-x)^2;x,a) \\
&\quad +nT_{n,\rho,c}(r(t,x)(t-x)^2;x,a)
\end{aligned}
$$

$$\begin{aligned}
\lim_{n\to\infty} n[T_{n,\rho,c}(f;x,a)-f(x)] &= f'(x)\lim_{n\to\infty} nT_{n,\rho,c}(t-x;x,a)\\
&+\frac{f''(x)}{2!}\lim_{n\to\infty}(x)T_{n,\rho,c}((t-x)^2;x,a)\\
&+\lim_{n\to\infty} nT_{n,\rho,c}(r(t,x)(t-x)^2;x,a)\\
&=\frac{cx+\rho}{\rho}f'(x)+\frac{x}{2\rho}\left(cx+\rho+1+\frac{\rho}{a-1}\right)f''(x)\\
&+\lim_{n\to\infty} nT_{n,\rho,c}\left(r(t,x)(t-x)^2;x,a\right)\\
&=\frac{cx+\rho}{\rho}f'(x)+\frac{x}{2\rho}\left(cx+\rho+1+\frac{\rho}{a-1}\right)f''(x)+E.
\end{aligned}$$

By Cauchy–Schwarz inequality, we have

$$|E|\le \lim_{n\to\infty} nT_{n,\rho,c}(r^2(t,x);x,a)^{1/2}T_{n,\rho,c}\left((t-x)^4;x,a\right)^{1/2}. \tag{3.4}$$

It is easy to show that $T_{n,\rho,c}\left((t-x)^4;x,a\right)^{1/2}$ is bounded for $x\in[0,A]$. Also, observe that $r^2(x,x)=0$ and $r^2(.,x)\in C_B[0,\infty)$. Then, it follows that

$$\lim_{n\to\infty} nT_{n,\rho,c}(r^2(t,x);x,a)=r^2(x,x)=0 \tag{3.5}$$

uniformly with respect to $x\in[0,A]$. Now from (3.4), (3.5) we obtain

$$\lim_{n\to\infty} nT_{n,\rho,c}(r(t,x)(t-x)^2;x,a)=0.$$

Hence, $E=0$. Thus, we have

$$\lim_{n\to\infty} n[T_{n,\rho,c}(f;x,a)-f(x)]=\frac{cx+\rho}{\rho}f'(x)+\frac{x}{2\rho}\left(cx+\rho+1+\frac{\rho}{a-1}\right)f''(x),$$

which completes the proof.

4 Weighted Approximation

Let $B_{x^2}[0,\infty)=\{f:$ for every $x\in[0,\infty)$, $|f(x)|\le M_f(1+x^2)$, M_f being a constant depending on $f\}$. By $C_{x^2}[0,\infty)$, we denote the subspace of all continuous functions belonging to $B_{x^2}[0,\infty)$. Also, $C^*_{x^2}[0,\infty)$ is subspace of all functions $f\in C_{x^2}[0,\infty)$ for which $\lim\limits_{x\to\infty}\dfrac{f(x)}{1+x^2}$ is finite. The norm on $C^*_{x^2}[0,\infty)$ is $\|f\|_{x^2}=\sup\limits_{x\in[0,\infty)}\dfrac{|f(x)|}{1+x^2}$.

Theorem 3 *For each* $f \in C^*_{x^2}[0,\infty)$, *we have*

$$\lim_{n\to\infty} \|T_{n,\rho,c}(f;.,a) - f\|_{x^2} = 0$$

Proof Using [4] we see that it is sufficient to verify the following conditions

$$\lim_{n\to\infty} \|T_{n,\rho,c}(t^\nu;x,a) - x^\nu\|_{x^2} = 0, \nu = 0,1,2. \tag{4.1}$$

Since $T_{n,\rho,c}(1;x,a) = 1$,therefore for $\nu = 0$ (4.1) holds.
By Lemma 2 for $n > \frac{c}{\rho}$, we have

$$\begin{aligned}\|T_{n,\rho,c}(t;x,a) - x\|_{x^2} &= \sup_{x\in[0,\infty)} \frac{|T_{n,\rho,c}(t;x,a) - x|}{1+x^2}\\ &\le \left(\frac{n\rho}{n\rho - c} - 1\right) \sup_{x\in[0,\infty)} \frac{x}{1+x^2} + \frac{\rho}{n\rho - c}\\ &\le \left(\frac{c+2\rho}{2(n\rho - c)}\right),\end{aligned}$$

the condition (4.1) holds for $\nu = 1$ as $n \to \infty$.
Again by Lemma 2 for $n > \frac{2c}{\rho}$, we have

$$\begin{aligned}\|T_{n,\rho,c}(t^2;x,a) - x^2\|_{x^2} &= \sup_{x\in[0,\infty)} \frac{|T_{n,\rho,c}(t^2;x,a) - x^2|}{1+x^2}\\ &\le \left|\frac{n^2\rho^2}{(n\rho-c)(n\rho-2c)} - 1\right| \sup_{x\in[0,\infty)} \frac{x^2}{1+x^2}\\ &+ \frac{n\rho}{(n\rho-c)(n\rho-2c)}\left(3\rho+1+\frac{\rho}{a-1}\right) \sup_{x\in[0,\infty)} \frac{x}{1+x^2}\\ &+ \left.\frac{\rho(2\rho+1)}{(n\rho-c)(n\rho-2c)}\right]\\ &\le \left|\frac{n^2\rho^2}{(n\rho-c)(n\rho-2c)} - 1\right|\\ &+ \frac{n\rho}{(n\rho-c)(n\rho-2c)}\left(3\rho+1+\frac{\rho}{a-1}\right) + \frac{\rho(2\rho+1)}{(n\rho-c)(n\rho-2c)},\end{aligned}$$

the condition (4.1) holds for $\nu = 2$ as $n \to \infty$.
Hence the theorem.

Corollary 1 *For each* $f \in C_{x^2}[0,\infty)$, $a > 1$ *and* $\alpha > 0$, *we have*

$$\lim_{n\to\infty} \sup_{x\in[0,\infty)} \frac{|T_{n,\rho,c}(f;x,a) - f(x)|}{(1+x^2)^\alpha} = 0.$$

Proof For any fixed $x_0 > 0$,

$$\sup_{x\in[0,\infty)} \frac{|T_{n,\rho,c}(f;x,a)-f(x)|}{(1+x^2)^{1+\alpha}} \leq \sup_{x\leq x_0} \frac{|T_{n,\rho,c}(f;x,a)-f(x)|}{(1+x^2)^{1+\alpha}} + \sup_{x\geq x_0} \frac{|T_{n,\rho,c}(f;x,a)-f(x)|}{(1+x^2)^{1+\alpha}}$$
$$\leq \|T_{n,\rho,c}(f;.,a)-f\|_{C[0,x_0]} + \|f\|_{x^2} \sup_{x\geq x_0} \frac{|T_{n,\rho,c}(1+t^2;x,a)|}{(1+x^2)^{1+\alpha}}$$
$$+ \sup_{x\geq x_0} \frac{|f(x)|}{(1+x^2)^{1+\alpha}}.$$

The first term of the above inequality tends to zero from Theorem 1. By Lemma 2 for any fixed x_0 it is easily seen that $\sup_{x\geq x_0} \frac{|T_{n,\rho,c}(1+t^2;x,a)|}{(1+x^2)^{1+\alpha}} \leq \frac{M}{(1+x_0^2)^{\alpha}}$ for some positive constant M independent of x. We can choose x_0 so large that the right-hand side of the former inequality and last part of above inequality can be made small enough.

Thus the proof is completed.

Acknowledgments Authors are thankful to the referees for valuable suggestions, leading to an overall better presentation in the paper.

References

1. De Vore, R.A., Lorentz, G.G.: Constructive Approximation. Springer, Berlin (1993)
2. Finta, Z.: On converse approximation theorems. J. Math. Anal. Appl. **312**(1), 159–180 (2005)
3. Finta, Z., Gupta, V.: Direct and inverse estimates for Phillips type operators. J. Math. Anal. Appl. **303**(2), 627–642 (2005)
4. Gadzjiv, A.D.: Theorems of the type of P.P.Korovkin type theorems, Math. Zametki, **20**(5), 781–786 (1976). English Translation, Math. Notes **20**(5–6), 996–998 (1976)
5. Govil, N.K., Gupta, V., Soyabos, D.: Certain New classes of Durrmeyer Type operators. Appl. Math. Comput. **225**, 195–203 (2013)
6. Gupta, V.: Direct estimates for a new general family of Durrmeyer type operators. Bollettino dell'Unione Matematica Italiana **7**, 279–288 (2015)
7. Gupta, V.: Rate of convergence by the Bézier variant of Phillips operators for bounded variation functions. Taiwanese J. Math. **8**(2), 183–190 (2004)
8. Gupta, V., Agrawal, R.P.: Convergence Estimates in Approximation Theory. Springer, Berlin (2014)
9. Gupta, V., Agrawal, R.P., Verma, D.K.: Approximation for a new sequence of summation-integral type operators. Adv. Math. Sci. Appl. **23**(1), 35–42 (2013)
10. Păltănea, R.: Modified Szász-Mirakjan operators of integral form. Carpathian J. Math. **24**(3), 378–385 (2008)
11. Phillips, R.S.: An inversion formula for Laplace transforms and semi-groups of linear operators. Ann. Math. **59**, 325–356 (1954)
12. Serhan, V., Fatma, T.: Szász type operators involving Charlier polynomials. Math. Comput. Modell. **56**, 118–122 (2012)

Identities of Symmetry for the Generalized Degenerate Euler Polynomials

Dae San Kim and Taekyun Kim

Abstract In this paper, we give some identities of symmetry for the generalized degenerate Euler polynomials attached to χ which are derived from the symmetric properties for certain fermionic p-adic integrals on $\mathbb{Z}_p$.

Keywords Identities of symmetry · Generalized degenerate Euler polynomial · Fermionic p-adic integral

2010 Mathematics Subject Classification 11B68 · 11B83 · 11C08 · 65D20 · 65Q30 · 65R20

1 Introduction and Preliminaries

Let p be a fixed odd prime. Throughout this paper, $\mathbb{Z}_p$, $\mathbb{Q}_p$ and $\mathbb{C}_p$ will be the ring of p-adic integers, the field of p-adic rational numbers and the completion of the algebraic closure of $\mathbb{Q}_p$, respectively.

The p-adic norm $|\cdot|_p$ in $\mathbb{C}_p$ is normalized as $|p|_p = \frac{1}{p}$. Let $f(x)$ be continuous function on $\mathbb{Z}_p$. Then the fermionic p-adic integral on $\mathbb{Z}_p$ is defined as

$$I_{-1}(f) = \int_{\mathbb{Z}_p} f(x)\, d\mu_{-1}(x) \tag{1.1}$$

D.S. Kim
Department of Mathematics, Sogang University, Seoul 121-742, Republic of Korea
e-mail: dskim@sogang.ac.kr

T. Kim (✉)
Department of Mathematics, Kwangwoon University, Seoul 139-701, Republic of Korea
e-mail: tkkim@kw.ac.kr

V.K. Singh et al. (eds.), *Modern Mathematical Methods and High Performance Computing in Science and Technology*, Springer Proceedings in Mathematics & Statistics 171, DOI 10.1007/978-981-10-1454-3_4

$$= \lim_{N\to\infty} \sum_{x=0}^{p^N-1} f(x)(-1)^x, \quad (\text{see [9]}).$$

From (1.1), we note that

$$I_{-1}(f_n) + (-1)^{n-1} I_{-1}(f) = 2\sum_{l=0}^{n-1} (-1)^{n-1-l} f(l), \quad (\text{see [7]}), \tag{1.2}$$

where $n \in \mathbb{N}$.

As is well known, the Euler polynomials are defined by the generating function

$$\int_{\mathbb{Z}_p} e^{(x+y)t} d\mu_{-1}(y) = \frac{2}{e^t+1} e^{xt} = \sum_{n=0}^{\infty} E_n(x) \frac{t^n}{n!}. \tag{1.3}$$

When $x = 0$, $E_n = E_n(0)$ are called the Euler numbers (see [1–19]).

For a fixed odd integer d with $(p, d) = 1$, we set

$$X = \lim_{\overleftarrow{N}} \mathbb{Z}/dp^N\mathbb{Z}, \quad X^* = \bigcup_{\substack{0<a<dp \\ (a,p)=1}} \left(a + dp\mathbb{Z}_p\right),$$

$$a + dp^N\mathbb{Z}_p = \left\{x \in X \mid x \equiv a \pmod{dp^N}\right\},$$

where $a \in \mathbb{Z}$ lies in $0 \leq a < dp^N$.

It is known that

$$\int_{\mathbb{Z}_p} f(x)\, d\mu_{-1}(x) = \int_X f(x)\, d\mu_{-1}(x), \quad (\text{see [7–9]}),$$

where f is a continuous function on $\mathbb{Z}_p$.

Let $d \in \mathbb{N}$ with $d \equiv 1 \pmod 2$ and let χ be a Dirichlet character with conductor d. Then the generalized Euler polynomials attached to χ are defined by the generating function

$$\left(\frac{2\sum_{a=0}^{d-1}(-1)^a \chi(a) e^{at}}{e^{dt}+1}\right) e^{xt} = \sum_{n=0}^{\infty} E_{n,\chi}(x) \frac{t^n}{n!}. \tag{1.4}$$

In particular, for $x = 0$, $E_{n,\chi} = E_{n,\chi}(0)$ are called the generalized Euler numbers attached to χ.

For $d \in \mathbb{N}$ with $d \equiv 1 \pmod 2$, by (1.2), we get

$$\begin{aligned}\int_X \chi(y)\, e^{(x+y)t}\, d\mu_{-1}(y) &= \frac{2\sum_{a=0}^{d-1}(-1)^a \chi(a)\, e^{at}}{e^{dt}+1} e^{xt} \\ &= \sum_{n=0}^{\infty} E_{n,\chi}(x)\frac{t^n}{n!}, \quad (\text{see [9–11]}).\end{aligned} \tag{1.5}$$

From (1.5), we have

$$\int_X \chi(y)(x+y)^n\, d\mu_{-1}(y) = E_{n,\chi}(x), \quad (n \geq 0). \tag{1.6}$$

Carlitz considered the degenerate Euler polynomials given by the generating function

$$\begin{aligned}\frac{2}{(1+\lambda t)^{\frac{1}{\lambda}}+1}(1+\lambda t)^{\frac{x}{\lambda}} &= \sum_{n=0}^{\infty} \mathcal{E}_n(x \mid \lambda)\frac{t^n}{n!}, \quad (\text{see [3]}).\end{aligned} \tag{1.7}$$

Note that $\lim_{\lambda \to 0} \mathcal{E}_n(x \mid \lambda) = E_n(x)$, $(n \geq 0)$.
From (1.2), we note that

$$\begin{aligned}\int_X (1+\lambda t)^{\frac{x+y}{\lambda}}\, d\mu_{-1}(y) &= \frac{2}{(1+\lambda t)^{\frac{1}{\lambda}}+1}(1+\lambda t)^{\frac{x}{\lambda}} \\ &= \sum_{n=0}^{\infty} \mathcal{E}_n(x \mid \lambda)\frac{t^n}{n!}.\end{aligned} \tag{1.8}$$

Thus, by (1.8), we get

$$\int_X (y+x \mid \lambda)_n\, d\mu_{-1}(y) = \mathcal{E}_n(x \mid \lambda), \quad (n \geq 0), \tag{1.9}$$

where $(x \mid \lambda)_n = x(x-\lambda)\cdots(x-(n-1)\lambda)$, for $n \geq 1$, and $(x \mid \lambda)_0 = 1$.

From (1.2), we can derive the following equation:

$$\begin{aligned}&\int_X \chi(y)(1+\lambda t)^{\frac{x+y}{\lambda}}\, d\mu_{-1}(y) \\ &= \frac{2\sum_{a=0}^{d-1}(-1)^a \chi(a)(1+\lambda t)^{\frac{a}{\lambda}}}{(1+\lambda t)^{\frac{d}{\lambda}}+1}(1+\lambda t)^{\frac{x}{\lambda}},\end{aligned} \tag{1.10}$$

where $d \in \mathbb{N}$ with $d \equiv 1 \pmod 2$.

In view of (1.5), we define the generalized degenerate Euler polynomials attached to χ as follows:

$$\frac{2\sum_{a=0}^{d-1}(-1)^a \chi(a)(1+\lambda t)^{\frac{a}{\lambda}}}{(1+\lambda t)^{\frac{d}{\lambda}}+1}(1+\lambda t)^{\frac{x}{\lambda}} = \sum_{n=0}^{\infty} \mathcal{E}_{n,\lambda,\chi}(x)\frac{t^n}{n!}. \tag{1.11}$$

When $x = 0$, $\mathcal{E}_{n,\lambda,\chi} = \mathcal{E}_{n,\lambda,\chi}(0)$ are called the generalized degenerate Euler numbers attached to χ.

Let n be an odd natural number. Then, by (1.2), we get

$$\begin{aligned}&\int_X \chi(x)(1+\lambda t)^{\frac{nd+x}{\lambda}}\, d\mu_{-1}(x) + \int_X \chi(x)(1+\lambda t)^{\frac{x}{\lambda}}\, d\mu_{-1}(x) \\ &= 2\sum_{l=0}^{nd-1}(-1)^l \chi(l)(1+\lambda t)^{\frac{l}{\lambda}}.\end{aligned} \tag{1.12}$$

Now, we set

$$R_k(n,\lambda \mid x) = 2\sum_{l=0}^{n}(-1)^l \chi(l)(l \mid \lambda)_k. \tag{1.13}$$

From (1.2) and (1.12), we have

$$\begin{aligned}&\int_X (1+\lambda t)^{\frac{x+dn}{\lambda}} \chi(x)\, d\mu_{-1}(x) + \int_X \chi(x)(1+\lambda t)^{\frac{x}{\lambda}}\, d\mu_{-1}(x) \\ &= \frac{2\int_X (1+\lambda t)^{\frac{x}{\lambda}}\chi(x)\, d\mu_{-1}(x)}{\int_X (1+\lambda t)^{\frac{ndx}{\lambda}}\, d\mu_{-1}(x)} \\ &= \sum_{k=0}^{\infty} R_k(nd-1,\lambda \mid \chi)\frac{t^k}{k!},\end{aligned} \tag{1.14}$$

where $n, d \in \mathbb{N}$ with $n \equiv 1 \pmod 2$, $d \equiv 1 \pmod 2$.

In this paper, we give some identities of symmetry for the generalized degenerate Euler polynomials attached to χ derived from the symmetric properties of certain fermionic p-adic integrals on $\mathbb{Z}_p$.

2 Identities of Symmetry for the Generalized Degenerate Euler Polynomials

Let w_1, w_2 be odd natural numbers. Then we consider the following integral equation:

$$\begin{aligned}
&\frac{\int_X \int_X (1+\lambda t)^{\frac{w_1 x_1 + w_2 x_2}{\lambda}} \chi(x_1)\chi(x_2)\, d\mu_{-1}(x_1)\, d\mu_{-1}(x_2)}{\int_X (1+\lambda t)^{\frac{d w_1 w_2 x}{\lambda}}\, d\mu_{-1}(x)} \\
&= \frac{2\left((1+\lambda t)^{\frac{d w_1 w_2}{\lambda}} + 1\right)}{\left((1+\lambda t)^{\frac{w_1 d}{\lambda}} + 1\right)\left((1+\lambda t)^{\frac{w_2 d}{\lambda}} + 1\right)} \\
&\quad \times \sum_{a=0}^{d-1} \chi(a)(1+\lambda t)^{\frac{w_1 a}{\lambda}} (-1)^a \\
&\quad \times \sum_{b=0}^{d-1} \chi(b)(1+\lambda t)^{\frac{w_2 b}{\lambda}} (-1)^b .
\end{aligned} \tag{2.1}$$

From (1.10) and (1.11), we note that

$$\int_X \chi(y)(x+y \mid \lambda)_n\, d\mu_{-1}(y) = \mathcal{E}_{n,\lambda,\chi}(x), \quad (n \geq 0). \tag{2.2}$$

By (1.14), we get

$$\int_X \chi(x)(x+dn \mid \lambda)_k\, d\mu_{-1}(x) + \int_X \chi(x)(x \mid \lambda)_k\, d\mu_{-1}(x) = R_k(nd-1, \lambda \mid x), \tag{2.3}$$

where $k \geq 0$.

Thus, by (2.2) and (2.3), we get

$$\mathcal{E}_{k,\lambda,\chi}(nd) + \mathcal{E}_{k,\lambda,\chi} = R_k(nd-1, \lambda \mid \chi), \tag{2.4}$$

where $k \geq 0$, $n, d \in \mathbb{N}$ with $n \equiv 1 \pmod 2$, $d \equiv 1 \pmod 2$.

Now, we set

$$I_\chi(w_1, w_2 \mid \lambda) = \frac{\int_X \int_X \chi(x_1)\chi(x_2)(1+\lambda t)^{\frac{w_1x_1+w_2x_2+w_1w_2x}{\lambda}}\, d\mu_{-1}(x_1)\, d\mu_{-1}(x_2)}{\int_X (1+\lambda t)^{\frac{dw_1w_2x}{\lambda}}\, d\mu_{-1}(x)}. \tag{2.5}$$

From (2.5), we have

$$\begin{aligned} & I_\chi(w_1, w_2 \mid \lambda) \\ &= \frac{2\left((1+\lambda t)^{\frac{dw_1w_2}{\lambda}}+1\right)(1+\lambda t)^{\frac{w_1w_2x}{\lambda}}}{\left((1+\lambda t)^{\frac{w_1d}{\lambda}}+1\right)\left((1+\lambda t)^{\frac{w_2d}{\lambda}}+1\right)} \\ &\quad \times \sum_{a=0}^{d-1} \chi(a)(-1)^a (1+\lambda t)^{\frac{w_1a}{\lambda}} \\ &\quad \times \sum_{b=0}^{d-1} \chi(b)(-1)^b (1+\lambda t)^{\frac{w_2b}{\lambda}}. \end{aligned} \tag{2.6}$$

Thus, by (2.6), we see that $I_\chi(w_1, w_2 \mid \lambda)$ is symmetric in w_1, w_2. By (1.12), (1.14), (2.2) and (2.5), we get

$$\begin{aligned} & 2I_\chi(w_1, w_2 \mid \lambda) \\ &= \sum_{l=0}^{\infty}\left(\sum_{i=0}^{l}\binom{l}{i}\mathcal{E}_{i,\frac{\lambda}{w_2},\chi}(w_1x)\, w_2^i w_1^{l-i} R\left(dw_2-1, \frac{\lambda}{w_1}\middle|\chi\right)\right)\frac{t^l}{l!}. \end{aligned} \tag{2.7}$$

From the symmetric property of $I_\chi(w_1, w_2 \mid \lambda)$ in w_1 and w_2, we have

$$\begin{aligned} & 2I_\chi(w_1, w_2 \mid \lambda) \\ &= 2I_\chi(w_2, w_1 \mid \chi) \\ &= \sum_{l=0}^{\infty}\left(\sum_{i=0}^{l}\binom{l}{i}\mathcal{E}_{i,\frac{\lambda}{w_1},\chi}(w_2x)\, w_1^i w_2^{l-i} R\left(dw_1-1, \frac{\lambda}{w_2}\middle|\chi\right)\right)\frac{t^l}{l!}. \end{aligned} \tag{2.8}$$

Therefore, by (2.7) and (2.8), we obtain the following theorem.

Theorem 1 *For $w_1, w_2, d \in \mathbb{N}$ with $w_1 \equiv w_2 \equiv d \equiv 1 \pmod 2$, let χ be a Dirichlet character with conductor d. Then, we have*

$$\sum_{i=0}^{l}\binom{l}{i}\mathcal{E}_{i,\frac{\lambda}{w_1},\chi}\left(w_2x\right)w_1^iw_2^{l-i}R\left(dw_1-1,\frac{\lambda}{w_2}\bigg|\chi\right)$$
$$=\sum_{i=0}^{l}\binom{l}{i}\mathcal{E}_{i,\frac{\lambda}{w_2},\chi}\left(w_1x\right)w_2^iw_1^{l-i}R\left(dw_2-1,\frac{\lambda}{w_1}\bigg|\chi\right),$$

where $l \geq 0$.

When $x = 0$, by Theorem 1, we get

$$\sum_{i=0}^{l}\binom{l}{i}\mathcal{E}_{i,\frac{\lambda}{w_1},\chi}w_1^iw_2^{l-i}R\left(dw_1-1,\frac{\lambda}{w_2}\bigg|\chi\right)$$
$$=\sum_{i=0}^{l}\binom{l}{i}\mathcal{E}_{i,\frac{\lambda}{w_2},\chi}w_2^iw_1^{l-i}R\left(dw_2-1,\frac{\lambda}{w_1}\bigg|\chi\right),\quad (l\geq 0).$$

By (2.5), we get

$$2I_\chi\left(w_1,w_2\mid\lambda\right) \tag{2.9}$$
$$=\sum_{l=0}^{dw_2-1}(-1)^l\chi(l)\int_X(1+\lambda t)^{\frac{w_2}{\lambda}\left(w_2+w_1x+\frac{w_1}{w_2}l\right)}\chi(x_2)\,d\mu_{-1}(x)$$
$$=\sum_{n=0}^{\infty}\left(\sum_{l=0}^{dw_2-1}(-1)^l\chi(l)\,\mathcal{E}_{n,\frac{\lambda}{w_2},\chi}\left(w_1x+\frac{w_1}{w_2}l\right)w_2^n\right)\frac{t^n}{n!}.$$

On the other hand,

$$2I_\chi\left(w_2,w_1\mid\lambda\right)=2I_\chi\left(w_1,w_2\mid\lambda\right) \tag{2.10}$$
$$=\sum_{n=0}^{\infty}\left(\sum_{l=0}^{dw_1-1}(-1)^l\chi(l)\,\mathcal{E}_{n,\frac{\lambda}{w_1},\chi}\left(w_2x+\frac{w_2}{w_1}l\right)w_1^n\right)\frac{t^n}{n!}.$$

Therefore, by (2.9) and (2.10), we obtain the following theorem.

Theorem 2 *For* $w_1, w_2, d \in \mathbb{N}$ *with* $d \equiv 1 \pmod 2$, $w_1 \equiv 1 \pmod 2$ *and* $w_2 \equiv 1 \pmod 2$, *let* χ *be a Dirichlet character with conductor* d. *Then, we have*

$$w_2^n\sum_{l=0}^{dw_2-1}(-1)^l\chi(l)\,\mathcal{E}_{n,\frac{\lambda}{w_2},\chi}\left(w_1x+\frac{w_1}{w_2}l\right)$$
$$=w_1^n\sum_{l=0}^{dw_1-1}(-1)^l\chi(l)\,\mathcal{E}_{n,\frac{\lambda}{w_1},\chi}\left(w_2x+\frac{w_2}{w_1}l\right),\quad (n\geq 0).$$

To derive some interesting identities of symmetry for the generalized degenerate Euler polynomials attached to χ, we used the symmetric properties for certain fermionic p-adic integrals on $\mathbb{Z}_p$. When $w_2 = 1$, from Theorem 2, we have

$$\sum_{l=0}^{d-1} (-1)^l \chi(l) \mathcal{E}_{n,\lambda,\chi}(w_1 x + w_1 l) = w_1^n \sum_{l=0}^{dw_1-1} (-1)^l \chi(l) \mathcal{E}_{n,\frac{\lambda}{w_1},\chi}\left(x + \frac{1}{w_1} l\right).$$

In particular, for $x = 0$, we get

$$\sum_{l=0}^{d-1} (-1)^l \chi(l) \mathcal{E}_{n,\lambda,\chi}(w_1 l) = w_1^n \sum_{l=0}^{dw_1-1} (-1)^l \chi(l) \mathcal{E}_{n,\frac{\lambda}{w_1},\chi}\left(\frac{1}{w_1} l\right).$$

References

1. Araci, S., Bagdasaryan, A., Özel, C., Srivastava, H.M.: New symmetric identities involving q-zeta type functions. Appl. Math. Inf. Sci. **8**(6), 2803–2808 (2014). MR 3228678
2. Bayad, A., Chikhi, J.: Apostol-Euler polynomials and asymptotics for negative binomial reciprocals. Adv. Stud. Contemp. Math. (Kyungshang) **24**(1), 33–37 (2014). MR 3157406
3. Carlitz, L.: Degenerate Stirling, Bernoulli and Eulerian numbers. Utilitas Math. **15**(80i:05014), 51–88 (1979). MR 531621
4. Dangi, R., Tiwari, M., Parihar, C.L.: Generalized Euler polynomials and their properties. J. Rajasthan Acad. Phys. Sci. **12**(4), 385–392 (2013). MR 3299610
5. Jiu, L., Moll, V.H., Vignat, C.: Identities for generalized Euler polynomials. Integral Transforms Spec. Funct. **25**(10), 777–789 (2014). MR 3230659
6. Kim, D.S.: Identities associated with generalized twisted Euler polynomials twisted by ramified roots of unity. Adv. Stud. Contemp. Math. (Kyungshang) **22**(3), 363–377 (2012). MR 2976595
7. Kim, D.S., Lee, N., Na, J., Park, K.H.: Identities of symmetry for higher-order Euler polynomials in three variables (I). Adv. Stud. Contemp. Math. (Kyungshang) **22**(1), 51–74 (2012). MR 2931605
8. Kim, T.: Symmetry identities for the twisted generalized Euler polynomials. Adv. Stud. Contemp. Math. (Kyungshang) **19**(2), 151–155 (2009). MR 2566912 (2010j:11041)
9. Kim, T.: Symmetry of power sum polynomials and multivariate fermionic $\mathbb{Z}_p$. Russ. J. Math. Phys. **16**(1), 93–96 (2009). MR 2486809 (2010c:11028)
10. Kim, T.: An identity of symmetry for the generalized Euler polynomials. J. Comput. Anal. Appl. **13**(7), 1292–1296 (2011). MR 2791956 (2012d:11050)
11. Kozuka, K.: On a p-adic interpolating power series of the generalized Euler numbers. J. Math. Soc. Jpn. **42**(1), 113–125 (1990). MR 1027544 (90j:11020)
12. Luo, Q.-M.: q-analogues of some results for the Apostol-Euler polynomials. Adv. Stud. Contemp. Math. (Kyungshang) **20**(1), 103–113 (2010). MR 2597996 (2011e:05031)

13. Luo, Q.-M., Qi, F.: Relationships between generalized Bernoulli numbers and polynomials and generalized Euler numbers and polynomials. Adv. Stud. Contemp. Math. (Kyungshang) **7**(111), 11–18 (2003). MR 1981601
14. Rim, S.-H., Jeong, J.-H., Lee, S.-J., Moon, E.-J., Jin, J.-H.: On the symmetric properties for the generalized twisted Genocchi polynomials. Ars Combin. **105**, 267–272 (2012). MR 2976377
15. Ryoo, C.S.: On the generalized Barnes type multiple q-Euler polynomials twisted by ramified roots of unity. Proc. Jangjeon Math. Soc. **13**(2), 255–263 (2010). MR 2676690 (2011e:11043)
16. Şen, E.: Theorems on Apostol-Euler polynomials of higher order arising from Euler basis. Adv. Stud. Contemp. Math. (Kyungshang) **23**(2), 337–345 (2013). MR 3088764
17. Simsek, Y., Yurekli, O., Kurt, V.: On interpolation functions of the twisted generalized Frobenius–Euler numbers. Adv. Stud. Contemp. Math. (Kyungshang) **15**(2), 187–194 (2007). MR 2356176 (2008g:11193
18. Yang, S.L., Qiao, Z.K.: Some symmetry identities for the Euler polynomials. J. Math. Res. Exposition **30**(3), 457–464 (2010). MR 2680613 (2011d:11045)
19. Zhang, Z., Yang, H.: Some closed formulas for generalized Bernoulli–Euler numbers and polynomials. Proc. Jangjeon Math. Soc. **11**(2), 191–198 (2008). MR 2482602 (2010a:11036)

Using MathLang to Check the Correctness of Specifications in Object-Z

David Feller, Fairouz Kamareddine and Lavinia Burski

Abstract The importance of thoroughly checking software specifications is widely recognised in the software industry, particularly for software involved in dealing with safety critical systems. The MathLang project has been successfully used to check large mathematical texts for correctness in a stepwise fashion. Currently MathLang is being tested for checking the correctness of formal specifications written in Z. Since object-orientation is a vital concept for software specification, it is important that the tools available for thoroughly checking specifications can be used with a language powerful enough to express specifications for object-oriented software. This paper aims to test the usefulness of MathLang for the computerisation of formal specifications written in Object-Z.

Keywords Software specification and correctness · Object-oriented design · MathLang · Object-Z

1 Introduction

Inadequate checking of software is a serious problem in the software industry. According to Frentiu [1]:

> Experience shows that more than 75 % of finished software products have errors during maintenance, and deadlines are missed and cost overruns are a rule not an exception. It was estimated that more than 50 % of the development effort was spent on testing and debugging. Nevertheless, some errors are not detected by testing, and some of them are never detected. More, there are projects that have never been finished. And it is not an exception; it is estimated that from each six large projects two of them are never finished.

Rigorous checking of software systems could help with these issues. This is of particular importance to the designers of safety critical systems who cannot afford to find bugs in their software by testing it on users. This is because such a bug could

D. Feller · F. Kamareddine (✉) · L. Burski
School of Mathematical and Computer Sciences, Heriot Watt University,
Edinburgh EH14 4AS, UK
e-mail: f.d.kamareddine@hw.ac.uk

V.K. Singh et al. (eds.), *Modern Mathematical Methods and High Performance Computing in Science and Technology*, Springer Proceedings in Mathematics & Statistics 171, DOI 10.1007/978-981-10-1454-3_5

cause injury or death in the course of being found. According to MacKenzie [2], the total number of computer-related accidental deaths, worldwide, to the end of 1992, can be expressed, in conventional format, as 1,100 +/− 1,000. One can easily imagine other cases where a high degree of confidence that a piece of software will always function as intended is needed before it can be used (e.g., software dealing with sensitive information). This paper is concerned with creating software which aids in the formal proof of software correctness.

1.1 Why Formally Prove Software Correctness?

Formally proving that a piece of software is correct can give us a high degree of confidence that it will function as intended; checking the validity of that proof, even more so. Testing and debugging can fail because of some condition that the software developers forgot to check. As Dijkstra [3] observed: program testing can be used to show the presence of bugs, but never to show their absence. Testing and debugging can take much longer than expected to locate those errors that are hard to isolate and hence fixing those errors can only happen late in the development cycle. However, a specification that has been proven correct should function as defined, provided that said proof is correct.

1.2 The Difficulties of Formally Proving Correctness

For large systems, formally proving correctness can be repetitive and labour intensive. It is not guaranteed that software developers have a great deal of experience with formal proof and it is certainly not guaranteed that every developer working on a large piece of software could aid in the formal checking of software correctness. As [4] states, many software engineers reject the use of formal methods for software validation, arguing that it is too complex and time-consuming a process for most programmers. Further it is still possible for such proofs to be subject to human error. As such, it is important that we have good tools to aid in formally checking the correctness of software specifications.

Contributions This paper presents the first step in the development of a new tool to aid in a *stepwise easy to use fashion* in the formal checking of the correctness of software specifications written in the specification language Object Z [5, 6]. This first step allows for the type checking and grammatical correctness of documents written in Object Z. We present a development path for expansion of the tool to aid in more complete checking of specifications in Object Z where also logical and rhetorical correctness can be checked. We also explain why our proposed method might provide a basis for development of tools for checking correctness in other specification languages.

1.3 Related Work

One usually formalises a Z [7] specification into a complete proof right away [8–11], as shown by arrow **e** in Fig. 1. The thickness of the arrow here represents the level of difficulty, the huge expertise needed, and the amount of work necessary to take that path. Our proposal is to carry out the correctness checking in smaller steps, each of which is more focused and very simple to carry out. These smaller steps are based on MathLang [12]. MathLang is a system for computerising mathematical texts which aims to reduce the complexity of checking the correctness of a text by breaking down the process into more manageable steps which can be easily completed with the aid of a computer. MathLang starts by separating out the work that needs to be done in computerising a mathematical text into three main aspects. These are the Core Grammatical aspect (CGa), the Document Rhetorical aspect (DRa) and the Text and Symbol aspect (TSa).

The CGa checks the internal grammatical structure of a text is correct by capturing the structures and common concepts with a finite set of concepts which are derived from weak type theory. The TSa captures the mathematical relations which hold between the parts of the text as represented by the CGa. The DRa captures the logical roles that are held by chunks of text that tell us where they feature in an argument or proof. This information can be used to generate a proof skeleton in a theorem prover making the move from document to formal checking much simpler. Further it allows simple checking of the grammar and general structure of the document to be performed automatically.

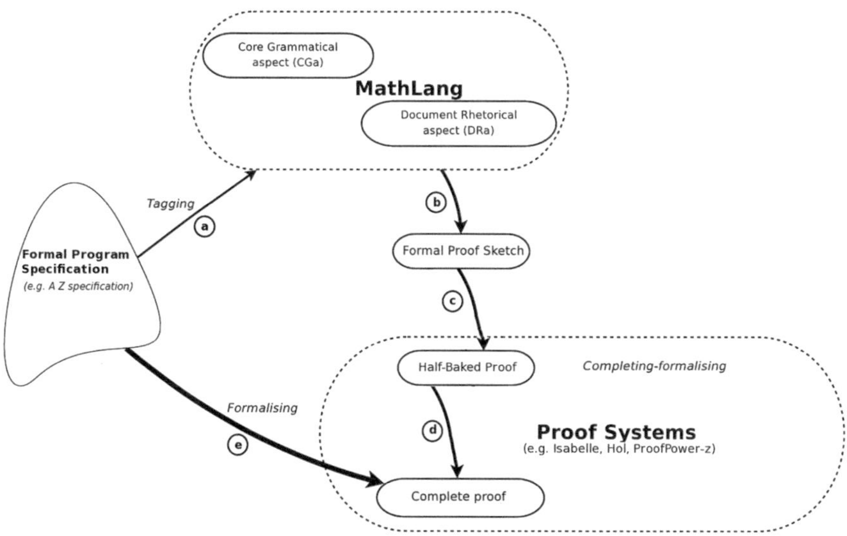

Fig. 1 The different steps taken to achieve a full proof using the ZMathLang method

MathLang for Z (ZMathLang) [13], divides **e** into a number of smaller paths via **a**, **b**, **c** and **d**. Following this path the user would apply the Z core grammatical aspect (ZCGa) and the Z document rhetorical aspect (ZDRa), to show that the specification is weakly type checked and is also rhetorically correct (i.e. no loops in the reasoning). Then the user would take the ZCGa and ZDRa annotated specification into a general proof skeleton, then a 'half-baked' proof and ultimately a complete proof. Breaking this down would allow a user with minimal theorem prover expertise to obtain a fully proved Z specification.

Unlike MathLang for mathematics, ZMathLang does not require a Text-Symbolic aspect (TSa) as the mathematical relations in Z are already formal.

Object-Z Object-Z [14] is an extension of the Z language for writing formal specifications that has added functionality for dealing with object-oriented concepts. Both Z and Object-Z have been designed to make formally proving the correctness of specifications relatively easy. They each have a standard notation which can be easily manipulated in a mathematical fashion, allowing for proof of correctness using standard mathematical methods.

Z allows specifications to be split up into different schema boxes—each of which represent individual functions within the software. The input, output and manipulation of data is expressed through a mathematical notation based around set theory. Object Z introduces class boxes, which allow a specification to be identified with a class of objects and standard notation for creating instances of objects and for initialising and running methods within a specification.

Why Choose Object-Z over Z as a Specification Language for Correctness Checking? Object-oriented programming allows developers to separate out programs into modules whose functions and interactions are (relatively) easy for developers to understand and whose contents are easier to alter without needing to change too much of the rest of the code. This is especially important for large pieces of software with multiple developers whose code can quickly grow unmanageable and difficult for humans to interpret. Most popular languages in use today are object oriented. Examples include Java, Python, C++, Visual Basic .NET and Ruby. It is important, therefore, that our tool for checking formal correctness works for specifications of object-oriented languages. Smith [14] noted the following benefits of an object-oriented specification language:

- The modularity it brings to system design. Modularity increases the clarity of specifications by allowing a reader to focus on one part at a time.
- It provides a precise methodology for system design. This methodology involves the specification of a system by first specifying the behaviour of its constituent objects by classes, and by utilising inheritance and polymorphism where appropriate
- Seamless development—the use of common concepts and system structuring at each stage of system development: from the specification right through to the implementation. This is possible when using an object-oriented approach to specification and then implementing in an object-oriented programming language. It makes the specification more accessible to the programmer, who may not be a formalist, and facilitates his or her task of transforming the specification to implementation.

Another reason to prefer Object-Z over Z when using MathLang to check the correctness of formal specifications is that checking the correctness of Object-Z gives us much more confidence that MathLang is well suited to checking the correctness of software specifications than checking the correctness of Z alone does, as object-oriented models might present unique problems for conversion into a framework like that of MathLang due to the subtyping of objects.

Tools for Object-Z TOZE [15] is a graphical editor for Object-Z documents which allows syntax and type checking without demanding experience with LaTeX or requiring the user to save the specification and use tools outside the text editor. Kimber [16] gives a tool which maps 80 % of Object-Z to perfect developer [17] allowing the direct verification of the soundness of simple specifications.

By checking Object-Z in MathLang, one allows some flexibility in whether to perform syntax and weak type checking, or check that simple dependencies are fulfilled, or provide a full proof. Another benefit especially when the soundness of an Object-Z specification is difficult to verify directly, consists of the MathLang automated layers which are crucial to embed the entire specification into a theorem prover. This could particularly be useful in large software projects where no or very few individuals are familiar with a theorem prover, and it would take a long time to translate the full specification into a theorem prover by hand.

1.4 Overview

In Sect. 2 we create a new weak typing system for MathLang to incorporate Object-Z specifications. We give the weak types and the rules in which we check specifications. In Sect. 2.2 we give a step by step example of how we can check for weak typing errors in an Object-Z specification. We explain how to label the specification and how to run the weak type checker on the specification. Section 3 goes on to explain how we implemented the Object-Z Core Grammatical aspect (OZCGa), and how it has been tested. The problems encountered and how they were dealt with is described in Sect. 3.4. Finally, our conclusion along with benefits and limitations is described in Sect. 4.

2 Adapting MathLang to Include Object-Z

We look at the first aspect of the ZMathLang framework (CGa) extended to weak type check Object-Z specifications.

OZCGa includes 7 weak types **Spec**, $\mathbf{\Gamma}$, $\mathcal{T}$, $\mathbb{S}$, $\mathcal{Z}$, $\mathcal{E}$, $\mathcal{D}$, $\mathbb{O}$, $\mathcal{M}$ corresponding to *specification*, *schematext*, *term*, *set*, *declaration*, *expression*, *definition*, *object* and *method* respectively. We categorise the parts of the Z syntax using these types in order to define the core grammatical aspect OZCGa.

We have three types of variables in our syntax

- $V = V^{\mathcal{T}}$, Variables giving terms.
- $V = V^{\mathbb{S}}$, Variables giving sets.
- $V = V^{\mathbb{O}}$, Variables giving objects.

We have three types of constants in our syntax:

- $C = C^{\mathcal{T}}$, Constants for terms.
- $C = C^{\mathbb{S}}$, Constants for sets.
- $C^{\mathcal{E}}$, Constants giving expressions. These can be further broken down into the following:
  - C^{bool}, constants $\subseteq$, $=$,$\neq$ taking expressions as parameters.
  - C^{termop}, constant term operators $<, \leq, =, >, \geq, \neq$ taking two terms as parameters.
  - C^{setop}, constant set operators $\subseteq$, $=$, $\neq$, *partition* taking a set and a sequence of sets as parameters.
  - $C^{termsetop}$, constant term/set operators $\in$, $\notin$ taking one term and one set as parameters.
  - C^{objop}, constant object operators $=$, $\neq$ taking two objects as parameters.
  - $C^{objsetop}$, constant object/set operators $\in$, $\notin$ taking one object and one set as parameters.

We have three types of constants for our Object-Z syntax:

- $C = C^{\mathbb{O}}$, Constants for objects.
- $C = C^{\mathcal{M}}$, Constants for methods.
  - $C = C^{obop}$, constant for object operator taking an object as a parameter.
  - $C = C^{methop}$, constant for method operator $\fatsemi$ taking two methods as parameters.

We have three types of binders in Z

- $B^{\mathbb{S}}$, the binder $\cup$ giving sets and taking sets as parameters.
- $B^{\mathcal{E}}$, binders $\exists$, $\forall$ giving expressions and taking expressions as parameters.
- $\downarrow$, gives an object and takes an object as its parameter.

Definitions in Object-Z can define constants including those of the form $C^{\mathbb{O}}$ which take a specification as its parameter.

Declarations express the relationship between something and its type. In the ZCGa we have two kinds of declarations, these can be *SET* (the type of all sets) or a particular set. We write either $V^{\mathbb{S}} : SET$, $V^{\mathcal{T}} : \mathbb{S}$ or $V^{\mathbb{O}} : \mathbb{S}$.

Expressions, terms and sets in Z are given as described in our rules for constants variables and binders.

A schematext can be empty or it can contain a declaration, expression or method. A declaration in a schematext represents the introduction of a new variable of a known type.

A specification is either empty or it consists of schematext and definitions where the parts of the schematext which are not defined inside the schematext itself have a corresponding definition in the specification.

2.1 A Formalisation of These Typing Rules

If we formally represent these typing rules we see that they are a subset of the typing rules of MathLang. The only differences [13] are that we change book to specification, context becomes schematext and statements become expressions. We eliminate nouns and adjectives and only have one syntax for definition.

We use the notation **::** for typing between an entity and its weak type and $\vdash$ to denote derivability. Here are some examples (we only state the meaning of the first 3 and leave the rest as obvious):

1. spec is a weakly typed specification:
   $\vdash$ spec **:: Spec**
2. Γ is a weakly well typed paragraph relative to specification spec:
   spec $\vdash \Gamma$ **::** $\mathbf{\Gamma}$
3. t is a weakly typed term, relative to specification spec and schematext Γ:
   spec; $\Gamma \vdash$ t **::** $\mathcal{T}$
4. spec; $\Gamma \vdash$ s **::** $\mathbb{S}$
5. spec; $\Gamma \vdash$ Z **::** $\mathcal{Z}$
6. spec; $\Gamma \vdash$ e **::** $\mathcal{E}$
7. spec; $\Gamma \vdash$ D **::** $\mathcal{D}$
8. spec; $\Gamma \vdash$ o **::** $\mathbb{O}$
9. spec; $\Gamma \vdash$ m **::** $\mathcal{M}$

The next definition is crucial for analysing the grammatical correctness of the specification since it collects the defined constants and declared variables of specifications and paragraphs:

Definition 1 1. Let $\theta \in$ spec be a definition paragraph $\Gamma \lhd D$ where D is of the form $c(x_1, \ldots, x_n) := A$. We define defcons(D) $= c$.
2. defcons(spec) $= \{$defcons(D) $| \Gamma \lhd D$ is a paragraph of spec for some $\Gamma\}$.
3. Internal constants defined using $==$ are noted as defcons(Γ).
4. We define for parameter P the weak type of P with respect to spec and Γ as: $\mathrm{wt}_{spec;\,\Gamma}$(P) $=$ W if and only if spec; $\Gamma \vdash$ P :: W.
5. Predefined constants such as $\mathbb{P}$, dom and ran are noted as prefcons(spec).
6. We use OK(spec;Γ) to denote $\vdash$ spec **:: Spec** and spec $\vdash \Gamma$ **::** Γ.
7. We define dvar(Γ) as follows:
   (a) if $\Gamma = \emptyset$, then dvar(Γ) $= \emptyset$.
   (b) if $\Gamma = \Gamma', x : A$ and $x \notin$ dvar(Γ), then dvar(Γ) $=$ dvar(Γ')$, x$.
   (c) Otherwise, if $\Gamma = \Gamma', e$, then dvar(Γ) $=$ dvar(Γ').

The next definition gives the typing rules that deal with type-orientedness:

Definition 2 1. Derivation rule for variables

$$\frac{OK(spec, \Gamma), x \in V^{\mathcal{T}/\mathbb{S}/\mathbb{O}}, x \in dvar(\Gamma)}{spec; \Gamma \vdash x :: \mathcal{T}/\mathbb{S}/\mathbb{O}}(var)$$

2. Derivation rule for internal constants

$$\frac{\begin{array}{c}OK(spec,\Gamma), \Gamma \lhd D \in spec, dvar(\Gamma')=(x_1,\ldots,x_n), defcons(D)=c \in C^{\mathcal{T}/\mathbb{S}/\mathbb{O}/\mathcal{E}/\mathcal{M}},\\ wt_{spec;\Gamma}(P_i)=wt_{spec;\Gamma'}(x_i), \text{for all} i=1,\ldots,n\end{array}}{spec; \Gamma \vdash c(P_1, \ldots, P_n) :: \mathcal{T}/\mathbb{S}/\mathbb{O}/\mathcal{E}/\mathcal{M}}(int-cons)$$

3. Derivation rule for external constants

$$\frac{\begin{array}{c}OK(spec,\Gamma), \text{ c external to spec}, c :: k_1,\ldots,k_n \to k,\\ spec;\Gamma \vdash P_i :: k_i (i=1,\ldots,n)\end{array}}{spec; \Gamma \vdash c(P_1, \ldots, P_n) :: k}(ext-cons)$$

4. Derivation rule for binders:

$$\frac{OK(spec; \Gamma; Z), b \in B, b :: k_1 \to k_2, spec; \Gamma, Z \vdash E :: k_1}{spec; \Gamma \vdash b_z(E) :: k_2}(bind)$$

5. Derivation rule for definitions:

$$\frac{\begin{array}{c}spec,\Gamma \vdash o::\mathbb{O}, OK(spec',\Gamma'),\\ c \in C^{\mathbb{O}}, c \notin prefcons(spec) \cup defcons(spec)\end{array}}{spec; \Gamma \vdash classbox(c(spec', \Gamma'), o) :: D}(obj-def)$$

That is, an entity that consists of a set equalling a number of variables bound by a constant and that has not yet been defined has weak type definition.

6. Derivation rule for an empty schematext is

$$\frac{\vdash spec :: \mathbf{Spec}}{spec \vdash \emptyset :: \mathbf{\Gamma}}(emp-cont)$$

7. Derivation rule for adding a set declaration to a paragraph is

$$\frac{OK(spec; \Gamma), x \in V^{\mathbb{S}}, x \notin dvar(\Gamma)}{spec \vdash \Gamma, x : SET :: \mathbf{\Gamma}}(set-dec)$$

8. Derivation rule for adding a term declaration is

$$\frac{OK(spec; \Gamma), spec; \Gamma \vdash s :: \mathbb{S}, x \in V^{\mathcal{T}}, x \notin dvar(\Gamma)}{spec \vdash \Gamma, x : s :: \mathbf{\Gamma}}(term-dec)$$

9. Derivation rule for adding an object declaration to a paragraph

$$\frac{OK(spec;\Gamma),\ spec;\Gamma \vdash s :: \mathbb{S}/\mathbb{O},\ x \in V^{\mathbb{O}},\ x \notin dvar(\Gamma)}{spec \vdash \Gamma, x : s :: \mathbf{\Gamma}}(obj-dec)$$

10. Derivation rule for adding an expression is

$$\frac{OK(spec;\Gamma),\ spec;\Gamma \vdash e :: \mathcal{E}}{spec \vdash \Gamma, x : e :: \mathbf{\Gamma}}(assump)$$

11. Derivation rule for adding methods

$$\frac{OK(spec;\Gamma),\ spec;\Gamma \vdash m :: \mathcal{M}}{spec \vdash \Gamma, x : m :: \mathbf{\Gamma}}(meth)$$

12. Derivation rule for an empty specification

$$\frac{}{\vdash \emptyset :: \boldsymbol{Spec}}(emp-spec)$$

13. Derivation rule for extending a specification is

$$\frac{spec \vdash \Gamma :: \mathbf{\Gamma}}{\vdash spec, \Gamma :: \mathbf{Spec}}(spec-ext)$$

2.2 An Example of an Object-Z Class with Weak Types Labelled

We have implemented the above syntax and type derivation rules to obtain an OZCGa type checker that checks whether specifications written in Object-Z are grammatically correct. To use this type checker, we need first to annotate the Object-Z specification with our weak types. For this, we create commands within a LaTeX package (see Table 1) where each of the weak types, is associated to a LaTeX command and the colour in which the contents appears.

We use an example of an Object-Z specification 'TwoCards' which describes an action where the balance is the first card plus the second card. Money is allowed to be withdrawn on both these cards. An example of the specification is shown in Fig. 5 (source file in Fig. 2). We use the commands from Table 1 to annotate this specification, giving us the source code shown in Fig. 3 (Fig. 4).

Table 1 The LaTeX commands to annotate an Object-Z specification with

Weak Type	Command	Colour
specification	\specification{...}	...
schematext	\text{...}	...
term	\term{...}	...
set	\set{...}	...
declaration	\declaration{...}	...
expression	\expression{...}	...
definition	\definition{...}	...
object	\object{...}	...
method	\method{...}	...

Fig. 2 Part of an Object-Z specification source code

```
\begin{schema}{TwoCards}
c1,c2:CreditCard \\
totalbal:\num
\where
c1 \neq c2\\
totalbal = c1.balance + c2.balance
\end{schema}

\begin{schema}{withdraw1}
\where
c1.withdraw
\end{schema}

\begin{schema}{withdraw2}
\where
c2.withdraw
\end{schema}

\begin{schema}{tranferAvail}
\where
c1.withdrawAvail \semi c2.deposit
\end{schema}
```

Colours now appear around each grammatical part of the specification (Fig. 5). These colours can be used to reduce the complexity of the specification and can also assist beginners in learning the syntax of Object-Z. More examples of the labelling and weakly typed Object-Z specification can be found in Appendix D. Other examples of weakly typed Z specifications are also given in Appendixes A, B and C.

After the specification has been labelled using the ‘ozcga’ package in LaTeX, our weak type checker goes through the specification an checks it for grammatical correctness. This weak type checking is run in a terminal through a program implemented in python.

```
\begin{class}{\object{TwoCards}}
\also
\specification{
\begin{schema}{TwoCards}
\text{\declaration{\object{c1},\object{c2}:
\expression{CreditCard}}\\
\declaration{\term{totalbal}:\expression{\num}}}
\where
\text{\expression{\object{c1}\neq\object{c2}}\\
\expression{\term{totalbal}=
\term{\object{c1}.\term{balance}+\object{c2}.
\term{balance}}}}
\end{schema}
\begin{schema}{withdraw1}
\where
\text{\method{\object{c1}.withdraw}}
\end{schema}
\begin{schema}{withdraw2}
\where
\text{\method{\object{c2}.withdraw}}
\end{schema}
\begin{schema}{tranferAvail}
\where
\text{\method{\method{
\object{c1}.withdrawAvail}\semi\method{\object{c2}.
deposit}}}
\end{schema}}
\end{class}
```

Fig. 3 Part of an Object-Z specification labelled in OZCGa source code. The full version can be found in appendices

TwoCards

$c1, c2 : CreditCard$
$totalbal : \mathbb{Z}$

$c1 \neq c2$
$totalbal = c1.balance + c2.balance$

withdraw1

$c1.withdraw$

withdraw2

$c2.withdraw$

tranferAvail

$c1.withdrawAvail \fatsemi c2.deposit$

Fig. 4 Part of an Object-Z specification

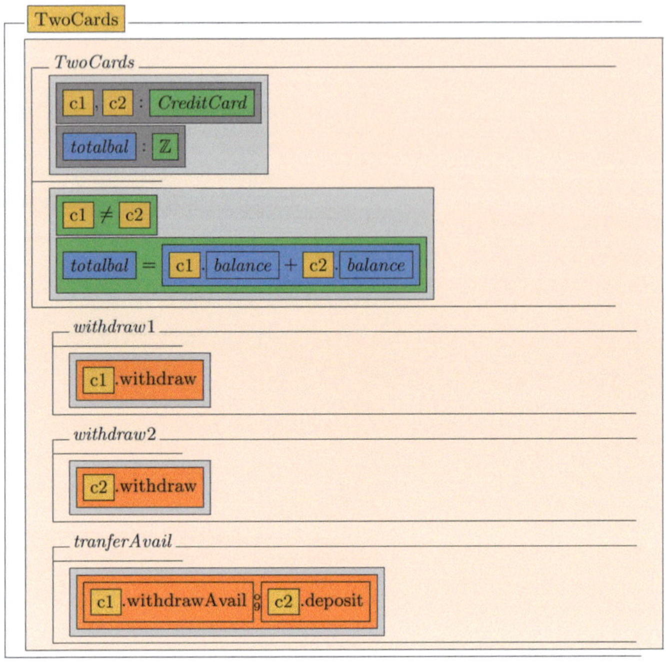

Fig. 5 Part of an Object-Z specification labelled in OZCGa and compiled with pdflatex. The full version can be found in Appendix D

3 Implementation of the OZCGa

In this section we look at the specific implementation of the Core Grammatical aspect of MathLang for Object-Z (OZCGa), that we created. We go over the implementation and the specific examples we used to test it.

3.1 Expansion of Existing Functional Software

The LaTeX style file used for labelling Object-Z specifications to be checked by the OZCGa.py is built around the oz style file [18], which is the file the Community Z Tools website [19] suggests for typesetting Object-Z documents in LaTeX, so it can be easily applied to existing Object-Z documents. The software for the OZCGa itself is built around the software used for the ZCGa—so we can be fairly confident that the functionality and reliability of the ZCGa has been preserved in the adaptation to the OZCGa. In addition it means that Z specifications which can be checked by the ZCGa do not need changing to be checked with the OZCGa.

3.2 The Style File

The zcga.sty style file used for labelling Z specifications for the ZCGa imports the zed.sty style file. For the ozcga.sty style file we chose to import the oz.sty style file to deal with specifications written in Object-Z. We chose oz.sty because it contained all the commands used in the zed.sty style file so ZCGa specifications written in LaTeX with the ozcga.sty style file in place of the zcga.sty style file are compatible with the ZCGa. We have also kept the original weak type labels and used the same format for labelling the two additional weak types object and method.

3.3 The Code Structure

We have retained all the original code for the ZCGa python file (though we have added to several functions within it as well as incorporating new functions). This makes us much more confident in the backwards compatibility of the OZCGa. It also gives us a very high degree of confidence that the structures present in both Z and Object-Z are type checked correctly.

Where possible we have copied the parts of each typing rule for terms as closely as possible when expanding the rules to deal with objects. Most the expression constants between objects and sets and between objects and other objects are the same as those between terms and sets and those between terms and other terms respectively. Similarly the declaration rule for objects closely resembles the declaration rules for terms. This allows us to have a high degree of confidence that these rules are well implemented—as they follow the same structure as well implemented code.

3.4 Problems Encountered

While the typing rules of Z are perfectly compatible with the object definition rule of Object-Z the way in which they are realised in the zcga.py file makes the object declaration rule difficult to implement without extensive restructuring of the code. The reason for this is that the ZCGa assumes that only one specification is checked at a time so the checking of specifications is not separable from the checking of documents. There are two problems with this approach when checking Object-Z specifications which require that the specifications contained within them are themselves well typed.

- A document can easily contain two specifications which are individually badly typed, but together well typed (if types only defined in one specification are used in the other).

- A document can easily contain two well typed specifications which would be badly typed if treated like they were one specification (if the same type is declared in each specification).

Our Solution Since we wanted to retain as much of the structure of the ZCGa as possible our solution is to ask the user to break down their specification so that only one class is checked at once. In order to do this all data from other classes, apart from the declarations of terms called within the class being checked using object.term should be deleted. An example of how this breakdown is done can be found in Appendix E.

4 Conclusion

In this paper we described a new translation path from an Object-Z specification into a theorem prover in a stepwise fashion. We have derived and implemented a new weak typing system for Object Z and thus completing the first step in the ZMath-Lang path. This weak typing system is defined by weak types and derivation rules for object-orientedness and shows that the system can weakly type check Object-Z specifications. The style file which was produced also acts as a clear visual reference for the typing of an Object-Z specification, many errors can be caught at the stage of adding these labels.

This paper is concerned with the Core Grammatical aspect of Object-Z. The next step is taking Object-Z specification through to the Document Rhetorical aspect to check the document rhetorical correctness (e.g. loops in the reasoning). Using the DRa we can automatically produce dependencies graphs and the proof skeleton. Other work which might be of interest is to create the MathLang path for other languages such as SysML or for specifications which are written non-formally in natural language.

Limitations to the ZDRa are that specifications need to be labelled by hand and the program runs on a terminal. Perhaps having a user interface to label the specification and run the program may make things more friendly to use.

Appendices

A Credit Card ZCGa

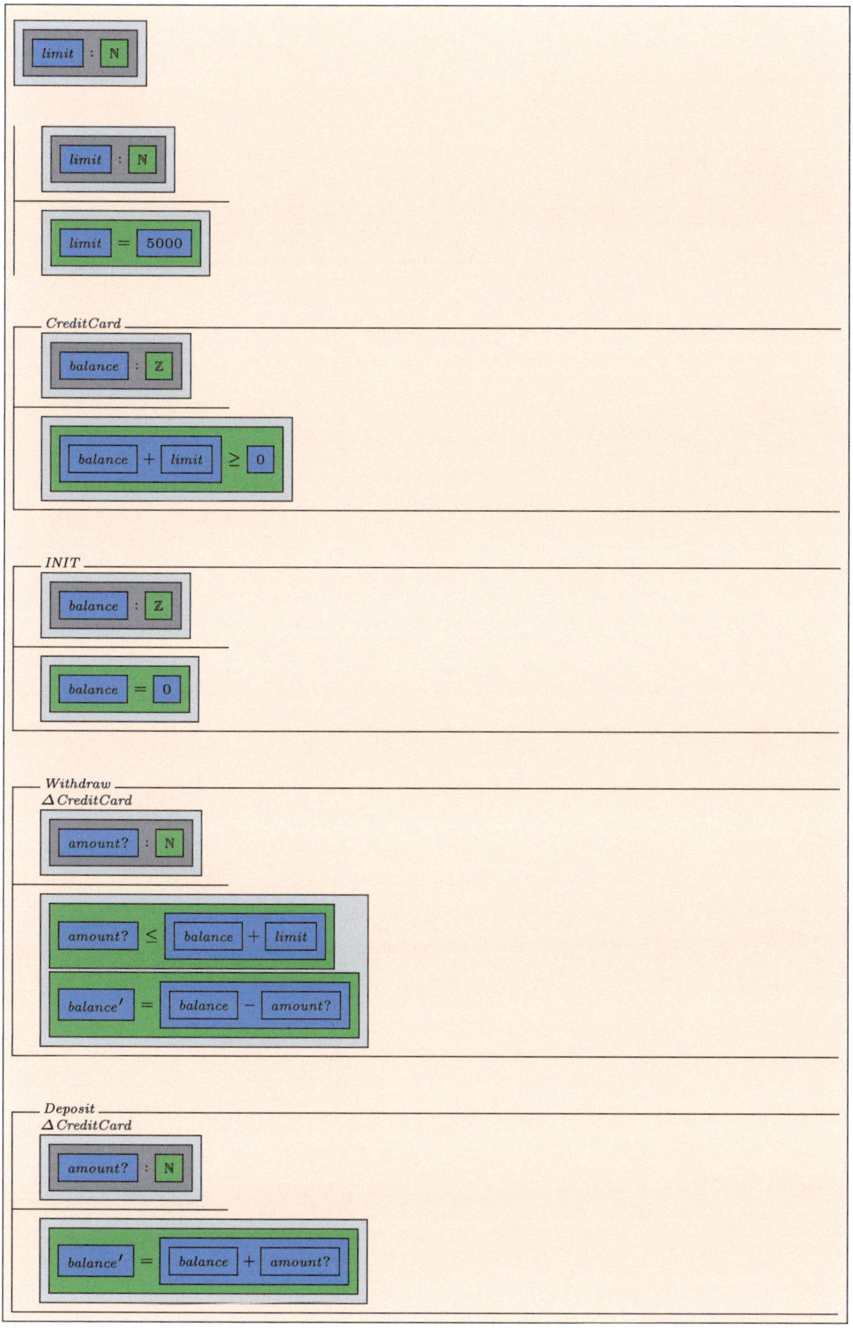

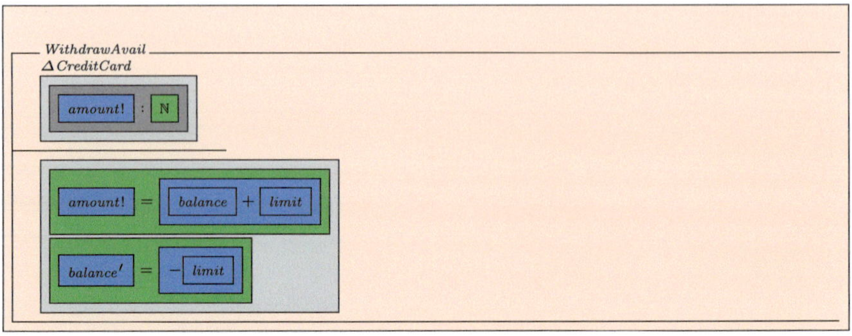
WithdrawAvail
ΔCreditCard
amount! : ℕ
amount! = balance + limit
balance′ = −limit

B CheckIn ZCGa

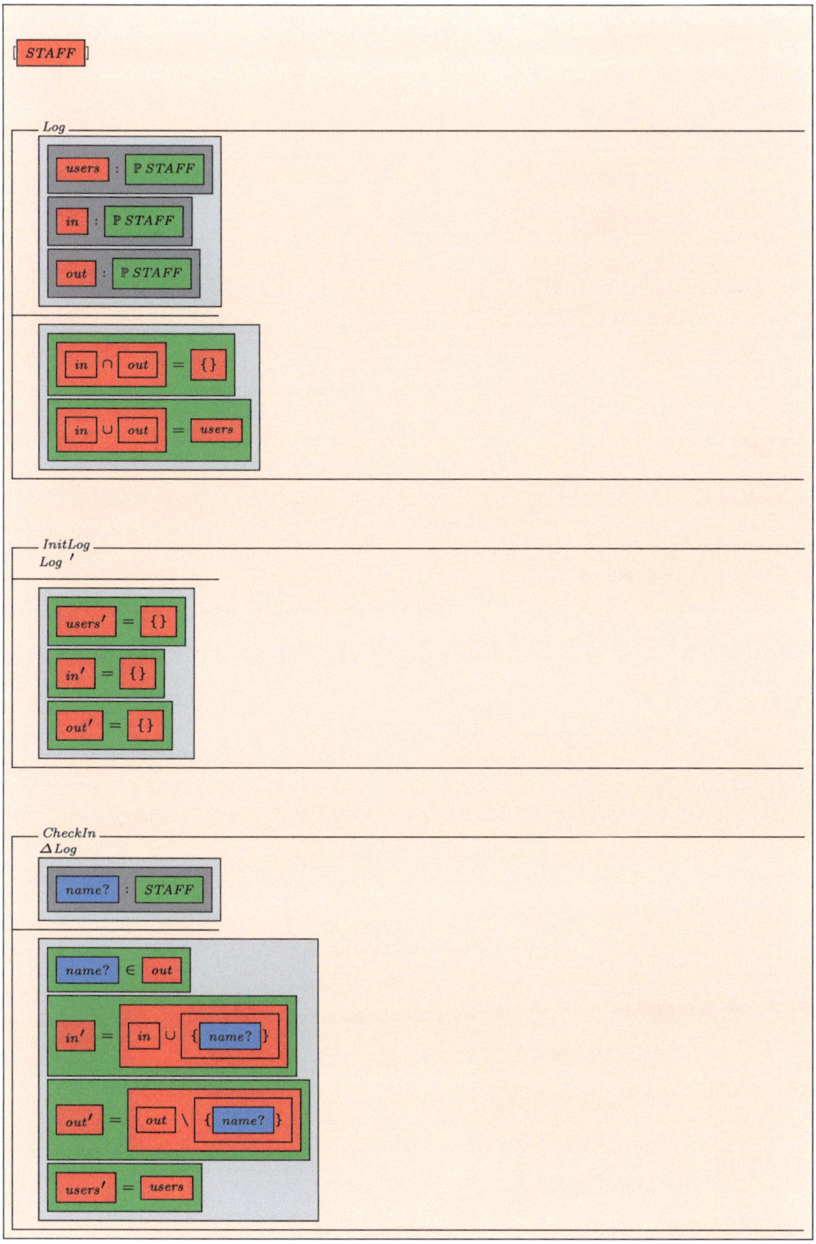

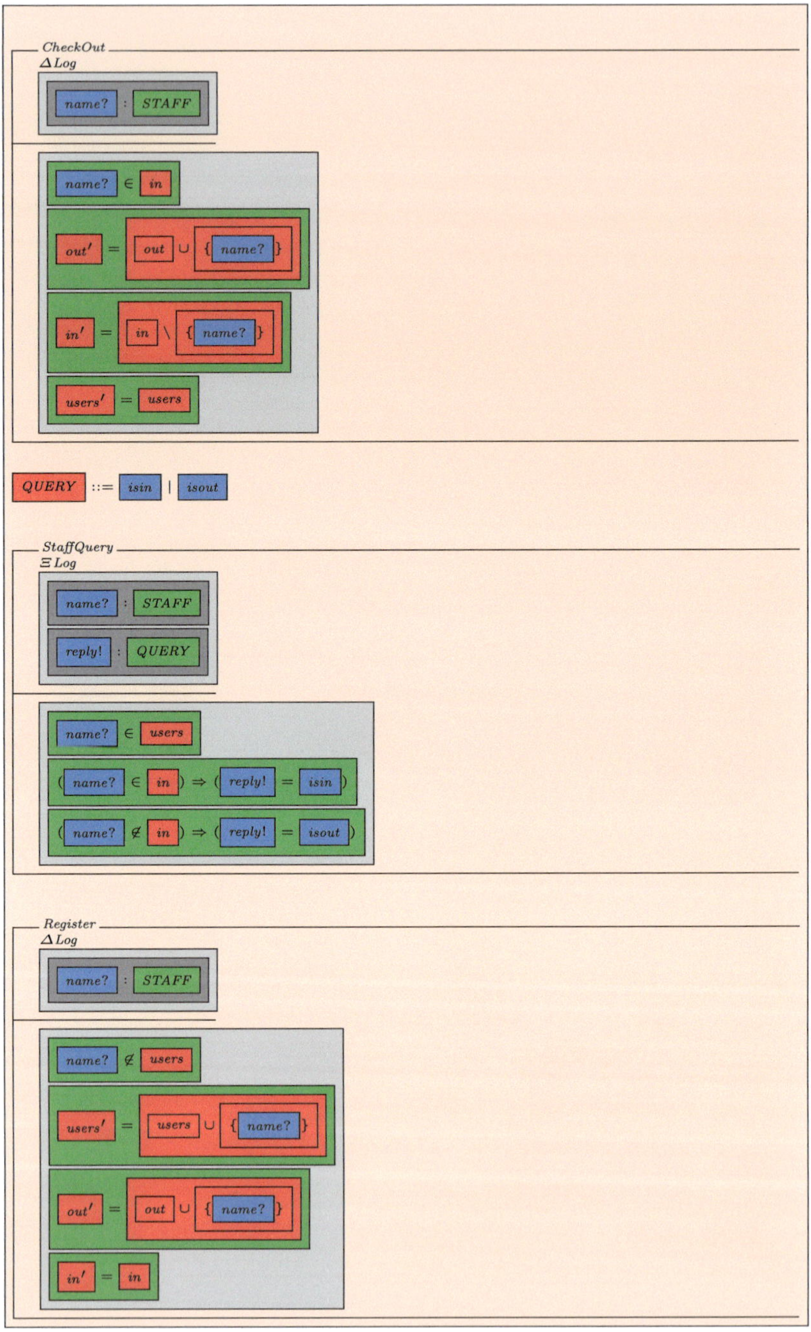
CheckOut
ΔLog
name? : STAFF
name? ∈ in
out′ = out ∪ {name?}
in′ = in \ {name?}
users′ = users
QUERY ::= isin | isout
StaffQuery
ΞLog
name? : STAFF
reply! : QUERY
name? ∈ users
(name? ∈ in) ⇒ (reply! = isin)
(name? ∉ in) ⇒ (reply! = isout)
Register
ΔLog
name? : STAFF
name? ∉ users
users′ = users ∪ {name?}
out′ = out ∪ {name?}
in′ = in

C Rich Tea ZCGa

$[INGREDIENTS]$

$TEMP ::= hot \mid cold$

$CHOICE ::= earlgrey \mid englishbreakfast$

TeaState

$cup : \mathbb{P}(INGREDIENTS \cup CHOICE)$

$kettle : \mathbb{P}\,INGREDIENTS$

$plate : \mathbb{P}\,INGREDIENTS$

$ingrTemp : INGREDIENTS \twoheadrightarrow TEMP$

$service : \mathbb{P}(\mathbb{P}\,INGREDIENTS)$

$\mathrm{dom}\ ingrTemp \subseteq cup \cup kettle$

$cup \cap kettle = \{\}$

InitTeaState

$cup : \mathbb{P}(INGREDIENTS \cup CHOICE)$

$kettle : \mathbb{P}\,INGREDIENTS$

$ingrTemp : INGREDIENTS \twoheadrightarrow TEMP$

$service : \mathbb{P}(\mathbb{P}\,INGREDIENTS)$

$cup = \{\}$

$kettle = \{\}$

$plate = \{\}$

$ingrTemp = \{\}$

$service = \{cup\}$

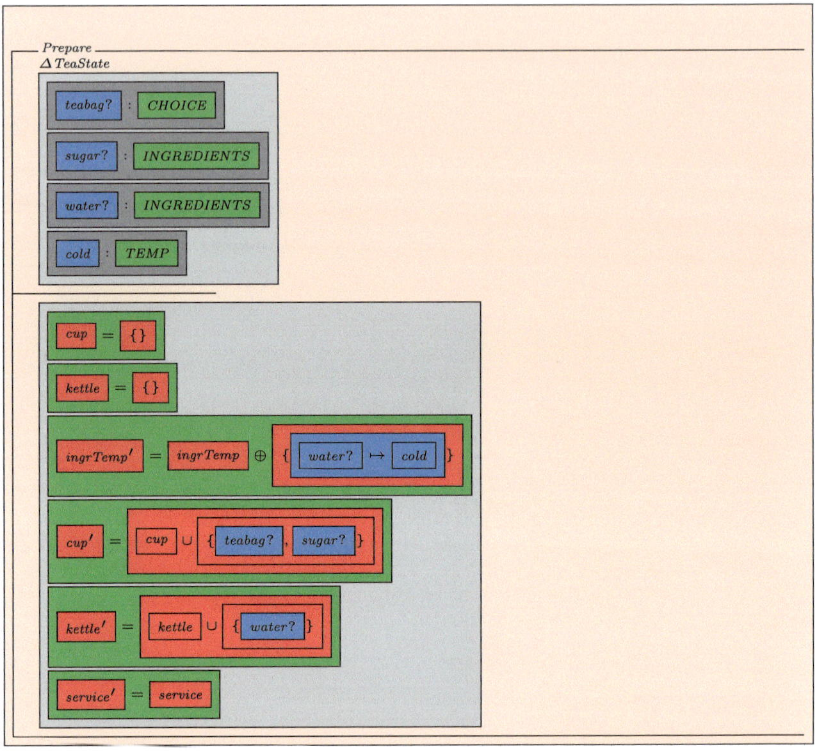
Prepare
Δ TeaState
teabag? : CHOICE
sugar? : INGREDIENTS
water? : INGREDIENTS
cold : TEMP
cup = { }
kettle = { }
ingrTemp′ = ingrTemp ⊕ { water? ↦ cold }
cup′ = cup ∪ { teabag?, sugar? }
kettle′ = kettle ∪ { water? }
service′ = service

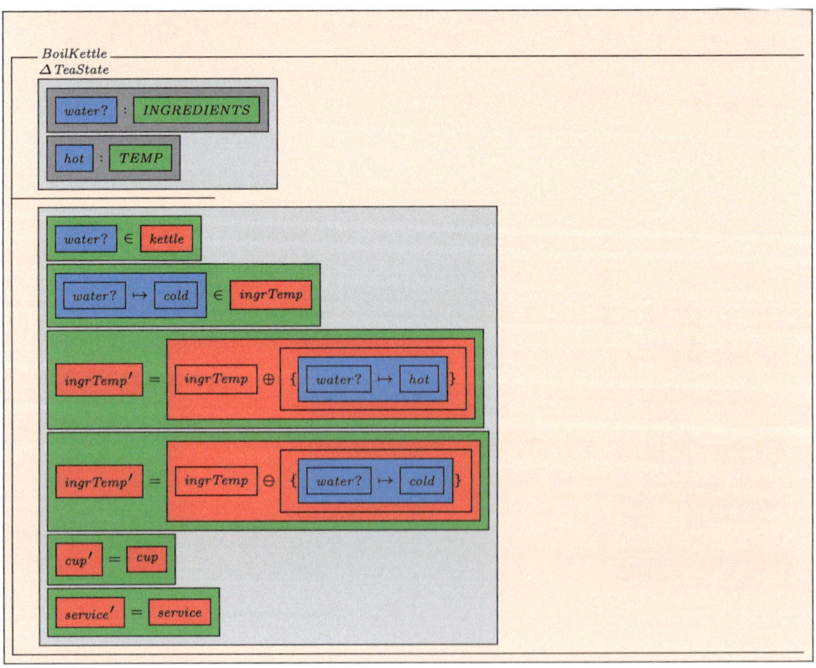
BoilKettle
Δ TeaState
water? : INGREDIENTS
hot : TEMP
water? ∈ kettle
water? ↦ cold ∈ ingrTemp
ingrTemp′ = ingrTemp ⊕ { water? ↦ hot }
ingrTemp′ = ingrTemp ⊖ { water? ↦ cold }
cup′ = cup
service′ = service

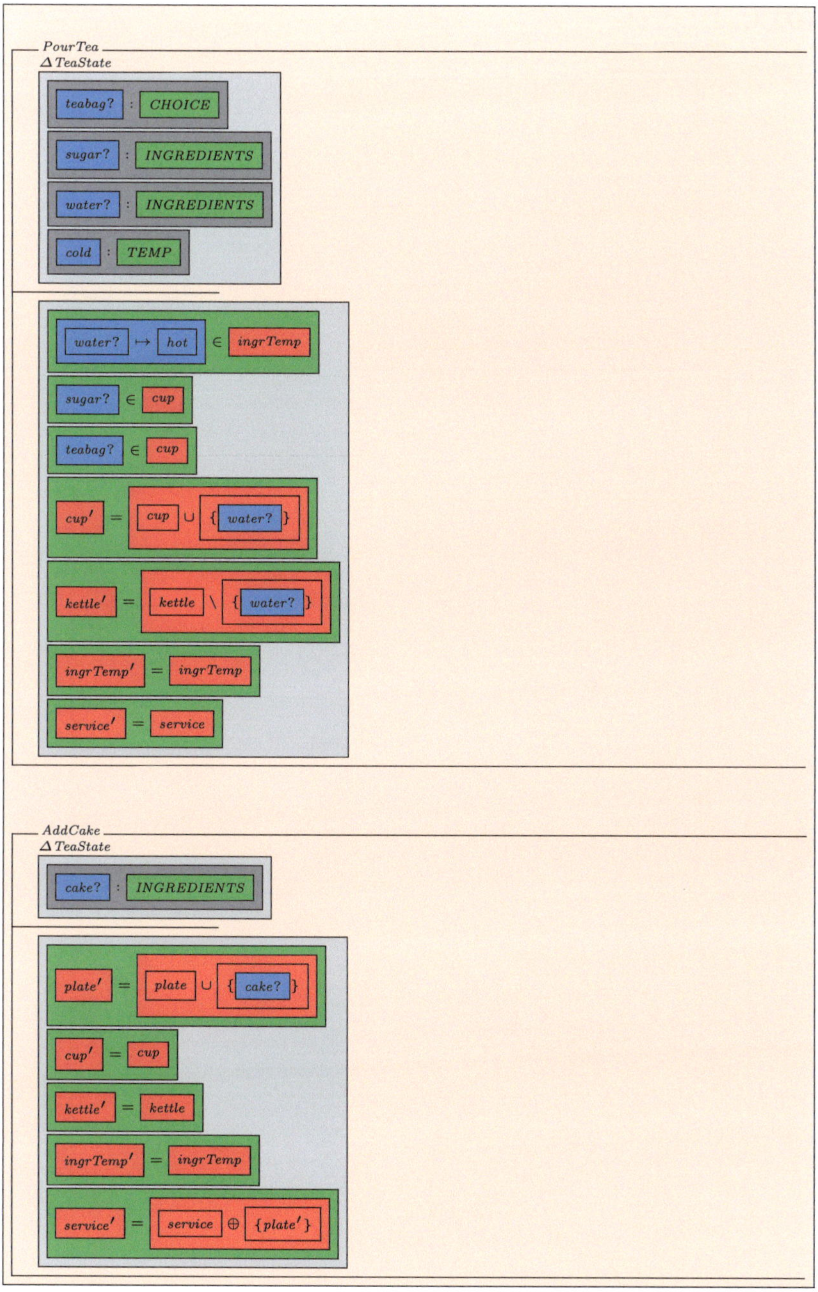
PourTea
Δ TeaState
teabag? : CHOICE
sugar? : INGREDIENTS
water? : INGREDIENTS
cold : TEMP
water? ↦ hot ∈ ingrTemp
sugar? ∈ cup
teabag? ∈ cup
cup′ = cup ∪ {water?}
kettle′ = kettle \ {water?}
ingrTemp′ = ingrTemp
service′ = service
AddCake
Δ TeaState
cake? : INGREDIENTS
plate′ = plate ∪ {cake?}
cup′ = cup
kettle′ = kettle
ingrTemp′ = ingrTemp
service′ = service ⊕ {plate′}

D OZCGa LaTeX file

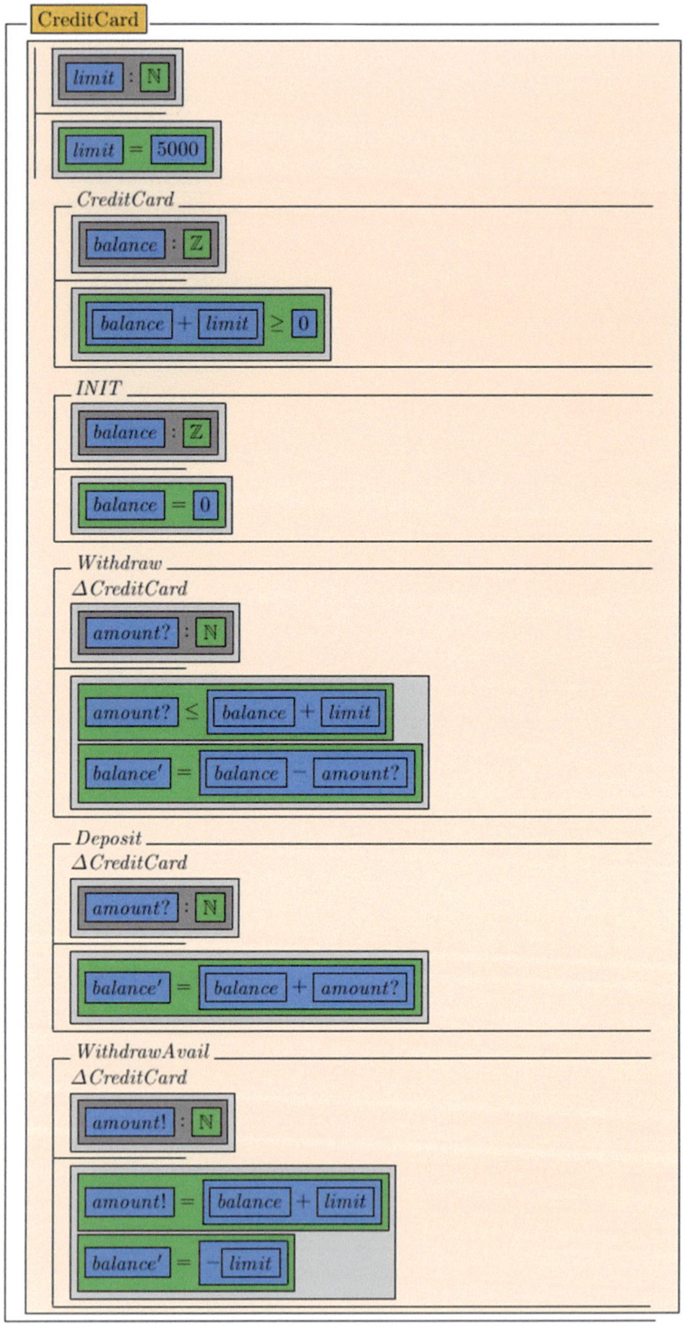

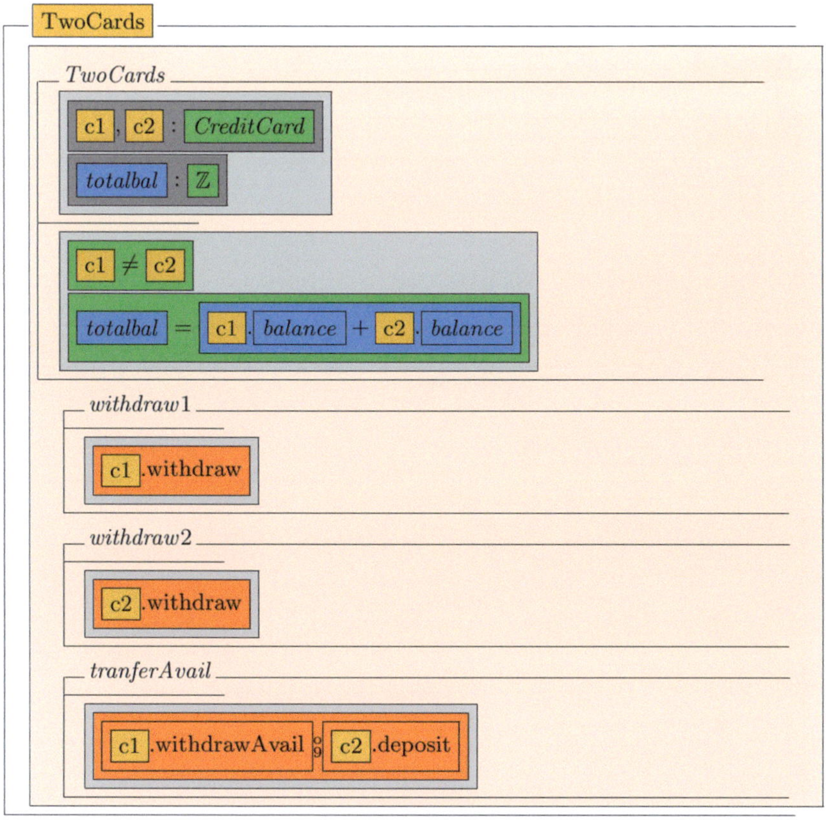
TwoCards
TwoCards
c1, c2 : CreditCard
totalbal : ℤ
c1 ≠ c2
totalbal = c1.balance + c2.balance
withdraw1
c1.withdraw
withdraw2
c2.withdraw
tranferAvail
c1.withdrawAvail ⨾ c2.deposit

E A Breakdown of the Specification for Checking

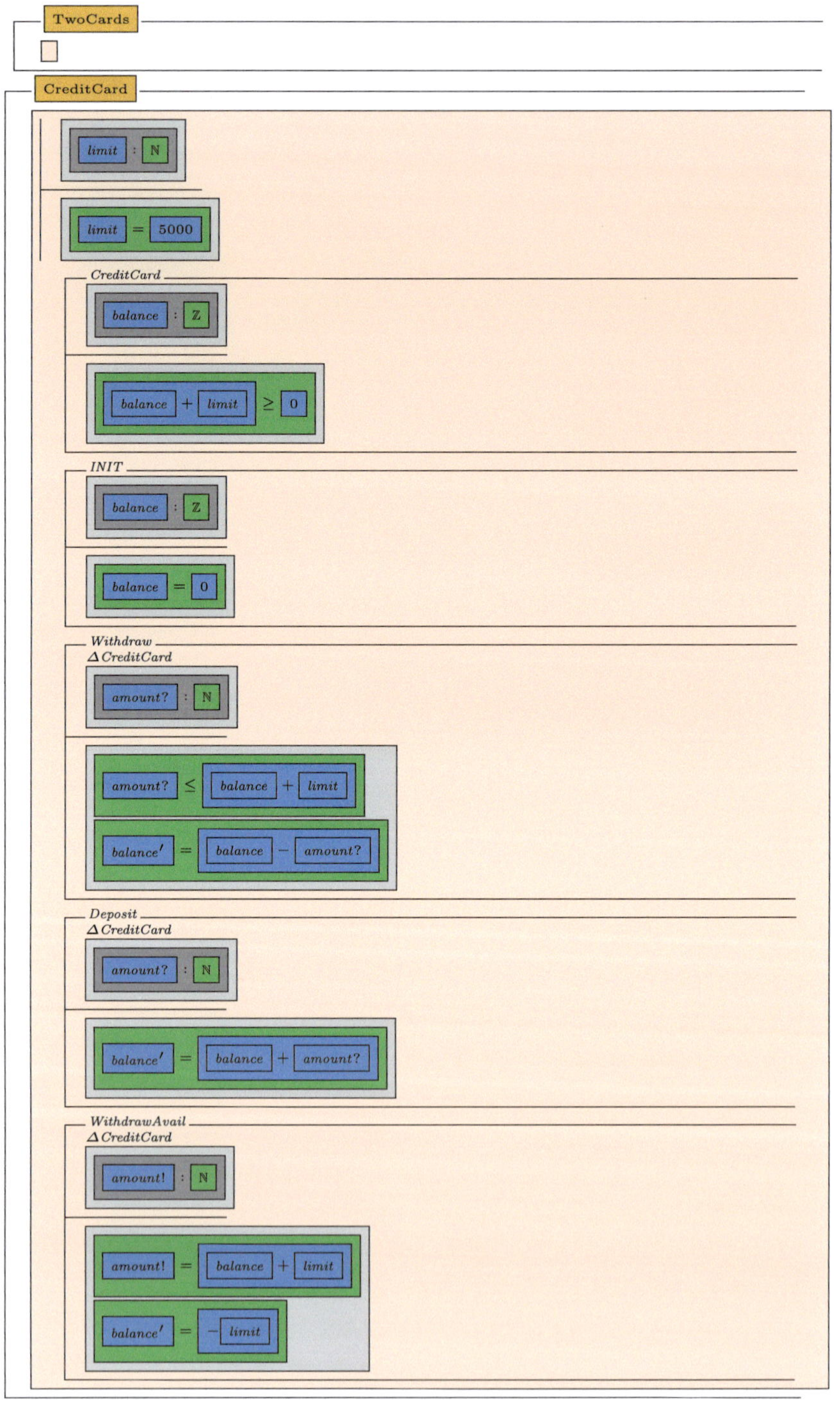

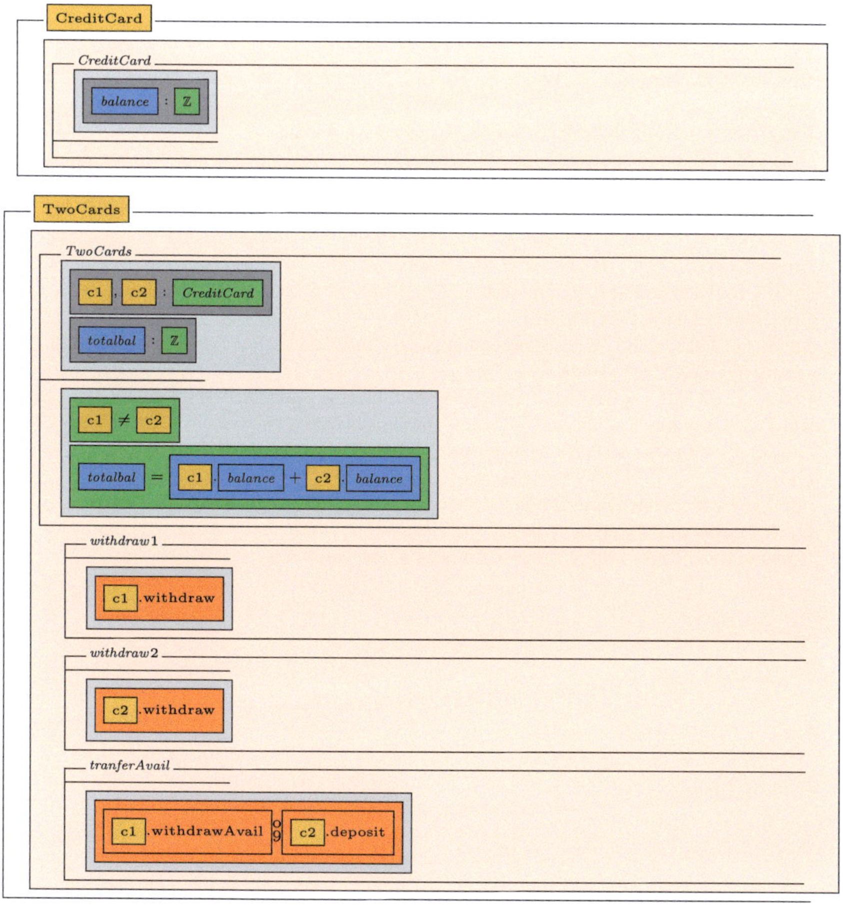

References

1. Frentiu, M., Correctness, a very important quality factor in programming. Stud. Univ. Babes-Bolyai Ser. Inf. **L**(1), 11–20 (2005)
2. MacKenzie, D.: Computer-related accidental death: an empirical exploration. Sci. Public Policy **21**(5), 233–248 (1994)
3. Dijkstra, E.W.: Notes on Structured Programming. Technological University Eindhoven, Department of Mathematics (1970)
4. DeMillo, R.A., Lipton, R.J., Perlis, A.J.: Social processes and proofs of theorems and programs. Commun. ACM **22**(5), 271–280 (1979)
5. Li, Y., Pan, X., Hu, T., Sung, S.Y., Yuan, H.: Specifying complex systems in Object-Z: a case study of petrol supply systems. J. Softw. **9**(7), 11 (2014)
6. Preibusch, S., Kamller, F.: Checking the TWIN Elevator System by Translating Object-Z to SMV. Formal Methods for Industrial Critical Systems: 12th International Workshop, FMICS, pp 38–57 (2007)

7. ISO/IEC 13568, Information technology Z formal specification notation Syntax, type system and semantics (2002)
8. Brucker, A.D., Rittinger, F., Wolff, B.: HOL-Z 2.0: a proof environment for Z-specifications. J. Univers. Comput. Sci. **9**(2), 152–172 (2003)
9. https://www.brucker.ch/projects/hol-z/. Accessed April 2015
10. Jones, R.: Methods and tools for the verification of critical properties. In: 5th Refinement Workshop, Springer Workshops in Computing (2004)
11. Arthan, R.: On Formal Specification of a Proof Tool. Lemma 1 Ltd
12. Kamareddine, F., Wells, J., Zengler, C., Barendregt, H.: Computerising mathematical text. Computational Logic. Handbook of the History of Logi, vol. 9, pp. 343–396. Elsevier (2014)
13. Burski, L., Kamareddine, F.: Translating Z into Isabelle Syntax using MathLang. Heriot Watt University, ULTRA Group (2015)
14. Smith, G.: The Object-Z Specification Language. Software Verification Research Centre, University of Queensland (1999)
15. Parker, T.: TOZE - A Graphical Editor for the Object-Z Specification Language with Syntax and Type Checking Capabilities. Masters Thesis, University of Wisconsin-La Crosse (2008)
16. Kimber, T.: Object-Z to Perfect Developer. Masters Thesis. Imperial College London, London (2007)
17. http://www.eschertech.com/products/perfect_developer.php. Accessed April 2015
18. Oz Style File. http://web.mit.edu/tex/stuff/latex-dist/psnfss/oz.sty. Accessed August (2015)
19. Community Z Tools. http://czt.sourceforge.net/. Accessed August (2015)

Ultimate Numerical Bound Estimation of Chaotic Dynamical Finance Model

Dharmendra Kumar and Sachin Kumar

Abstract This paper has investigated the boundedness of a 3D chaotic Dynamical Finance Model. We have discussed two bounds of this model. First by Lagrange multiplier method and second by optimization method. It was verified by using fmincon solver. Lyapunov Exponent calculated using Wolf algorithm and presented graphically in this paper. Lyapunov dimension of Dynamic Finance Model also discussed. Numerical simulations are presented to show the effectiveness of the proposed scheme.

Keywords Ultimate bounds · Chaotic finance model · Lyapunov stability theory · Positively invariant set · Fmincon · Optimization

1 Introduction

Lorenz first studied the chaotic dynamics in 1963 [9]. Dynamical diagnosis of financial models observed in recent research on financial chaotic dynamics. To understand the highly complex dynamics of real economic and financial systems, we need to study its global dynamical properties. Chaotic behaviour is not stochastic or random. On the contrary, a chaotic system is one that is completely deterministic, yet appears as if it were purely random, even to the extent of satisfying standard tests of randomness. Such systems are not predictable. Further, they do not necessarily require systems of complex equations to describe them. Remarkably, chaos may be generated from the simplest of nonlinear equations where, unlike in linear systems, the smallest changes can lead to extreme variability.

D. Kumar (✉)
Department of Mathematics, SGTB Khalsa College,
University of Delhi, Delhi 110007, India
e-mail: dhku06@gmail.com

S. Kumar
Department of Mathematics, University of Delhi, Delhi 110007, India
e-mail: sachinambariya@gmail.com

V.K. Singh et al. (eds.), *Modern Mathematical Methods and High Performance Computing in Science and Technology*, Springer Proceedings in Mathematics & Statistics 171, DOI 10.1007/978-981-10-1454-3_6

Chaotic phenomenon in economics was first found in 1985 [1]. The instability and complexity make the precise economic prediction greatly limited, and the reasonable prediction behaviour has become complicated as well. In the fields of finance, stocks and social economics, with all kinds of economic problems being more and more complicated. So it has become more and more important to make a systematic and deep study in the internal structural characteristics in such a complicated economic system. Chaos is about the irregular behaviour of solutions to deterministic equations of motion, and has received much attention from mathematicians and physicists over recent years. The equations must be nonlinear to generate chaotic solutions, but apart from that it can be remarkably simple. A nonlinear difference equation in one variable can generate chaos. Even in an ordinary differential equation in three variables can generate chaos. Chaotic solutions are only accurate for a length of time governed by the errors on initial conditions and the Lyapunov exponent of the system, which quantifies the exponential divergence of trajectories in chaotic systems. However, when considered in the underlying state space, in many cases chaotic solutions relax onto a strange attractor, which has a fractal structure and typically a non-integral dimension.

As we know, though a chaotic system is bounded, it is not an easy work to estimate and examine its bound. Therefore, an interesting fundamental question is how to estimate the bound of strange attractor. This objective can be achieved using four methods viz

1. the hyper plane oriented method [3],
2. the iteration theorem and the first order extremum theorem [23],
3. Lyapunov stability theory combined with the comparison principle method [8],
4. the optimization method [4, 19].

In recent papers, optimization theory is used for estimating the ultimate bounds of a class of High Dimensional Quadratic Autonomous Dynamical System [20]. The composition operators on Lorentz–Karamata–Bochner spaces and characterization of the properties like boundedness, closedness and essential range of these operators on the space has been discussed in [13, 14]. By using Boyd and Wong fixed point theorem, some existence, and uniqueness theorems of solutions and iterative approximation for solving these class of functional equations are established in [2, 15, 16]. In [21, 24], author used the Lagrange multiplier method to find two kinds of explicit ultimate bound sets and estimates the Hausdorff dimension of the novel hyperchaotic system. An estimate through the Lyapunov function of the upper bound of a threshold is precisely the threshold itself [17]. Using optimization method and the comparison principle, ultimate bounds and positively invariant sets of the hyperchaotic Lorenz Stenflo system found in [19]. Four-dimensional ellipsoidal ultimate bound and two-dimensional parabolic bound of Lorenz Haken system discussed in [7]. Ultimate bound and positively invariant set for the Lorenz system and the unified chaotic systems was studied in [6]. The discussion on ellipsoidal ultimate bound for unified chaotic system and two dimensional bound for the Chen system, Lu system, and unified system can be found in [5]. In [10], unification of the Lorenz and the Chen system using the unified system. Partial bound for the Chen system using suitable Lyapunov function [18] was discussed.

Chaos can be defined on bounded-state behaviour that is not equilibrium solution or a periodic solution or a quasi-periodic solution. This article is focused on analysis of dynamic properties and possible occurrence of chaotic behaviour. In this article, we found the Lyapunov Exponents and two types of bounds one from Lagrange method of multiplier and another by method of optimization with the help of toolbox in MATLAB namely fmincon solver.

2 Dynamic Finance Model

In [11, 12], author reported a dynamic model of finance, composed of three first-order differential equations. The model describes the time-variation of three state variables: the interest rate, x, the investment demand, y, and the price index, z. The factors that influence changes in x mainly come from two aspects: first, contradictions from the investment market, i.e. the surplus between investment and savings, and second, structural adjustment from good prices. The changing rate of y is in proportion to the rate of investment, and in proportion to an inversion with the cost of investment and interest rates. Changes in z, on the one hand, are controlled by a contradiction between supply and demand in commercial markets, and on the other hand, are influenced by inflation rates. By choosing, an appropriate coordinate system and setting appropriate dimensions for every state variable, [11, 12] offer the simplified finance model as

$$\begin{aligned} \dot{x} &= z + (y - a)x \\ \dot{y} &= 1 - by - x^2 \\ \dot{z} &= -x - cz \end{aligned} \tag{1}$$

where a is the saving amount, b is the cost per investment, and c is the elasticity of demand of commercial markets. It is obvious that all three constants a, b, and c, are non-negative. The parameters were chosen to be

$$a = 3, \quad b = \frac{1}{10}, \quad c = 1.0 \tag{2}$$

with an initial state $(x_0, y_0, z_0) = (2, 3, 2)$. In this case, the system has Lyapunov Exponents:

$$L_1 = 0.7848, \quad L_2 = 0.2260, \quad L_3 = -1.3332$$

It can be seen that the largest Lyapunov exponent is positive, indicating that the system has chaotic characteristics. Two L_1, L_2 are positive Lyapunov exponent, and the third one is negative. Thus, the system is chaotic. The time histories, phase

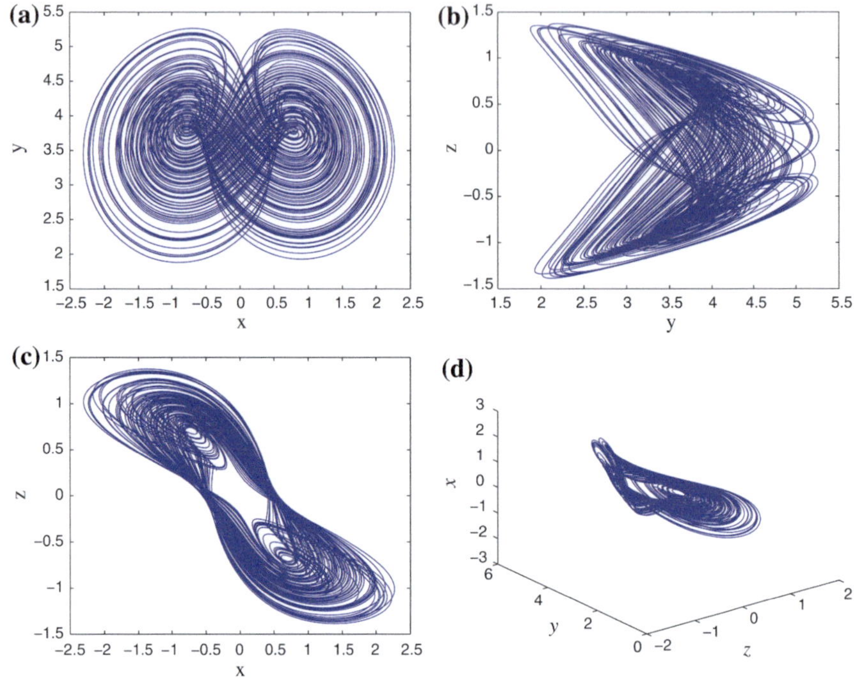

Fig. 1 Phase diagrams for the dynamic finance model. **a** Phase Portrait in x-y plane. **b** Phase Portrait in y-z plane. **c** Phase Portrait in x-z plane. **d** Phase Portrait in x-y-z plane

diagrams, and the largest Lyapunov Exponent were used to identify the dynamics of the system. The largest Lyapunov Exponent were calculated using the scheme proposed by Wolf [22].

Therefore, the Lyapunov dimension of the new chaotic (1) is given by

$$D_L = j + \frac{1}{|L_{j+1}|}\sum_{i=1}^{j} L_i = 2 + \frac{L_1 + L_2}{|L_3|} = 2.7582 \tag{3}$$

So, the chaos in this system (1) is very obvious. Thus the corresponding attractors are shown in Fig. 1.

3 Dynamical Behaviors of the Financial Chaotic System

3.1 Symmetry and Invariance

The system is invariant under the transformation (Fig. 2)

$$(x(t), y(t), z(t)) \to (-x(t), y(t), -z(t))$$

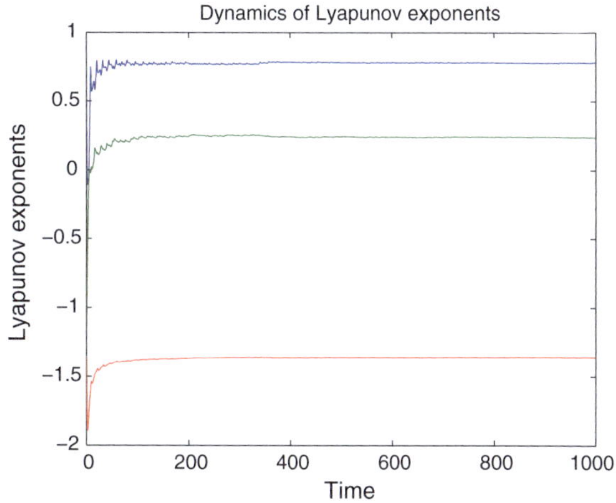

Fig. 2 Lyapunov exponent of the dynamic finance model

Thus, if $(x(t), y(t), z(t))$ is a solution, so is $(-x(t), y(t), -z(t))$. We see from Eq. (1) that is if $x(0) = 0$ and $z(0) = 0$ thus x and z remain zero for all t

$$\dot{y} = 1 - by$$

which is linear in $y(t)$. Solution is given by

$$y(t) = \frac{1}{b} + \left(y(0) - \frac{1}{b}\right) e^{-bt} \tag{4}$$

Unlike Lorenz equations instead of z-axis, the y-axis is always a part of the stable manifold for the equilibrium at the origin.

3.2 Dissipativity and Existence of Attractor

For the (1), it can be observed that

$$\nabla V = \frac{\partial \dot{x}}{\partial x} + \frac{\partial \dot{y}}{\partial y} + \frac{\partial \dot{z}}{\partial z} = -(a + b + c) \tag{5}$$

So, when $(a + b + c) > 0$, $\nabla V < 0$, (1) is dissipative, with an exponential contraction rate:

$$\frac{dV}{dt} = -(a + b + c)V \tag{6}$$

That is, a volume element V_0 is contracted by the flow into a volume element

$$V(t) = V_0 e^{-(a+b+c)t} \text{ in time t.}$$

With our canonical values of 3 for a and 1/10 for b and 1 for c, this is

$$V(t) = V_0 e^{-4.1t}$$

This means that each volume containing the system trajectory shrinks to zero as $t \to \infty$ at an exponential rate, $(a + b + c)$. Therefore, all system orbits are ultimately confined to a specific subset of zero volume, and the asymptotic motion settles onto an attractor.

4 Main Result

Theorem 4.1 ([21]) *All solution of system with parameters are globally bounded for time t.*

Proof 4.2 Define the Lyapunov function

$$V(x, y, z) = \frac{1}{2}(x^2 + y^2 + (z - a)^2) \tag{7}$$

Computing the derivative of $V(x, y, z)$ along the trajectory of (1) gives

$$\begin{aligned} \dot{V} &= x\dot{x} + y\dot{y} + (z - a)\dot{z} \\ &= x(z + (y - a)x) + y(1 - by - x^2) + (z - a)(-x - cz) \\ &= -ax^2 + ax - by^2 + y - cz^2 + caz \end{aligned}$$

Hence, one may take d_0 sufficiently large such that

$$a\left(x - \frac{1}{2}\right)^2 + b\left(y - \frac{1}{2b}\right)^2 + c(z - a)^2 > \frac{1}{4b} + \frac{a}{4} + ca^2 \tag{8}$$

That is equivalent to say (8) provided that (x, y, z) satisfies $V(x, y, z) = d$ with $d > d_0$. Consequently, on the surface

$$\{(x, y, z) | V(x, y, z) = d\}$$

where $d > d_0$, one has $\dot{V}(x, y, z) < 0$, which implies that the set

$$\{(x, y, z) | V(x, y, z) \leq d\}$$

is a trapping region, so that the solutions of (1) are globally bounded. In particular, when the system parameters are specified, it is easy to obtain *Max V*. For example, when $a = 3$, $b = 0.1$, $c = 1.0$.

4.1 Langrange Multiplier Method

Consider the following problem, which is constructed using Lyapunov function theory

$$Max\ V(x,y,z) = \frac{1}{2}(x^2 + y^2 + (z-a)^2) \tag{9}$$

Subject to constraint:

$$\Gamma : 3\left(x - \frac{1}{2}\right)^2 + \frac{1}{10}(y-5)^2 + (z-3)^2 = \frac{11}{4} \tag{10}$$

In order to solve the above maximization problem using Lagrange multiplier, we let

$$L = \frac{1}{2}(x^2 + y^2 + (z-3)^2) + \mu\left(3\left(x - \frac{1}{2}\right)^2 + \frac{1}{10}(y-5)^2 + (z-3)^2 - \frac{11}{4}\right) \tag{11}$$

and

$$L_x = x + 6\mu\left(x - \frac{1}{2}\right) = 0 \tag{12}$$

$$L_y = y + \frac{\mu}{5}(y-5) = 0 \tag{13}$$

$$L_z = (z-3) + 2\mu(z-3) = 0 \tag{14}$$

$$L_\mu = 3\left(x - \frac{1}{2}\right)^2 + \frac{1}{10}(y-5)^2 + (z-3)^2 - \frac{11}{4} = 0 \tag{15}$$

Here, we will discuss for $\mu = -9.7675$:

$$x = 0.5086, \quad y = 10.2438, \quad z = 2.9999 \tag{16}$$

Max V = 52.5971.
Therefore, the estimate of ultimate bound for (1) is

$$\Omega = \left\{(x,y,z) | \frac{1}{2}(x^2 + y^2 + (z-3)^2) \le 52.5971\right\} \tag{17}$$

is called positively invariant set for $a = 3$, $b = 0.1$, $c = 1$ and $(x_0, y_0, z_0) = (2, 3, 2)$. Based on these parameters, maximization problem gives $R_{\max} = 52.5971$. Therefore, the corresponding ultimate bound of system is achieved.

4.2 Optimization Method Using Matrix Analysis

Theorem 4.3 ([20]) *Suppose that there exits a real symmetric matrix* $P > 0$ *and a vector* $\mu \in \mathbb{R}^3$ *such that*

$$Q = A^T P + PA + 2(B_1^T P u^T, B_2^T P u^T, \ldots, B_n^T P u^T)^T < 0$$

and $\sum x_i X^T (B_i^T P + PB_i)X = 0$ *for any* $X = (x_1, x_2, \ldots, x_n)^T \in \mathbb{R}^3$ *and* $u = (u_1, u_2, \ldots, u_n) = 2\mu^T P$ *then (1) is bounded and has the following ultimate bound set also called positively invariant set:*

$$\Omega = \left\{ X \in \mathbb{R}^3 | (X + \mu)^T P (X + \mu) \le R_{\max} \right\} \tag{18}$$

where $R_{\max}$ *is a real number to be determined by*

$$Max\ (X + \mu)^T P (X + \mu) \tag{19}$$

subject to

$$X^T Q X + 2(\mu^T P A + C^T P) X + 2C^T P \mu = 0. \tag{20}$$

$$A = \begin{pmatrix} -a & 0 & 1 \\ 0 & -b & 0 \\ -1 & 0 & -c \end{pmatrix}; X = \begin{pmatrix} x_1 \\ x_2 \\ x_3 \end{pmatrix}; B_1 = \begin{pmatrix} 0 & 1/2 & 0 \\ -1 & 0 & 0 \\ 0 & 0 & 0 \end{pmatrix}; B_2 = \begin{pmatrix} 1/2 & 0 & 0 \\ 0 & 0 & 0 \\ 0 & 0 & 0 \end{pmatrix};$$

$$B_3 = \begin{pmatrix} 0 & 0 & 0 \\ 0 & 0 & 0 \\ 0 & 0 & 0 \end{pmatrix}; \mathrm{C} = \begin{pmatrix} 0 \\ 1 \\ 0 \end{pmatrix};$$

$$P = \begin{pmatrix} p_{11} & 0 & 0 \\ 0 & p_{11} & 0 \\ 0 & 0 & p_{33} \end{pmatrix}; \mu = \begin{pmatrix} 0 \\ -cp_{11} \\ 0 \end{pmatrix}; Q = \begin{pmatrix} -2ap_{11} - 2cp_{11}^2 & 0 & p_{11} - p_{33} \\ 0 & -2bp_{11} & 0 \\ p_{11} - p_{33} & 0 & -2cp_{33} \end{pmatrix}$$

Theorem 4.4 *Suppose that* $a > 0, b > 0, c > 0, p_{11} > 0, p_{33} > 0$. *Denote*

$$\Omega = \left\{ X \in \mathbb{R}^3 | p_{11} x_1^2 + p_{11} \left(-a + x_2\right)^2 + p_{33} x_3^2 \le R_{\max} \right\} \tag{21}$$

Then Ω is the ultimate bound of system (1), where $R_{\max}$ can be derived from the following optimization problem (Fig. 3 and Table 1):

$$Maximize : p_{11}x_1^2 + p_{11}\left(cp_{11} + x_2\right)^2 + p_{33}x_3^2$$

subject to constraint:

$$(-2ap_{11} - 2cp_{11}^2)x_1 + (p_{11} - p_{33})x_3 \; -2bp_{11}x_2 \; (p_{11} - p_{33})x_1 - 2cp_{33}x_3$$

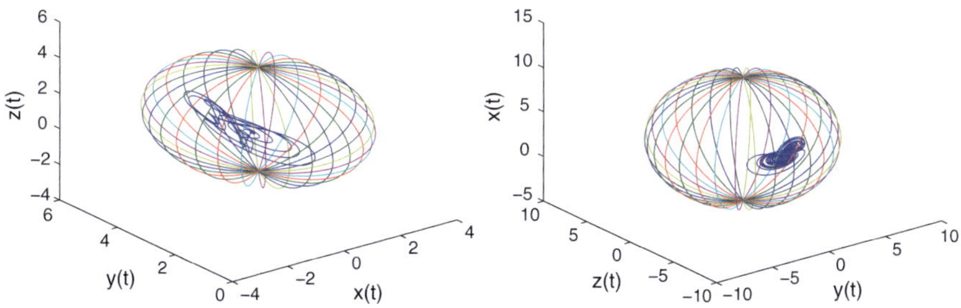

Fig. 3 The bound of the chaotic attractor of system with $a = 3, b = 0.1$ and $c = 1, p_{11} = 1,$ $p_{33} = 2$

Table 1 Simulation results using MATLAB

Parameters values	Initial condition (x_0, y_0, z_0)	$R_{\max}$
$p_{11} = 1, p_{33} = 2$	(2, 3, 2)	281560.49
$p_{11} = 0.1, p_{33} = 0.2$	(2, 3, 2)	129247.49
$p_{11} = 0.001, p_{33} = 0.001$	(2, 3, 2)	56796
$p_{11} = 0.001, p_{33} = 0.002$	(2, 3, 2)	79519.34
$p_{11} = 1000, p_{33} = 2$	(2, 3, 2)	1.2E48
$p_{11} = 1, p_{33} = 2000$	(2, 3, 2)	1.2E12
$p_{11} = 0.001, p_{33} = 0.001$	(2, 3, 2)	56796
$p_{11} = 1, p_{33} = 2$	(0, 0, 0)	165775.15
$p_{11} = 1, p_{33} = 1$	(2, 3, 2)	639152.0

5 Conclusion

Research on the ultimate bound for dynamics system is very important in both control theory and synchronization. It is very difficult to estimate the ultimate bounds of some typical chaotic systems. Even for the classical Chen system, there does not exist any effective bound estimation method reported over the last decade. In this paper, we have investigated the ultimate bound and positively invariant set for a financial dynamic model. To the best of our knowledge, we discussed first time Lyapunov Exponent and it is proved that positive value of Lyapunov Exponent shows chaotic nature of the dynamic finance model. Further, using MATLAB bounds are numerically calculated and observed for different parameters. In this paper, we have shown the ellipsoidal boundedness of the financial dynamic model. Numerical simulations show the effectiveness and advantage of our methods.

References

1. Baumol, W.J., Quandt, R.E.: Chaos models and their implications for forecasting. East. Econ. J. **11**, 3–15 (1985)
2. Deepmala, Das, A.K.: On solvability for certain functional equations arising in dynamic programming, mathematics and computing, Springer Proceedings in Mathematics and Statistics, vol. 139, pp. 79–94 (2015)
3. Leonov, G.A.: Bound for attractors and the existence of homoclinic orbit in Lorenz system. J. Appl. Math. Mech. **65**, 19–32 (2001)
4. Li, D.M., Lu, J.A., Wu, X.Q., Chen, G.R.: Estimating the bounds for the Lorenz family of chaotic systems. Chaos Solitons Fractals **23**, 529–534 (2005)
5. Li, D., Lu, J., C, Wu: Estimating the bounds for the Lorenz family of chaotic systems. Chaos Solitons Fractals **23**, 529–534 (2005)
6. Li, D., Lu, J., Wu, X., Chen, G.: Estimating the ultimate bound and positively invariant set for the Lorenz system and a unified chaotic system. J. Math. Anal. Appl. **323**, 844–853 (2006)
7. Li, D., Wu, X., Lu, J.: Estimating the ultimate bound and positively invariant set for the hyper chaotic LorenzHaken system. Chaos Solitons Fractals **39**, 1290–1296 (2009)
8. Liao, X.X.: On the global basin of attraction and positively invariant set for the Lorenz chaotic system and its application in chaos control and synchronization. Sci. China Ser. E **34**, 404–419 (2004)
9. Lorenz, E.N.: Deterministic non-periodic flow. J. Atmos. Sci. **20**, 130–141 (1963)
10. Lu, J., Chen, G., Cheng, D., Celikovsky, S.: Bridge the gap between the Lorenz system and the Chen system. Int. J. Bifurc. Chaos **12**, 2917–2926 (2002)
11. Ma, J.H., Chen, Y.S.: Study for the bifurcation topological structure and the global complicated character of a kind of nonlinear finance system (I). Appl. Math. Mech. **22**, 1240–1251 (2001)
12. Ma, J.H., Chen, Y.S.: Study for the bifurcation topological structure and the global complicated character of a kind of nonlinear finance system (II). Appl. Math. Mech. **22**, 1375–1382 (2001)
13. Mishra, V.N.: Some problems on approximations of functions in banach spaces, Ph.D. Thesis, Indian Institute of Technology, Roorkee - 247667, Uttarakhand, India (2007)
14. Mishra, V.N., Mishra, L.N.: Trigonometric approximation of signals (functions) in L_p ($p1$) norm. Int. J. Contemp. Math. Sci. **7**(19), 909–918 (2012)
15. Pathak, H.K., Deepmala: Existence and uniqueness of solutions of functional equations arising in dynamic programming. Appl. Math. Comput. **218**(13), 7221–7230 (2012)

16. Pathak, H.K., Deepmala: Some existence theorems for solvability of certain functional equations arising in dynamic programming, Bull. Calcutta Math. Soc. **104**(3), 237–244 (2012)
17. Pogromsky, A., Santoboni, G., Nijmeijer, H.: An ultimate bound on the trajectories of the Lorenz system and its applications. Nonlinearity **16**, 1597–1605 (2003)
18. Qin, W.X., Chen, G.: Chen G. On the boundedness of the solutions of the Chen system. J. Math. Anal. Appl. **329**, 445–451 (2007)
19. Wang, P., Li, D.M., Hu, Q.: Bounds of the hyper-chaotic LorenzStenflo system. Commun. Nonlinear Sci. Numer. Simul. **15**, 2514–2520 (2010)
20. Wang, P., Li, D., Wu, X., Lu, J., Yu, X.: Ultimate bound estimation of a class of high dimensional quadratic autonomous dynamical systems. Int. J. Bifur. Chaos Appl. Sci. Eng. **21**, 2679–2694 (2011)
21. Wang, P., Zhang, Y., Shaolin, T., Wan, L.: Explicit ultimate bound sets of a new hyperchaotic system and its application in estimating the Hausdorff dimension. Int. J. Nonlinear Dyn. **74**, 133–142 (2013)
22. Wolf, A., Swift, J.B., Swinney, H.L., Vastano, J.A.: Determining Lyapunov exponents from a time series. Phys. D **16**, 285–317 (1985)
23. Zhang, F.C., Shu, Y.l., Yang, H.l.: Bounds for a new chaotic system and its application in chaos synchronization, Commun. Nonlinear Sci. Numer. Simul. **16**(13), 1501–1508 (2011)
24. Zhang, F.C., Mu, C., Xiaowu, L.: On the boundness of some solutions of the Lu system. Int. J. Bifur. Chaos Appl. Sci. Eng. **22**(1), 1–5 (2012)

Basic Results on Crisp Boolean Petri Nets

Gajendra Pratap Singh and Sangita Kansal

Abstract The concept of Petri net as a discrete event-driven dynamical system was invented by Carl Adam Petri in his doctoral thesis 'Communication with Automata' in 1962. Petri nets are one of the best defined approach to modeling of discrete and concurrent systems. The dynamics of Petri nets represent the long-term behavior of the modeled system. Petri nets combine mathematical concepts with a pictorial representation of the dynamical behavior of the modeled systems. A Petri net is a bipartite directed graph consisting of two type of nodes, namely, place nodes and transition nodes. Directed arcs connect places to transitions and transitions to places to represent flow relation. In this paper, we present some basic results on 1-safe Petri net that generates every binary n-vector exactly once as marking vectors in its reachability tree, known as crisp Boolean Petri net. These results can be used for characterizing crisp Boolean Petri nets.

Keywords 1-safe Petri net · Binary n-vectors · Hamming distance · Reachability tree · Digraphs

1 Introduction

Petri nets are the graphical tool invented by Carl Adam Petri [8]. They are very reliable tool to model and study the structure of those discrete event-driven systems that are complex and tricky in nature. Petri nets are frequently used in many fields such as manufacturing processes, logistics, supply chain management, inventory, marketing, optimization, telecommunication systems, traffic systems, biological system

G.P. Singh (✉)
School of Computational and Integrative Sciences, Jawaharlal Nehru University,
New Mehrauli Road, New Delhi 110067, India
e-mail: gajendraresearch@gmail.com

S. Kansal
Department of Applied Mathematics, Delhi Technological University,
New Delhi 110042, Delhi, India
e-mail: sangita_kansal15@rediffmail.com

V.K. Singh et al. (eds.), *Modern Mathematical Methods and High Performance Computing in Science and Technology*, Springer Proceedings in Mathematics & Statistics 171, DOI 10.1007/978-981-10-1454-3_7

[2, 3], etc. Of all existing models, Petri nets and their extensions are of undeniable fundamental interest because they define easy graphical support for the representation and the understanding of basic mechanism and behaviors. It has changed the outlook of researchers of mathematics and computer science due to its applicability into real-life modeling problems related to engineering, computer science, technology, etc. The structure of Petri net is a bipartite directed graph. As a graphical tool, Petri net can be used for planning and designing a system with given objectives more practically effective than flowchart and block design diagrams. As a mathematical tool, it enables one to set up state equations, algebraic equations, and other mathematical models which govern the behavior of discrete dynamical systems. Kansal et al. [6] proposed a 1-safe *star Petri net* S_n, with $|P| = n$ and $|T| = n + 1$, having a central transition. This Petri net S_n generates all the binary n-vectors, as its marking vectors. We call such Petri nets as *Boolean Petri nets* [5]. In this paper, some basic results on crisp Boolean Petri nets have been presented. By a crisp Boolean Petri net, we mean a 1-safe Petri net which generates all the binary n-vectors as its marking vectors exactly once [7]. These marking vectors containing binary bits '0' and '1' are very much used in the design of generalized cyclic multiswitches such as those used to control automatic washing machine [1] and many interconnection networks with many alternatives and well-known properties such as regularity, symmetry, star networks, etc.

2 Preliminaries

For standard terminology and notation on Petri nets, we refer the reader to Peterson [9].

A *Petri net* is a 5-tuple $C = (P, T, I^-, I^+, \mu^0)$, where
(a) P is a nonempty set of 'places',
(b) T is a nonempty set of 'transitions',
(c) $P \cap T = \emptyset$,
(d) $I^-, I^+ : P \times T \longrightarrow N$, where N is the set of nonnegative integers, are called the *negative* and the *positive* 'incidence functions' (or, 'flow functions') respectively,
(e) $\forall p \in P, \exists t \in T : I^-(p, t) \neq 0$ or $I^+(p, t) \neq 0$ and
$\forall t \in T, \exists p \in P : I^-(p, t) \neq 0$ or $I^+(p, t) \neq 0$,
(f) $\mu^0 : P \rightarrow N$ is the *initial marking*.

In fact, $I^-(p, t)$ and $I^+(p, t)$ represent the number of arcs from p to t and t to p respectively. I^-, I^+ and μ^0 can be viewed as matrices of size $|P| \times |T|$, $|P| \times |T|$ and $|P| \times 1$, respectively.

A *marking* μ is a mapping $\mu : P \longrightarrow N$. A marking μ can hence be represented as a vector $\mu \in N^n$, $n = |P|$, such that the ith component of μ is the value $\mu(p_i)$.

A transition $t \in T$ is said to be *enabled* at μ if and only if $I^-(p, t) \leq \mu(p), \forall p \in P$. An enabled transition may or may not 'fire' (depending on whether or not the event actually takes place). After firing at μ, the new marking μ' is given by the rule

$$\mu'(p) = \mu(p) - I^-(p, t) + I^+(p, t), \text{ for all } p \in P.$$

We say that *t fires at* μ to yield μ' (or, that *t fires* μ to μ'), and we write $\mu \xrightarrow{t} \mu'$, whence μ' is said to be *directly reachable* from μ. Hence, it is clear, what is meant by a sequence like

$$\mu^0 \xrightarrow{t_1} \mu^1 \xrightarrow{t_2} \mu^2 \xrightarrow{t_3} \mu^3 \ldots \xrightarrow{t_k} \mu^k,$$

which simply represents the fact that the transitions $t_1, t_2, t_3, \ldots, t_k$ have been successively fired to transform the marking μ^0 into the marking μ^k. The whole of this sequence of transformations is also written in short, $\mu^0 \xrightarrow{\sigma} \mu^k$, where $\sigma = t_1, t_2, t_3, \ldots, t_k$.

A marking μ is said to be *reachable from* μ^0, if there exists a sequence of transitions which can be successively fired to obtain μ from μ^0. The set of all markings of a Petri net C reachable from a given marking μ is denoted by $M(C, \mu)$ and, together with the arcs of the form $\mu^i \xrightarrow{t_r} \mu^j$, represents what in standard terminology called the *reachability graph* $R(C, \mu)$ of the Petri net C. If the reachability graph has no cycle then it is called *reachability tree* of the Petri net C.

A place in a Petri net is *safe* if the number of tokens in that place never exceeds one. A Petri net is *safe* if all its places are safe.

The *preset* of a transition t is the set of all input places to t, i.e., ${}^\bullet t = \{p \in P : I^-(p, t) > 0\}$. The *postset* of t is the set of all output places from t, i.e., $t^\bullet = \{p \in P : I^+(p, t) > 0\}$. Similarly, $p's$ preset and postset are ${}^\bullet p = \{t \in T : I^+(p, t) > 0\}$ and $p^\bullet = \{t \in T : I^-(p, t) > 0\}$, respectively.

A transition without any input place is called a *source transition*, i.e., ${}^\bullet t = \emptyset$. A source transition is unconditionally enabled that represents an event or operation which can occur in every step. A transition without any output place is called a *sink transition*, i.e., $t^\bullet = \emptyset$. The firing of a sink transition consumes tokens but does not generate new token in the net.

Let $C = (P, T, I^-, I^+, \mu^0)$ be a Petri net with $|P| = n$ and $|T| = m$, the incidence matrix $I = [a_{ij}]$ is an $n \times m$ matrix of integers, $|P| = n$ and $|T| = m$ and its entries are given by $a_{ij} = a_{ij}^+ - a_{ij}^-$ where $a_{ij}^+ = I^+(p_i, t_j)$ is the number of arcs from transition t_j to its output place p_i, known as positive incidence matrix and $a_{ij}^- = I^-(p_i, t_j)$ is the number of arcs from place p_i to its output transition t_j, known as negative incidence matrix. In other words, $I = I^+ - I^-$.

The *Hamming distance* between two bit-strings $u = u_1, u_2, \ldots, u_n$, $v = v_1, v_2, \ldots, v_n \in \{0, 1\}^n$ of length n is the number of bit positions in which u and v differ: $d_H^n(u, v) = |\{i \in \{1, 2, \ldots, n\} : u_i \neq v_i\}|$. The Hamming distance between 1011101 and 1001001 is two.

Let $C = (P, T, I^-, I^+, \mu^0)$ be a Petri net and Z be a subnet of C. Then Z is called a *strong chain cycle* (SCC) of C or C is said to have a strong chain cycle (SCC) Z, if $|{}^\bullet t| = 2$, $|p^\bullet| = 2$ and $|t^\bullet| = 1\ \forall\, p, t \in Z$ [10]. If an SCC Z contains all the places of C then C is said to have a strong chain cycle covering all the places as shown in the following Fig. 1. Note that an SCC containing k places, where $k \leq n = |P|$ will always have k self-loops.

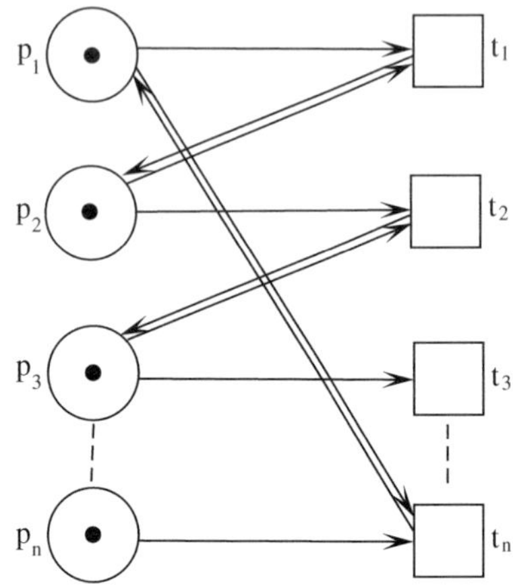

Fig. 1 Petri net having an SCC covering all the places

3 Results on Crisp Boolean Petri Nets

In this section, some results on crisp Boolean Petri net have been discussed.

Lemma 1 *In a crisp Boolean Petri net* $C = (P, T, I^-, I^+, \mu^0)$, $|P| = n$, *there exists exactly one sink transition.*

Proof Let $C = (P, T, I^-, I^+, \mu^0)$ be a crisp Boolean Petri net, i.e., a 1-safe Petri net that generates every binary n-vector exactly once. We shall prove this result by contradiction. Suppose C has two sink transitions, say t_i and t_j. Therefore $^\bullet t_i \neq \emptyset$ and $^\bullet t_j \neq \emptyset$. Since C is Boolean, $\mu^0(p) = 1, \ \forall\, p \in P$ ([7]). Therefore, all the transitions are enabled and fire. Now three cases arise.

Case 1: $|^\bullet t_i| = |^\bullet t_j| = n$. Then both the transitions t_i and t_j after firing produce zero marking vector, i.e., $(0, 0, 0, \ldots, 0)$. This leads to the contradiction that C is crisp Boolean.

Case 2: $|^\bullet t_i| = |^\bullet t_j| = 1$. This case has two subcases;

subcase 1: $^\bullet t_i = {}^\bullet t_j = \{p_k\}$, after firing of t_i and t_j, same binary n-vector will be generated which has 0 at kth place and 1 at other places, again a contradiction.

subcase 2: $^\bullet t_i = \{p_i\}$ and $^\bullet t_j = \{p_j\}, \forall\, i \neq j$. If t_i fires, it generates the marking vector $(1, 1, \ldots 0, 1 \ldots, 1)$, i.e., 0 at ith place. Now at this marking vector t_j is enabled and after firing it generates $(1, 1, \ldots, 0, 1, \ldots, 0, \ldots, 1)$, i.e., 0 at ith and jth places and 1 at other places. The same vector will be generated if t_j fires first at μ^0 and then t_i, again a contradiction.

Case 3: $1 < |^\bullet t_i| < n$ and $1 < |^\bullet t_j| < n$. Let $|^\bullet t_i| = m$ and $|^\bullet t_j| = r$, $1 < m; r < n$. After firing of t_i and t_j at μ^0, we get the marking vectors of Hamming distance m and r from μ^0, respectively, at the first stage only. So not all the binary n-vector of Hamming distance less than m and r will be generated after the first stage of firing (Remark 3.4.2 [10]), which is a contradiction that C is crisp Boolean.

Hence C cannot have more than one sink transition.

Lemma 2 *If t be the only sink transition in a crisp Boolean Petri net $C = (P, T, I^-, I^+, \mu^0)$, $|P| = n$, then $|{}^\bullet t|=1$ or n.*

Proof Let $|{}^\bullet t| = m$ where $1 < m < n$. Let ${}^\bullet t = \{p_1, p_2, p_3, \ldots, p_m\}$. Since $\mu^0(p) = 1, \ \forall p \in P$, t is enabled and fires. After firing, it generates the marking vector having 0 at m places and 1 at other places. That means, we cannot get the zero marking vector. This leads to the contradiction to the assumption.

Lemma 3 *If $C = (P, T, I^-, I^+, \mu^0)$, $|P|=n > 2$ be a crisp Boolean Petri net, then ${}^\bullet t_i \neq {}^\bullet t_j$ and $t_i^\bullet \neq t_j^\bullet, \forall\, i \neq j$.*

Proof Suppose $C = (P, T, I^-, I^+, \mu^0)$, $|P|=n > 2$ be a crisp Boolean Petri net and let ${}^\bullet t_i = {}^\bullet t_j$ and $t_i^\bullet = t_j^\bullet, \forall\, i \neq j$. In this case, repetition of binary n-vectors occur, contradiction to the fact that C is crisp Boolean. Hence, ${}^\bullet t_i \neq {}^\bullet t_j$ and $t_i^\bullet \neq t_j^\bullet$, $\forall\, i \neq j$.

4 Conclusions and Scope

In this paper, we have discussed some basic results on crisp Boolean Petri net. It has several properties that makes it attractive, e.g., the situation where the decision can be made at once. Further, one can think about to develop an algorithm for the embedding of a 1-safe Petri net into a crisp Boolean Petri net. Embedding [11, 12] is very useful concept in computer science to embed any subsystem into the connected networks. At presently, a computationally good characterization of crisp Boolean Petri net is still open.

References

1. Acharya, B.D.: Set-indexers of a graph and set-graceful graphs. Bull. Allahabad Math. Soc. **16**, 1–23 (2001)
2. Blateke, M.A., Heiner, M., Marwan, W.: Petri nets in system biology, Ist edn. Ottovon Guericke Universitat, Magdeburg, Germany (2011)
3. Chaouiya, C.: Petri net modelling of biological networks. Brief. Bioinform. **8**(4), 210–219 (2007)
4. Jensen, K.: Coloured Petri Nets. Lecture Notes in Computer Science, vol. 254, pp. 248–299. Springer-Verlag, Berlin (1986)
5. Kansal, S., Acharya, M., Singh, G.P.: Boolean petri nets. In: Pawlewski, P. (ed.) Petri nets—Manufacturing and Computer Science, Chap.17, pp. 381–406. In-Tech Global Publisher, Morn Hill (2012). ISBN 978-953-51-0700-2

6. Kansal, S., Singh, G.P., Acharya, M.: On Petri nets generating all the binary n-vectors, Scientiae Mathematicae Japonicae, 71(2), 209–216:e-2010, 113–120 (2010)
7. Kansal, S., Singh, G.P., Acharya, M.: 1-Safe Petri nets generating every binary n-vector exactly once, Scientiae Mathematicae Japonicae, 74(1), 29–36:e-2011, 127–137 (2011)
8. Petri, C.A.: Kommunikation mit automaten. Schriften des Institutes fur Instrumentelle Mathematik, Bonn (1962)
9. Peterson, J.L.: Petri net Theory and the Modeling of Systems, Englewood Cliffs. Prentice-Hall Inc, New Jersey (1981)
10. Singh, G.P.: Some Advances in the theory of Petri nets, Ph.D. thesis submitted to The Faculty of Technology, University of Delhi, Delhi, India (2013)
11. Singh, G.P., Kansal, S., Acharya, M.: Embedding an arbitrary 1-safe Petri net in a Boolean Petri net. Int. J. Comput. Appl. Found. Comput. Sci. New York, USA **70**(6), 7–9 (2013)
12. Singh, G.P., Kansal, S., Acharya, M.: Construction of a crisp Boolean Petri net from a 1-safe Petri net. Int. J. Comput. Appl. Found. Comput. Sci. New York, USA **73**(17), 1–4 (2013)

The Properties of Multiple Orthogonal Polynomials with Mathematica

Galina Filipuk

Abstract In this paper some computational aspects of studying various properties of multiple orthogonal polynomials are presented. The results were obtained using the symbolic and numerical computations in *Mathematica* (www.wolfram.com). This paper is mainly based on papers Filipuk et al., J Phys A: Math Theor 46:205–204, 2013, [1], Van Assche et al., J Approx Theory 190:1–25, 2015, [2], Zhang and Filipuk, Symmetry Integr Geom Methods Appl 10:103, 2014, [3] (joint with W. Van Assche and L. Zhang). We also perform the Painlevé analysis of certain nonlinear differential equation related to multiple Hermite polynomials and show the existence of two types of polar expansions, which might be useful to obtain relations for zeros of these polynomials.

Keywords Multiple orthogonal polynomials · Recurrence coefficients · Wronskians · Linear differential equations · Multiple Hermite polynomials · Symbolic computations in Mathematica

1 Introduction

The theory of orthogonal polynomials began in the nineteenth century in the papers of Hermite, Laguerre, Chebyshev, Jacobi, Bessel, and others. Nowadays the theory of orthogonal polynomials is a vivid, dynamic and an important part of the modern theory of special functions and mathematical analysis. Orthogonal polynomials also appear in quantum mechanics, number theory, approximation theory, stochastic processes, and other areas of mathematics and mathematical physics.

G. Filipuk (✉)
Institute of Mathematics of the Polish Academy of Sciences,
Śniadeckich str. 8, 00-956 Warsaw, Poland
e-mail: G.Filipuk@mimuw.edu.pl

G. Filipuk
Faculty of Mathematics, Informatics and Mechanics,
University of Warsaw, Banacha 2, 02-097 Warsaw, Poland

V.K. Singh et al. (eds.), *Modern Mathematical Methods and High Performance Computing in Science and Technology*, Springer Proceedings in Mathematics & Statistics 171, DOI 10.1007/978-981-10-1454-3_8

Let μ be a positive measure on the real line for which all the moments $\mu_n = \int x^n d\mu(x)$ are finite. It is well known (see, for example, [4–6]) that there exists a sequence of orthogonal polynomials $(p_n)_{n\in\mathbb{N}}$ (with a positive leading coefficient) such that

$$\int p_n(x)p_k(x)\,d\mu(x) = \delta_{n,k}, \tag{1}$$

where $\delta_{n,k}$ is the Kronecker delta. These orthonormal polynomials satisfy the so-called three-term recurrence relation:

$$xp_n(x) = a_{n+1}p_{n+1}(x) + b_n p_n(x) + a_n p_{n-1}(x), \tag{2}$$

where $p_{-1} = 0$ and the recurrence coefficients are given by the following integrals:

$$a_n = \int xp_n(x)p_{n-1}(x)\,d\mu(x), \qquad b_n = \int xp_n^2(x)\,d\mu(x). \tag{3}$$

Monic polynomials $p_n(x) = x^n + ...$ satisfy a similar three-term recurrence relation:

$$xp_n(x) = p_{n+1}(x) + b_n p_n(x) + a_n^2 p_{n-1}(x).$$

There exist various generalizations of orthogonal polynomials, for instance, multiple orthogonal polynomials. The topics connected to multiple orthogonal polynomials have become popular in recent years. This field of research is relatively new and has many open problems. Multiple orthogonal polynomials originated from the Hermite-Padé approximation in the context of irrationality and transcendence proofs in number theory (see, for instance, [5, 7, 8]). They also play a significant role in other areas of modern mathematics, for instance, in random matrix theory and certain models of mathematical physics [9].

2 Multiple Orthogonal Polynomials

Multiple orthogonal polynomials are polynomials of one variable orthogonal with respect to r different measures $\mu_1, \mu_2, \ldots, \mu_r$, where r is a natural number and $r > 1$. Let $\vec{n} = (n_1, n_2, \ldots, n_r) \in \mathbb{N}^r$ be a multi-index of size $|\vec{n}| = n_1 + n_2 + \ldots + n_r$, and suppose that $\mu_1, \mu_2, \ldots, \mu_r$ are positive measures on the real line which are absolutely continuous with respect to the Lebesgue measure μ.

Multiple orthogonal polynomials of type I are given by a vector $(A_{\vec{n},1}, \ldots, A_{\vec{n},r})$, where $A_{\vec{n},j}$ are polynomials of degree $\leq n_j - 1$, for which

$$\int x^k \sum_{j=1}^{r} A_{\vec{n},j}(x)w_j(x)\,d\mu(x) = 0, \qquad k = 0, 1, \ldots, |\vec{n}| - 2, \tag{4}$$

where w_j is the Radon–Nikodym derivative $d\mu_j/d\mu$, and the following normalization condition holds:

$$\int x^{|\vec{n}|-1} \sum_{j=1}^{r} A_{\vec{n},j}(x) w_j(x)\, d\mu(x) = 1. \tag{5}$$

Multiple orthogonal polynomials of type II are monic polynomials $P_{\vec{n}}(x) = x^{|\vec{n}|} + \cdots$, of degree $|\vec{n}|$, for which

$$\begin{aligned} \int P_{\vec{n}}(x) x^k \, d\mu_1(x) &= 0, \qquad k = 0, 1, \ldots, n_1 - 1, \\ &\vdots \\ \int P_{\vec{n}}(x) x^k \, d\mu_r(x) &= 0, \qquad k = 0, 1, \ldots, n_r - 1. \end{aligned} \tag{6}$$

The type I and II multiple orthogonal polynomials satisfy certain biorthogonality condition [5]. Multiple orthogonal polynomials of type I are much less studied than the polynomials of type II.

For instance, let $r = 2, \vec{n} = (n, m)$ and $c_1 \neq c_2$. The multiple Hermite polynomials of type II are defined by

$$\begin{aligned} \int_{-\infty}^{\infty} x^k H_{n,m}(x) e^{-x^2 + c_1 x} \, dx &= 0, \qquad k = 0, 1, \ldots, n - 1, \\ \int_{-\infty}^{\infty} x^k H_{n,m}(x) e^{-x^2 + c_2 x} \, dx &= 0, \qquad k = 0, 1, \ldots, m - 1. \end{aligned}$$

There exists an explicit formula for multiple Hermite polynomials using the usual Hermite polynomials which is useful to study them numerically.

3 Recurrence Relations for Multiple Orthogonal Polynomials

There exist recurrence relations similar to (2) for multiple orthogonal polynomials [10]. However, they are more complicated then in the case $r = 1$. For instance, in case $r = 2$ the recurrence relations are given by

$$\begin{aligned} x P_{n,m}(x) &= P_{n+1,m}(x) + c_{n,m} P_{n,m}(x) + a_{n,m} P_{n-1,m}(x) + b_{n,m} P_{n,m-1}(x), \\ x P_{n,m}(x) &= P_{n,m+1}(x) + d_{n,m} P_{n,m}(x) + a_{n,m} P_{n-1,m}(x) + b_{n,m} P_{n,m-1}(x) \end{aligned}$$

with $a_{0,m} = 0$ and $b_{n,0} = 0$ for all $n, m \geq 0$. For general $r > 1$ one has the following nearest-neighbor recurrence relations:

$$x P_{\vec{n}}(x) = P_{\vec{n}+\vec{e}_1}(x) + b_{\vec{n},1} P_{\vec{n}}(x) + \sum_{j=1}^{r} a_{\vec{n},j} P_{\vec{n}-\vec{e}_j}(x),$$
$$\vdots$$
$$x P_{\vec{n}}(x) = P_{\vec{n}+\vec{e}_r}(x) + b_{\vec{n},r} P_{\vec{n}}(x) + \sum_{j=1}^{r} a_{\vec{n},j} P_{\vec{n}-\vec{e}_j}(x),$$

where $\vec{e}_j = (0, \ldots, 0, 1, 0, \ldots, 0)$ is the jth standard unit vector with 1 on the jth entry and $(a_{\vec{n},1}, \ldots, a_{\vec{n},r})$, $(b_{\vec{n},1}, \ldots, b_{\vec{n},r})$ are the recurrence coefficients. For $r = 2$ the following notation for the recurrence coefficients is used:

$$a_{\vec{n},1} = a_{n,m}, \quad a_{\vec{n},2} = b_{n,m}, \quad b_{\vec{n},1} = c_{n,m}, \quad b_{\vec{n},2} = d_{n,m}.$$

4 Linear Ordinary Differential Equations for Multiple Orthogonal Polynomials

Paper [1] is devoted to the derivation of differential equations for multiple orthogonal polynomials (of the first and second type). The ladder operators are found for multiple orthogonal polynomials ([1, Theorems 2.1 and 2.2]) by using the Riemann–Hilbert problem and the Christoffel–Darboux formula. The ladder operators allow one to factorize the differential equations for multiple orthogonal polynomials with very few assumptions on the weight function. Several cases of weights of multiple orthogonal polynomials are considered (Hermite, Laguerre of the first and second kinds and with the cubic exponential function). Differential equations are computed for these weights to illustrate the general method. To give an example, the following theorem is proved in this paper.

Theorem 1 ([1, Theorem 2.1]) *Let $\mu_1, \ldots, \mu_r$ be absolutely continuous measures with the weights $w_1, \ldots, w_r$ and let the weights w_i vanish at the endpoints of the support of the measure μ_i. Suppose that all the indices $\vec{n} = (n_1, \ldots, n_r) \in \mathbb{N}^r$ are normal (see [1, p. 3] for the definition) and the functions*

$$\{w_1, xw_1, \ldots, x^{n_1-1}w_1, w_2, xw_2, \ldots, x^{n_2-1}w_2, \ldots, w_r, xw_r, \ldots, x^{n_r-1}w_r\}$$

are linearly independent. Then the type II multiple orthogonal polynomials satisfy the following equations:

$$P'_{\vec{n}}(x) = P_{\vec{n}}(x) \int P_{\vec{n}}(t) \sum_{k=1}^{r} A_{\vec{n},k}(t) \frac{v'_k(t) - v'_k(x)}{x-t} w_k(t)\, dt$$
$$- \sum_{j=1}^{r} a_{\vec{n},j} P_{\vec{n}-\vec{e}_j}(x) \int P_{\vec{n}}(t) \sum_{k=1}^{r} A_{\vec{n}+\vec{e}_j,k}(t) \frac{v'_k(t) - v'_k(x)}{x-t} w_k(t)\, dt,$$

$$P'_{\vec{n}-\vec{e}_i}(x) = P_{\vec{n}}(x) \int P_{\vec{n}-\vec{e}_i}(t) \sum_{k=1}^{r} A_{\vec{n},k}(t) \frac{v'_k(t) - v'_k(x)}{x-t} w_k(t)\, dt$$
$$- \sum_{j=1}^{r} \left(a_{\vec{n},j} \int P_{\vec{n}-\vec{e}_i}(t) \sum_{k=1}^{r} A_{\vec{n}+\vec{e}_j,k}(t) \frac{v'_k(t) - v'_k(x)}{x-t} w_k(t)\, dt - v'_i(x)\delta_{i,j} \right) P_{\vec{n}-\vec{e}_j}(x),$$

where $v_k(x) := -\ln w_k(x)$, *and* $a_{\vec{n},j}$ *are the recurrence coefficients for multiple orthogonal polynomials.*

A similar result [1, Theorem 2.2] is proved for multiple orthogonal polynomials of type I. In case $r = 1$ (ordinary orthogonal polynomials) this result coincides with the results in [11].

For multiple Hermite polynomials $H_{n,m}$ the recurrence coefficients are known explicitly:

$$2a_{n,m} = n, \quad 2b_{n,m} = m, \quad 2c_{n,m} = c_1, \quad 2d_{n,m} = c_2.$$

The following lowering and raising equations for $H_{n,m}$ hold:

$$H'_{n,m}(x) = nH_{n-1,m}(x) + mH_{n,m-1}(x),$$
$$H'_{n-1,m}(x) = -2H_{n,m}(x) + (2x - c_1)H_{n-1,m}(x),$$
$$H'_{n,m-1}(x) = -2H_{n,m}(x) + (2x - c_2)H_{n,m-1}(x).$$

Ladder operators allow one to find linear equations for multiple orthogonal polynomials ([1, Sect. 2.1]). For instance, the type II multiple orthogonal polynomials $H_{n,m}$ satisfy the following linear differential equation of order 3:

$$p'''(x) + (c_1 + c_2 - 4x)p''(x) + \left(c_1(c_2 - 2x) + 2(m + n - 1 - c_2x + 2x^2)\right) p'(x)$$
$$+ 2(c_1m + c_2n - 2(m + n)x)p(x) = 0. \tag{7}$$

Similar results hold for multiple orthogonal polynomials of the first type.

We can actually search for the multiple Hermite polynomials symbolically in *Mathematica*. The following code produces the type II multiple Hermite polynomials from the differential equation:

```
f[m_, n_, p_] := D[p,{x, 3}] + (c1 + c2 - 4 x) D[p,{x, 2}]
+(c1 (c2- 2 x) + 2 (m + n - 1 - c2 x + 2 x^2)) D[p,x]
+ 2 (c1 m + c2 n - 2 (m + n) x) p;
g[m_, n_] := Module[{k = m + n}, ((Sum[a[i] x^i,{i, 0, k}] //.
Solve[(# == 0) & /@ (CoefficientList[ f[m, n,
Sum[a[i] x^i,{i, 0, k}] /. a[k] -> 1], x]),
Table[a[i], {i, 0, k - 1}]]) /. a[k] -> 1)[[1]]]
```

Similarly, for multiple Laguerre polynomials of the first kind $L_{n,m}$ defined by the orthogonality conditions

$$\int_0^\infty x^k L_{n,m}(x) x^{\alpha_1} e^{-x}\, dx = 0, \qquad k = 0, 1, \dots, n-1,$$
$$\int_0^\infty x^k L_{n,m}(x) x^{\alpha_2} e^{-x}\, dx = 0, \qquad k = 0, 1, \dots, m-1,$$

where $\alpha_1, \alpha_2 > 0$ and $\alpha_1 - \alpha_2 \notin \mathbb{Z}$, the ladder operators have the following form:

$$\begin{aligned} xL'_{n,m}(x) &= (n+m)L_{n,m}(x) + \left(\frac{n(n+\alpha_1)(n+\alpha_1-\alpha_2)}{n-m+\alpha_1-\alpha_2}\right) L_{n-1,m}(x) \\ &\quad + \left(\frac{m(m+\alpha_2)(m+\alpha_2-\alpha_1)}{m-n+\alpha_2-\alpha_1}\right) L_{n,m-1}(x), \\ xL'_{n-1,m}(x) &= -L_{n,m}(x) - (n+\alpha_1-x)\, L_{n-1,m}(x), \\ xL'_{n,m-1}(x) &= -L_{n,m}(x) - (m+\alpha_2-x)\, L_{n,m-1}(x). \end{aligned}$$

Furthermore, the third order differential equation satisfied by $L_{n,m}$ is given by

$$\begin{aligned} &x^2 p'''(x) + \left(-2x^2 + (\alpha_1+\alpha_2+3)x\right) p''(x) \\ &\quad + \left(x^2 - x(\alpha_1+\alpha_2-n-m+3) + (\alpha_1+1)(\alpha_2+1)\right) p'(x) \\ &\quad - (x(n+m) - (n+m+nm+\alpha_1 m+\alpha_2 n))\, p(x) = 0. \end{aligned} \tag{8}$$

5 Wronskians with Multiple Orthogonal Polynomials

Karlin and Szegő developed an interesting and general theory regarding the determinants whose entries are orthogonal polynomials. Let

$$Q_n(x) = k_n(-x)^n + \dots, \quad k_n > 0, \quad n \in \mathbb{N} = \{0, 1, 2, 3, \dots\}$$

be a sequence of polynomials. The Wronskian of these polynomials is defined by

$$\begin{aligned} W(n, l; x) &:= W\left(Q_n(x), Q_{n+1}(x), \dots, Q_{n+l-1}(x)\right) \\ &= \det \begin{pmatrix} Q_n(x) & Q_{n+1}(x) & \cdots & Q_{n+l-1}(x) \\ Q'_n(x) & Q'_{n+1}(x) & \cdots & Q'_{n+l-1}(x) \\ \vdots & \vdots & \vdots & \vdots \\ Q_n^{(l-1)}(x) & Q_{n+1}^{(l-1)}(x) & \cdots & Q_{n+l-1}^{(l-1)}(x) \end{pmatrix}. \end{aligned}$$

It is known that, for l even,

$$(-1)^{l/2} W(n, l; x) > 0, \qquad x \in \mathbb{R},$$

i.e., the Wronskian keeps a constant sign for all real x; if l is odd, then $W(n, l; x)$ has exactly n simple real zeros and the zeros of $W(n, l; x)$ and $W(n+1, l; x)$ strictly interlace.

Another important class of determinants is the Hankel determinant

$$\begin{aligned} T(n, l; x) &:= T\left(Q_n(x), Q_{n+1}(x), \ldots, Q_{n+l-1}(x)\right) \\ &= \det \begin{pmatrix} Q_n(x) & Q_{n+1}(x) & \cdots & Q_{n+l-1}(x) \\ Q_{n+1}(x) & Q_{n+2}(x) & \cdots & Q_{n+l}(x) \\ \vdots & \vdots & \vdots & \vdots \\ Q_{n+l-1}(x) & Q_{n+l}(x) & \cdots & Q_{n+2l-2}(x) \end{pmatrix}, \end{aligned}$$

which is called the Turánian. Karlin and Szegő showed that, if l is even, $T(n, l; x)$ has the sign $(-1)^{l/2}$ on a certain interval I for specific three classical systems of orthogonal polynomials. Note that, if $l = 2$, then one has

$$T(n, 2; x) = Q_{n+1}^2(x) - Q_n(x) Q_{n+2}(x) > 0.$$

This inequality is called the Turán inequality, which was first proved for the Legendre polynomials $P_n(x) = P_n^{(1/2)}(x)$.

It is known that the Turán type inequalities are true for the majority of classical orthogonal polynomials. There have been numerous studies and generalizations of the classical results. Also Turán type inequalities have recently been found for many special functions and their q-analogues (numerous papers by A. Baricz et al.).

In paper [3] the Wronskians with the multiple orthogonal polynomials are studied. It is proved that depending on the size of the determinant some determinants have constant sign and others may have zeros ([3, Theorems 1.1–1.3]). These results generalize similar statements of Karlin and Szegő for ordinary orthogonal polynomials. The Túran type inequalities are proved for multiple Hermite and Laguerre polynomials ([3, Theorems 4.1 and 4.2]).

To give more details from [3], we suppose that $w_1, w_2, \ldots, w_r$ are r weights with supports on the real axis. Define $\vec{n}_0, \vec{n}_1, \ldots, \vec{n}_{l-1}$ to be a path from $\vec{n}$ to an arbitrary multi-index $\vec{n}_{l-1}$, where in each step the multi-index $\vec{n}_k$, for $k = 0, \ldots, l-1$, is increased by one at exactly one direction. For instance, for $r = 2$ we have in each step either $(n, m) \to (n+1, m)$ or $(n, m) \to (n, m+1)$. For every such kind of fixed path, we define the associated Wronskian by

$$W(\vec{n}, l; x) := W\left(P_{\vec{n}_0}(x), P_{\vec{n}_1}(x), \ldots, P_{\vec{n}_{l-1}}(x)\right)$$
$$= \det \begin{pmatrix} P_{\vec{n}_0}(x) & P_{\vec{n}_1}(x) & \cdots & P_{\vec{n}_{l-1}}(x) \\ P'_{\vec{n}_0}(x) & P'_{\vec{n}_1}(x) & \cdots & P'_{\vec{n}_{l-1}}(x) \\ \vdots & \vdots & \vdots & \vdots \\ P^{(l-1)}_{\vec{n}_0}(x) & P^{(l-1)}_{\vec{n}_1}(x) & \cdots & P^{(l-1)}_{\vec{n}_{l-1}}(x) \end{pmatrix},$$

where $P_{\vec{n}}$ is the type II multiple orthogonal polynomial. Clearly, $W(\vec{n}, l; x)$ is a polynomial in x depending on the parameters $\vec{n}$, l and the path starting from $\vec{n}$.

It is proved in [3] that if l is even, then

$$W(\vec{n}, l; x) > 0, \qquad x \in \mathbb{R}.$$

If l is odd, then for each fixed multi-index $\vec{n}$ the polynomials $W(\vec{n}, l; x)$ have exactly $|\vec{n}|$ simple zeros on the real axis. Furthermore, given two paths with l multi-indices, if the last $l-1$ multi-indices of one path coincide with the first $l-1$ multi-indices of another path, then the real zeros of two associated Wronskians strictly interlace [3].

The Turán type inequalities for multiple Hermite polynomials are given by

$$H_{\vec{n}+\vec{e}_k}(x) H_{\vec{n}+\vec{e}_j}(x) - H_{\vec{n}}(x) H_{\vec{n}+\vec{e}_k+\vec{e}_j}(x) > 0, \quad x \in \mathbb{R},$$

for $j, k = 1, \ldots, r$, where $\vec{e}_k = (0, \ldots, 0, 1, 0, \ldots, 0)$ denotes the kth standard unit vector with 1 on the kth entry. In particular, by taking $j = k$, we have

$$H^2_{\vec{n}+\vec{e}_k}(x) - H_{\vec{n}}(x) H_{\vec{n}+2\vec{e}_k}(x) > 0, \quad x \in \mathbb{R}.$$

Multiple Laguerre polynomials of the first kind for general r are defined by

$$\int_0^\infty x^k L^{\vec{\alpha}}_{\vec{n}}(x) x^{\alpha_j} e^{-x}\, dx = 0, \qquad k = 0, 1, \ldots, n_j - 1,$$

for $j = 1, \ldots, r$, where $\alpha_j > -1$ and $\alpha_i - \alpha_j \notin \mathbb{Z}$ whenever $i \neq j$. As in the case of multiple Hermite polynomials, there also exists an explicit formula for the multiple Laguerre polynomials of the first kind. We have the following Turán type inequalities:

$$L^{\vec{\alpha}}_{\vec{n}+\vec{e}_k}(x) L^{\vec{\alpha}-\vec{e}_j}_{\vec{n}+\vec{e}_j}(x) - L^{\vec{\alpha}}_{\vec{n}}(x) L^{\vec{\alpha}-\vec{e}_j}_{\vec{n}+\vec{e}_j+\vec{e}_k}(x) > 0, \qquad x > 0,$$

for $\vec{\alpha} > \vec{0}$ and $j, k = 1, \ldots, r$.

We can check the validity of the Turán type inequalities by using the function Reduce in *Mathematica*:

```
h[m_, n_] := g[m + 1, n]^2- g[m, n] g[m + 2, n] // Together
Reduce[h[3, 2] > 0]
```

The zero distribution on the complex plane of multiple orthogonal polynomials and their Wronskians can be studied numerically (using the program *Mathematica*) (see [3] for details) and it is shown that the zero distribution has a regular structure.

6 Multiple Orthogonal Polynomials Associated with an Exponential Cubic Weight

In paper [2] certain multiple orthogonal polynomials (and their properties), defined on some curves in the complex plane with the cubic exponential weight function, are studied ([2, Sect. 1.2]). The Rodrigues formula is found ([2, Sect. 1.2]); it is proved that the recurrence coefficients can be expressed in terms of the solutions of the discrete Painlevé equation ([2, Proposition 1.1, Theorem 1.3]); the ratio asymptotics of the polynomials is given ([2, Theorem 4.2]), asymptotics of the recurrence coefficients is studied ([2, Proposition 3.1]) and the zero distribution on the complex plane is presented (numerical computations are given in the computer program *Mathematica*).

To summarize the construction of multiple orthogonal polynomials associated with an exponential cubic weight function, we consider the three rays

$$\Gamma_k = \{z \in \mathbb{C} : \arg z = \omega^k\}, \quad \omega = e^{2\pi i/3}, \quad k = 0, 1, 2,$$

where the orientations are all taken from left to right.

We shall denote by $p_n^{(1)}$ the monic orthogonal polynomials satisfying

$$\int_\Gamma p_n(x)x^k e^{-x^3} dx = 0, \qquad k = 0, 1, \ldots, n-1, \tag{9}$$

with $\Gamma = \Gamma_0 \cup \Gamma_1$ and recurrence coefficients $\beta_n^{(1)}$ and $(\alpha_n^{(1)})^2$. In a similar manner, we set $p_n^{(2)}$ to be the polynomials satisfying (9) with $\Gamma = \Gamma_0 \cup \Gamma_2$, and denote by $\beta_n^{(2)}$ and $(\alpha_n^{(2)})^2$ the associated recurrence coefficients. The three-term recurrence relation is given by

$$xp_n(x) = p_{n+1}(x) + \beta_n p_n(x) + \alpha_n^2 p_{n-1}(x),$$

where

$$\beta_n = \frac{\int_\Gamma xp_n^2(x)e^{-x^3}dx}{\int_\Gamma p_n^2(x)e^{-x^3}dx}, \qquad \alpha_n^2 = \frac{\int_\Gamma xp_n(x)p_{n-1}(x)e^{-x^3}dx}{\int_\Gamma p_{n-1}^2(x)e^{-x^3}dx},$$

and the initial condition is taken to be $\alpha_0^2 p_{-1} = 0$. It can be shown (for instance, using ladder operators) that the recurrence coefficients β_n and α_n^2 satisfy the following string equations:

$$\alpha_{n+1}^2 + \beta_n^2 + \alpha_n^2 = 0, \quad 3\alpha_n^2(\beta_{n-1} + \beta_n) = n.$$

One can determine $(\beta_n^{(1),(2)}, (\alpha_n^2)^{(1),(2)})$ recursively from the string equations with initial condition $(\frac{\Gamma(2/3)}{\Gamma(1/3)}e^{\pi i/3}, 0)$ and $(\frac{\Gamma(2/3)}{\Gamma(1/3)}e^{-\pi i/3}, 0)$ respectively. Actually, one can prove that

$$\beta_n^{(1)} = b_n e^{\pi i/3}, \quad (\alpha_n^{(1)})^2 = a_n e^{-\pi i/3}, \quad \beta_n^{(2)} = b_n e^{-\pi i/3}, \quad (\alpha_n^{(2)})^2 = a_n e^{\pi i/3},$$

where

$$a_n + a_{n+1} = b_n^2, \quad 3a_{n+1}(b_n + b_{n+1}) = n + 1.$$

For $(k, l) \in \mathbb{N}^2$ one can define the multiple orthogonal polynomials $P_{k,l}$ of degree $k + l$ which satisfy the orthogonality conditions

$$\int_{\Gamma_0 \cup \Gamma_1} x^i P_{k,l}(x) e^{-x^3} dx = 0, \qquad i = 0, 1, \ldots, k-1,$$
$$\int_{\Gamma_0 \cup \Gamma_2} x^i P_{k,l}(x) e^{-x^3} dx = 0, \qquad i = 0, 1, \ldots, l-1.$$

If one of k and l is equal to zero, then $P_{k,l}$ reduce to the usual orthogonal polynomials with respect to the exponential cubic weight e^{-x^3}, i.e.,

$$P_{k,0}(x) = p_k^{(1)}(x), \qquad P_{0,k}(x) = p_k^{(2)}(x).$$

It can be shown that the following Rodrigues formula holds:

$$P_{n,n+m}(x) = \frac{(-1)^n}{3^n} e^{x^3} \frac{d^n}{dx^n}\left(e^{-x^3} P_{0,m}(x)\right),$$
$$P_{n+m,n}(x) = \frac{(-1)^n}{3^n} e^{x^3} \frac{d^n}{dx^n}\left(e^{-x^3} P_{m,0}(x)\right).$$

The recurrence coefficients of multiple orthogonal polynomials defined in such a way are also connected to the string equations.

7 The Painlevé Analysis of the Differential Equation Related to Multiple Hermite Polynomials

In [12] a method to find relations between zeros of orthogonal polynomials using a differential equation is presented. In short, given a differential equation, e.g., $y''(x) - 2xy'(x) + 2ny(x) = 0$ for the Hermite polynomials, we compute a differential equation for the logarithmic derivative $w(x) = y'(x)/y(x)$, which is the Riccati equation $w'(x) + w^2(x) - 2xw(x) + 2n = 0$ in our example. The equation for the function $w(x)$ is nonlinear and, hence, the Painlevé analysis can be performed, i.e., we search for the expansions of solutions of the form $\sum_{m=-r}^{\infty} c_m(x-x_0)^m,\ r>0$, in the neighborhood of the movable pole $x=x_0$ in the complex plane and find the coefficients c_m. On the other hand, the expansion of the function $w(x)$ in the Laurent series in the neighborhood of the zero x_j of the polynomial $y(x)$ will be of the form

$$w(x)=\frac{e_j}{x-x_j}-\sum_{m=0}^{\infty}\left(\sum_{k\neq i}\frac{e_k}{(x_k-x_j)^{m+1}}\right)(x-x_j)^m,$$

where e_j denotes the multiplicity of the zero. The identification of two expansions leads to relations between zeros. For instance, for the Hermite polynomials one has the famous relation $\sum_{k\neq j}^{n}(z_j-z_k)^{-1}-z_j=0,\ j=1,\dots,n$.

We can perform the Painlevé analysis for the second order nonlinear differential equation

$$\begin{aligned}&w''+w(3w'+2(m+n-1-c_2x+2x^2)+c_1(c_2+2x))+w'(c_1+c_2-4x)\\&\quad+w^3+(c_1+c_2-4x)w^2+2c_1m+2c_2n-4mx-4nx=0,\end{aligned}\tag{10}$$

for the function $w(x)=p'(x)/p(x)$, where $p(x)$ solves equation (7) for the multiple Hermite polynomials (with assumptions that $c_1\neq c_2$ and $m,n\geq 0$). This yields two types of expansions

$$w(x)=\frac{1}{x-x_0}+c_0-\frac{1}{3}q_1(x-x_0)+\dots,\tag{11}$$

where c_0 is arbitrary and $q_1=2(m+n-1)+(2x_0-c_1)(2x_0-c_2)+2(c_1+c_2-4x_0)c_0+3c_0^2$, and

$$w(x)=\frac{2}{x-x_0}+\frac{1}{3}(4x_0-c_1-c_2)+\frac{1}{18}q_2(x-x_0)+\dots,$$

where $q_2=c_1^2-c_1c_2+c_2^2-6(m+n-3)-2(c_1+c_2)x_0+4x_0^2$. Such expansions might be useful to further study the relations between zeros of multiple Hermite polynomials. A similar analysis of other differential equations related to multiple orthogonal polynomials will be published elsewhere.

Assume that the solutions of the Riccati equation $w'(x)=a(x)w(x)^2+b(x)w(x)+c(x)$ are simultaneously the solutions of equation (10). We get that $a(x)=-1$, $c(z)$ can be expressed in terms of $b(x)$, which, in turn, satisfies a certain second order nonlinear differential equation. In particular, if we assume that $b(x)$ is linear, i.e., $b(x)=Bx+B_1$, then $B=0$ and either $B_1=-c_2$, $n=0$ and $c(x)=-2m$ or $B_1=-c_1$, $m=0$ and $c(x)=-2n$ (compare to the nonlinear equation related to the Hermite polynomials). The Painlevé analysis of the nonlinear equation for $b(x)$ shows that there are also two types of polar expansions with residues 1 (with an arbitrary coefficient) and 2. Further, expansion (11) comes from the Riccati equation (with the expansion for $b(x)$ with the residue 2).

Acknowledgments The author is grateful to the organizers of the conference *Modern Mathematical Methods and High Performance Computing in Science and Technology* for their invitation and the opportunity to record a video lecture.

This paper is an extension of the video lecture. The author also acknowledges the support of the Alexander von Humboldt Foundation and the hospitality of the Catholic University Eichstätt-Ingolstadt.

References

1. Filipuk, G., Van Assche, W., Zhang, L.: Ladder operators and differential equations for multiple orthogonal polynomials. J. Phys. A: Math. Theor. **46**, 205204 (2013)
2. Van Assche, W., Filipuk, G., Zhang, L.: Multiple orthogonal polynomials associated with an exponential cubic weight. J. Approx. Theory **190**, 1–25 (2015)
3. Zhang, L., Filipuk, G.: On certain Wronskians of multiple orthogonal polynomials. Symmetry Integr. Geom. Methods Appl. **10**, 103 (2014)
4. Chihara, T.S.: An Introduction to Orthogonal Polynomials. Gordon and Breach, New York (1978)
5. Ismail, M.E.H.: Classical and Quantum Orthogonal Polynomials in One Variable. Encyclopedia of Mathematics and its Applications vol. 98. Cambridge University Press, Cambridge (2005)
6. Szegő, G.: Orthogonal Polynomials. AMS Colloquium Publications, vol. 23. American Mathematical Society, Providence (1975)
7. Nikishin, E.M., Sorokin, V.N.: Rational Approximations and Orthogonality. Translations of Mathematical Monographs vol. 92. American Mathematical Society, Providence (1991)
8. Van Assche, W.: Padé and Hermite-Padé approximation and orthogonality. Surv. Approx. Theory **2**, 61–91 (2006)
9. Kuijlaars, A.: Multiple orthogonal polynomials in random matrix theory. In: Bhatia, R. (ed.) Proceedings of the International Congress of Mathematicians, vol. 3, pp. 1417–1432. World Scientific, Singapore (2010)
10. Van Assche, W.: Nearest neighbor recurrence relations for multiple orthogonal polynomials. J. Approx. Theory **163**, 1427–1448 (2011)
11. Chen, Y., Ismail, M.E.H.: Ladder operators and differential equations for orthogonal polynomials. J. Phys. A: Math. G. **30**, 7817–7829 (1997)
12. Kudryashov, N.A., Demina, M.V.: Relations between zeros of special polynomials associated with the Painlevé equations. Phys. Lett. A **368**, 227–234 (2007)

The Problem of Soliton Collision for Non-integrable Equations

Georgy A. Omel'yanov

Abstract We describe an approach to construct multi-soliton asymptotics for non-integrable equations. The general idea is realized in the case of three waves and for the KdV-type equation with nonlinearity u^4.

Keywords Generalized Korteweg-de Vries equation · Soliton · Interaction · Weak asymptotics method · Weak solution · Non-integrability

2010 Mathematics Subject Classification 35Q53 · 35D30

1 Introduction

1.1 Statement of the Problem

The main question under consideration in this paper is the following: do the integrable equations form a compact cluster with a sharp frontier or there are non-integrable equations which preserve in a sense some properties of integrability? The standard approach to this problem is the investigation of small perturbations of integrable equations. On the contrary, we consider essentially non-integrable equations which cannot be reduced to an integrable case. Note that we study the waves of arbitrary amplitudes but assume the smallness of the dispersion ε. The last is equivalent to the consideration of large distances and time intervals. We restrict ourself to the generalized Korteweg-de Vries-4 equation,

$$\frac{\partial u}{\partial t} + \frac{\partial u^4}{\partial x} + \varepsilon^2 \frac{\partial^3 u}{\partial x^3} = 0, \quad x \in \mathbb{R}, \quad t > 0, \quad \varepsilon << 1, \tag{1}$$

and we consider the scenario of solitary wave collision only.

G.A. Omel'yanov (✉)
University of Sonora, Rosales y Encinas s/n, 83000 Hermosillo, Sonora, Mexico
e-mail: omel@mat.uson.mx

V.K. Singh et al. (eds.), *Modern Mathematical Methods and High Performance Computing in Science and Technology*, Springer Proceedings in Mathematics & Statistics 171, DOI 10.1007/978-981-10-1454-3_9

It is well known that an arbitrary number of solitary waves collide for integrable nonlinear equations in an remarkable manner: they pass through each other almost as linear waves. In particular, it is true for the KdV equation. Moreover, there exist explicit N-phase formulas which describe this collision. Conversely, for non-integrable equations there are neither exactly the same manner of interaction nor explicit N-phase formulas. However, it is possible to prove that in an asymptotic sense two waves interact almost elastically [1]. More in detail, let us consider the linear combination of two perturbed solitary waves

$$u = \sum_{i=1}^{2} G_i \omega\big(\beta_i(x - \varphi_i)/\varepsilon\big) \tag{2}$$

with variable amplitudes $G_i = G_i(t, \varepsilon)$ and nonlinear phases $\varphi_i = \varphi_i(t, \varepsilon)$. Here ω is a function such that $A\omega\big(\beta(x - \varphi_0)/\varepsilon\big)$ with $A = \text{const}$ and

$$\omega(\eta) = \cosh^{-2/3}(3\eta/2),\ A = c\beta^{2/3},\ c = (5/2)^{1/3},\ \varphi_0 = Vt,\ V = \beta^2 \tag{3}$$

represents the explicit solitary wave solution of (1) with the normalization condition $\omega(0) = 1$. Then there exist functions $G_i(t, \varepsilon), \varphi_i(t, \varepsilon), i = 1, 2$ such that (2) describes elastic interaction of two solitary waves in the weak asymptotic sense [1]:

Definition 1 A sequence $u(t, x, \varepsilon)$, belonging to $\mathcal{C}^\infty(0, T; \mathcal{C}^\infty(\mathbb{R}^1_x))$ for $\varepsilon = \text{const} > 0$ and belonging to $\mathcal{C}(0, T; \mathcal{D}'(\mathbb{R}^1_x))$ uniformly in $\varepsilon \geq 0$, is called a weak asymptotic mod $O_{\mathcal{D}'}(\varepsilon^2)$ solution of (1) if the relations

$$\frac{d}{dt}\int_{-\infty}^{\infty} u\psi dx - \int_{-\infty}^{\infty} u^4 \frac{\partial \psi}{\partial x} dx = O(\varepsilon^2), \tag{4}$$

$$\frac{d}{dt}\int_{-\infty}^{\infty} u^2\psi dx - 2\frac{4}{5}\int_{-\infty}^{\infty} u^5 \frac{\partial \psi}{\partial x} dx + 3\int_{-\infty}^{\infty} \left(\varepsilon\frac{\partial u}{\partial x}\right)^2 \frac{\partial \psi}{\partial x} dx = O(\varepsilon^2) \tag{5}$$

hold uniformly in t for any test function $\psi = \psi(x) \in \mathcal{D}(\mathbb{R}^1)$.

Here, the right-hand sides are $\mathcal{C}^\infty$-functions for $\varepsilon = \text{const} > 0$ and piecewise continuous functions uniformly in $\varepsilon \geq 0$. The estimates are understood in the $\mathcal{C}(0, T)$ sense:

$$g(t, \varepsilon) = O(\varepsilon^k) \leftrightarrow \max_{t \in [0,T]} |g(t, \varepsilon)| \leq c\varepsilon^k.$$

It turned out however that Definition 1 doesn't support asymptotics with three or more phases since it implies the appearance of ill-posed model equations for the parameters of the solutions (they are well-posed for the case of two phases). To overcome the obstacle it is necessary to change the viewpoint on the weak asymptotic solution: the analysis in [2] showed that for two-phase solutions the Definition 1 implies the fulfilment of two conservation laws in the weak sense. Moreover, the one-phase asymptotic theory for perturbed equations implies the fulfilment of a single "conservation law" again in the weak sense. Thus there appears the hypotheses that to construct N-phase asymptotics it is necessary to use N conservation laws. The first step in order to realize an appropriate construction has been done in [2]. The aim of this paper is to complete the analysis and to prove rigorously the existence of 3-phase asymptotic solution.

1.2 Weak Asymptotics Method: The Main Idea

The basic remark is very simple: rapidly varying solitary wave solutions (soliton or kink type) tend to distributions as the small parameter tends to zero. This allows us to treat the equation in the weak sense and, respectively, look for singularities instead of regular functions. Obviously, non-integrability implies that we cannot find neither classical nor weak exact solutions. However, we can construct an asymptotic weak solution considering the smallness of the remainder in the weak sense. In a sense, the situation here is similar to the shock waves: various regularization generates various profile for the wave, but in the limiting passage we obtain the same Rankine–Hugoniot conditions. For solitons the passage to the weak expansion results in the disappearance of the shape, but preserves the soliton's characteristics: amplitudes and phases. For the problem of interaction these parameters vary in a neighborhood of the time instant of the collision and stabilize ourself after that. Deriving uniform in time model equations for the parameters we can describe the scenario of the wave interaction.

Originally, such idea had been suggested by V. Danilov and V. Shelkovich for shock wave type solutions (1997, [3]), then generalized for soliton type solutions (V. Danilov and G. Omel'yanov 2003 [1]), and it has been developed and adapted later for many other problems (V. Danilov, G. Omel'yanov, V. Shelkovich, D. Mitrovic and others, see [4, 5] and references therein). Let us note finally that the treatment [2] of weak asymptotics as functions which satisfy some conservation or balance laws takes us back to the ancient Whitham's idea to construct one-phase asymptotic solution satisfying a Lagrangian. Now, for essentially non-integrable equations and multi-phase solutions, we use the appropriate number of the laws and satisfy them in the weak sense.

2 Asymptotics Construction

The gKdV-4 equation consists of three conservation laws, which we write in the differential form with the remainder:

$$\frac{\partial Q_j}{\partial t} + \frac{\partial P_j}{\partial x} = O_{\mathcal{D}'}(\varepsilon^2), \quad j = 1, 2, 3. \tag{6}$$

Here

$$Q_1 = u, \quad P_1 = u^4, \quad Q_2 = u^2, \quad P_2 = \frac{8}{5}u^5 - 3(\varepsilon u_x)^2, \tag{7}$$

$$Q_3 = (\varepsilon u_x)^2 - \frac{2}{5}u^5, \quad P_3 = 16u^3(\varepsilon u_x)^2 - u^8 - 3(\varepsilon^2 u_{xx})^2, \tag{8}$$

and we use the following definition of the smallness:

Definition 2 A function $v(t, x, \varepsilon)$ is said to be of the value $O_{\mathcal{D}'}(\varepsilon^k)$ if the relation

$$\int_{-\infty}^{\infty} v(t, x, \varepsilon)\psi(x)dx = O(\varepsilon^k)$$

holds uniformly in t for any test function $\psi \in \mathcal{D}(\mathbb{R}^1_x)$.

Let us consider three-phase asymptotic solution for the gKdV-4 Eq. (1) supplied by the initial condition

$$u|_{t=0} = \sum_{i=1}^{3} A_i \omega\big(\beta_i(x - x_{(i,0)})/\varepsilon\big). \tag{9}$$

Contrarily to Definition 1 we define the asymptotics in the following manner:

Definition 3 Let a sequence $u = u(t, x, \varepsilon)$ belong to the functional space indicated in Definition 1. Then u is called a 3-phase weak asymptotic mod $O_{\mathcal{D}'}(\varepsilon^2)$ solution of (1) if the relations (6) hold uniformly in t.

To construct the asymptotics we present the ansatz in the form similar to (2), that is

$$u = \sum_{i=1}^{3} G_i \omega\big(\beta_i(x - \varphi_i)/\varepsilon\big). \tag{10}$$

Here

$$G_i = A_i + S_i(\tau), \quad \varphi_i = \varphi_{i0}(t) + \varepsilon\varphi_{i1}(\tau), \quad \tau = \beta_1\big(\varphi_{30}(t) - \varphi_{10}(t)\big)/\varepsilon, \tag{11}$$

A_i are the original amplitudes and $\varphi_{i0} = V_i t + x_{(i,0)}$ describe the trajectories of the noninteracting waves; β_i, A_i, and V_i are connected by the equalities (3); the "fast time" τ characterizes the distance between the first and third trajectories. Next we set $A_1 < A_2 < A_3$, $x_{(i,0)} - x_{(i+1),0)} \geq \text{const} > 0$, $i = 1, 2$, and suppose the intersection of all trajectories $x = \varphi_{i0}(t)$ at the same point (x^*, t^*). We assume also that the amplitude and phase corrections $S_i(\tau)$, $\varphi_{i1}(\tau)$ are such that

$$S_i \to 0 \quad \text{as} \quad \tau \to \pm\infty, \tag{12}$$

$$\varphi_{i1} \to 0 \quad \text{as} \quad \tau \to -\infty, \quad \varphi_{i1} \to \varphi_{i1}^{\infty} = \text{const}_i \quad \text{as} \quad \tau \to +\infty \tag{13}$$

with an exponential rate.

To find $S_i(\tau)$, $\varphi_{i1}(\tau)$ we should calculate the weak expansions for Q_i and P_i. It is easy to check that

$$u = \varepsilon a_1 \sum_{i=1}^{3} \beta_i^{-1} G_i \delta(x - \varphi_i) + O_{\mathcal{D}'}(\varepsilon^3). \tag{14}$$

Here and in what follows we use the notation

$$a_k = \int_{-\infty}^{\infty} \omega^k(\eta)\, d\eta, \quad a_k^{(l)} = \int_{-\infty}^{\infty} \left(\frac{d^l \omega}{d\eta^l}\right)^k d\eta, \quad k \geq 1, \quad l \geq 1. \tag{15}$$

Next we take into account that $S_i(\tau)$ vanish exponentially fast as $|\varphi_1 - \varphi_3|$ grows, thus, the main contribution gives the point (x^*, t^*). We write

$$\varphi_{i0} = x^* + V_i(t - t^*) = x^* + \varepsilon \frac{V_i}{\dot{\psi}_0} \tau \text{ and } \varphi_i = x^* + \varepsilon \chi_i, \ \chi_i = V_i \tau / \dot{\psi}_0 + \varphi_{i1}, \tag{16}$$

where $\dot{\psi}_0 = \beta_1 (V_3 - V_1)$. Thus, we can modify (14) to the final form:

$$u = \varepsilon a_1 \sum_{i=1}^{3} \frac{A_i}{\beta_i} \delta(x - \varphi_i) + \varepsilon a_1 \sum_{i=1}^{3} \frac{S_i}{\beta_i} \big\{ \delta(x - x^*) - \varepsilon \chi_i \delta'(x - x^*) \big\} + O_{\mathcal{D}'}(\varepsilon^3). \tag{17}$$

Concerning nonlinear terms let us note that all the products of the waves with different phases are concentrated near the point (x^*, t^*). Thus, for any smooth function $F = F(z_0, \ldots, z_k)$ we write the corresponding weak expansion:

$$F\left(u, \ldots, \left(\varepsilon \frac{\partial}{\partial x}\right)^k u\right) = \varepsilon \left\{ \sum_{i=1}^{3} \frac{a_{F,i}^{(0)}}{\beta_i} \delta(x - \varphi_i) + \frac{1}{\beta_3} \mathcal{R}_F^{(0)} \delta(x - x^*) \right\}$$
$$- \varepsilon^2 \left\{ \sum_{i=1}^{3} \frac{a_{F,i}^{(1)}}{\beta_i^2} \delta'(x - \varphi_i) + \left(\frac{\chi_3}{\beta_3} \mathcal{R}_F^{(0)} + \frac{1}{\beta_3^2} \mathcal{R}_F^{(1)} \right) \delta'(x - x^*) \right\} + O_{\mathcal{D}'}(\varepsilon^3). \tag{18}$$

Here

$$a_{F,i}^{(n)} = \int_{-\infty}^{\infty} \eta^n F\big(A_i \omega(\eta), \ldots, A_i \beta_i^k \omega^{(k)}(\eta)\big) d\eta, \quad n = 0, 1, \tag{19}$$

$$\mathcal{R}_F^{(n)} = \int_{-\infty}^{\infty} \eta^n \left\{ F\left(\sum_{i=1}^{3} G_i \omega(\eta_{i3}), \ldots, \sum_{i=1}^{3} G_i \beta_i^k \omega^{(k)}(\eta_{i3}) \right) \right.$$
$$\left. - \sum_{i=1}^{3} F\Big(A_i \omega(\eta_{i3}), \ldots, A_i \beta_i^k \omega^{(k)}(\eta_{i3})\Big) \right\} d\eta, \tag{20}$$

where

$$\eta_{ij} = \theta_{ij}\eta - \sigma_{ij}, \quad \theta_{ij} = \beta_i/\beta_j, \quad \sigma_{ij} = \beta_i(\varphi_i - \varphi_j)/\varepsilon. \tag{21}$$

Calculating weak expansions for all terms of Definition 3 and substituting them into (6) we obtain linear combinations of $\delta'(x-\varphi_i), i = 1, 2, 3, \delta(x-x^*)$, and $\delta'(x-x^*)$. Therefore, we pass to the following system of model equations:

$$V_i\, a^{(0)}_{Q_j,i} - a^{(0)}_{P_j,i} = 0, \quad i = 1, 2, 3, \quad j = 1, 2, 3, \tag{22}$$

$$\mathcal{R}^{(0)}_{Q_j} = 0, \quad j = 1, 2, 3, \tag{23}$$

$$\dot{\psi}_0 \frac{d}{d\tau}\left\{\sum_{i=1}^{3} \varphi_{i1}\frac{a^{(0)}_{Q_j,i}}{\beta_i} + \frac{\chi_3}{\beta_3}\mathcal{R}^{(0)}_{Q_j} + \frac{1}{\beta_3^2}\mathcal{R}^{(1)}_{Q_j}\right\} - \frac{1}{\beta_3}\mathcal{R}^{(0)}_{P_j} = 0, \quad j = 1, 2, 3. \tag{24}$$

For each i the system (22) of three equation contains only two free parameters $A_i = A_i(\beta_i)$, $V_i = V_i(\beta_i)$. However, there is not any contradiction [2]:

Lemma 1 *Let $\omega(\eta)$, $A_i = A(\beta_i)$, and $V_i = V(\beta_i)$ be of the form (3). Then the equalities (22) are satisfied uniformly in $\beta_i > 0$.*

Let us simplify the Eq. (24). We take into account the equalities (23) and the following consequence of the definitions (16), (21) of φ_i and σ_{i3}:

$$\varphi_{i1} = \varphi_{31} + \beta_i^{-1}\sigma_{i3} - \dot{\psi}_0^{-1}\tau(V_i - V_3), \quad i = 1, 2. \tag{25}$$

Then the system (24) can be transformed to the form

$$\dot{\psi}_0 r_j \frac{d\varphi_{31}}{d\tau} + \dot{\psi}_0 \frac{d}{d\tau}\left\{\sum_{i=1}^{2} \frac{a^{(0)}_{Q_j,i}}{\beta_i^2}\sigma_{i3} + \frac{1}{\beta_3^2}\mathcal{R}^{(1)}_{Q_j}\right\} = f_j, \quad j = 1, 2, 3, \tag{26}$$

where

$$r_j = \sum_{i=1}^{3} \beta_i^{-1} a^{(0)}_{Q_j,i}, \quad f_j = \beta_3^{-1}\mathcal{R}^{(0)}_{P_j} + \sum_{i=1}^{2} \beta_i^{-1} a^{(0)}_{Q_j,i}(V_i - V_3). \tag{27}$$

Now it is clear that (26) contains three unknown functions, namely φ_{31}, σ_{13}, and σ_{23}. Moreover, (26) can be reduced easily to a system which contains σ_{i3}, $i = 1, 2$, only.

Taking into account our hypothesis (13) we supply (26) by the scattering-type condition

$$\varphi_{31} \to 0, \quad \sigma_{i3}/\tau \to \xi_{i3} \quad \text{as} \quad \tau \to -\infty, \quad i = 1, 2, \tag{28}$$

where $\xi_{i3} = \beta_i(V_i - V_3)/\dot{\psi}_0$.

The behavior of the solution of the problem (26), (28) describes the scenario of the wave collision: if φ_{31} remains bounded and $\sigma_{i3}/\tau \to \xi_{i3}$ when $\tau \to \infty$, then the waves interact elastically; each other behavior of the solution means another scenario of collision.

3 Analysis of the Model Equations

To simplify the further analysis let us assume that

$$\theta_{23} = \mu^3,\ \theta_{13} = \mu^{3(3+\alpha)/2},\ \text{where } \alpha \in [0, 1) \text{ and } \mu \text{ is a small parameter.} \tag{29}$$

Lemma 2 *Let $\sigma_{i3} \to \infty$ as $\tau \to \infty$, $i = 1, 2$. Then for sufficiently small μ the system (23) has a unique solution which satisfies the assumption (12).*

Proof We look for the asymptotic solution of the system (23) in the form:

$$S_1 = \frac{1}{2}\frac{c\beta_1}{\beta_3^{1/3}}\mu^\alpha(y - \mu^{2-\alpha}x),\ S_2 = -\frac{1}{2}\frac{c\beta_2}{\beta_3^{1/3}}\mu^\alpha(y + \mu^{2-\alpha}x),\ S_3 = c\beta_3^{2/3}\mu^2 x, \tag{30}$$

where x and y are free functions. Then the Eq. (23) for $j = 1$ is verified automatically. Furthermore, let us consider the monoids

$$M_m = u^m,\ M_2^{(k)} = (\varepsilon^k u_x^{(k)})^2,\ M_{3,2} = u^3(\varepsilon u_x)^2,\ \text{where } m \geq 2,\ k \geq 1. \tag{31}$$

Combining the general formula (18)–(20) with the representation (30) and omitting algebraic calculations we obtain the statement:

Lemma 3 *Under the assumption (29) the following relations hold*

$$\mathcal{R}^{(0)}_{M_m} = (c\beta_3^\gamma)^m m\mu^2 \Big\{a_m x + \lambda^{(0)}_{m-1,0,1}(\eta_{23}) - \frac{1}{2}\mu^{2(m-2)+\alpha}\big(a_m y - 2\lambda^{(0)}_{m-1,0,1}(\eta_{12})\big)$$
$$+ \frac{1}{2}\mu^{1+\alpha}\big(2\lambda^{(0)}_{m-1,0,1}(\eta_{13}) - y\lambda^{(0)}_{m-1,0,1}(\eta_{23})\big)\Big\} + O_{\mathcal{S}}(\mu^{2m-1+2\alpha}), \tag{32}$$

$$\mathcal{R}^{(0)}_{M_2^{(k)}} = c^2\beta_3^{2(\gamma+k)}\mu^2\{2a_2^{(k)}x + O_{\mathcal{S}}(\mu^2)\}, \tag{33}$$

$$\mathcal{R}^{(0)}_{M_{3,2}} = c^5\beta_3^{8\gamma}\mu^2\{5x\lambda^{(0)}_{3,2,0} + 3\lambda^{(0)}_{2,2,1}(\eta_{23}) + O_{\mathcal{S}}(\mu^{1+\alpha})\}. \tag{34}$$

Here and in what follows we use the notation

$$\lambda^{(n)}_{k,p,l}(\eta_{ij}) = \int_{-\infty}^{\infty} \eta^n\omega^k(\eta)\big(\omega'(\eta)\big)^p\omega^l(\eta_{ij})d\eta,\quad n = 0, 1, \tag{35}$$

and the following definition of the smallness:

Definition 4 A function $v(\sigma_{13}, \sigma_{23}, \mu)$ is said to be of the value $O_{\mathcal{S}}(\mu^\varkappa)$ if there exists a function $f(\tau) \geq 0$ from the Schwartz space $\mathcal{S}$ such that the inequality

$$|v(\sigma_{13}, \sigma_{23}, \mu)| \leq \mu^\varkappa f(\tau)$$

holds uniformly in $\tau \in \mathbb{R}$.

The formulas (32), (33) allow us to pass to the asymptotic representation of the Eq. (23) for $j = 2, 3$:

$$a_2\left(x - \frac{1}{2}\mu^\alpha y\right) = -\lambda^{(0)}_{1,0,1}(\sigma_{23}) - \mu^\alpha \lambda^{(0)}_{1,0,1}(\sigma_{12}) + O_{\mathcal{S}}(\mu^{1+\alpha}), \tag{36}$$

$$\big(2a_2^{(1)} - 5a_5\big)x = 5\lambda^{(0)}_{4,0,1}(\sigma_{23}) + 5\mu^{1+\alpha}\left(\lambda^{(0)}_{4,0,1}(\sigma_{13}) - \frac{1}{2}y\lambda^{(0)}_{4,0,1}(\sigma_{23})\right) + O_{\mathcal{S}}(\mu^2).$$

Obviously, the matrix in the left-hand side of (36) is degenerate $\mod O(\mu^\alpha)$. However, the right-hand side has the same rank.

Lemma 4 *For the function* ω *of the form (3) the relation*

$$\left(5a_5 - 2a_2^{(1)}\right)\lambda^{(0)}_{1,0,1}(\sigma_{ln}) = 5a_2\lambda^{(0)}_{4,0,1}(\sigma_{ln}) + O_{\mathcal{S}}(\theta_{ln}) \tag{37}$$

holds for all indices l, n.

Now we set

$$x = -\lambda^{(0)}_{1,0,1}(\sigma_{23}) + \mu^\alpha x_1/a_2 \tag{38}$$

and transform (36) to the final form:

$$x_1 - \frac{a_2}{2}y = -\lambda^{(0)}_{1,0,1}(\sigma_{12}) + O_{\mathcal{S}}(\mu), \tag{39}$$

$$x_1 = \frac{5}{4}\mu\left(2\lambda^{(0)}_{4,0,1}(\sigma_{13}) - y\lambda^{(0)}_{4,0,1}(\sigma_{23})\right) + O_{\mathcal{S}}(\mu^{2-\alpha}).$$

Since the matrix in the left-hand side of the system (39) is non-singular, we obtain the desired assertion of Lemma 2. □

The last step of the construction is the analysis of the problem (26), (28). To do it we use again the assumption (29) and the formulas (20), (32)–(34). However, since the calculation precision should be very high we omit the tedious algebra and present the final result only.

Lemma 5 *Let the assumption (29) be verified. Then the Eq. (26) can be reduced to the system*

$$\frac{d}{d\tau}\left\{\Delta_1(\sigma_{13} - \xi_{13}\tau) + \mu^{1+\alpha}Z_1\right\} = -2\mu^{(3+\alpha)/2}\mathcal{F}_1, \tag{40}$$

$$\frac{d}{d\tau}\left\{\Delta_2(\sigma_{23} - \xi_{23}\tau) - \mu^{(1+3\alpha)/2}Z_2\right\} = 2\mu^{(3+\alpha)/2}\mathcal{F}_2, \tag{41}$$

where $\Delta_i = 1 + O(\mu^{(1+\alpha)/2})$ *are constants,* ξ_{i3} *are defined in (28), and*

$$Z_2 = \frac{20}{a_5}\lambda^{(1)}_{4,0,1}(\eta_{23}) + O_{\mathcal{S}}(\mu), \quad Z_1 = Z_2 + \mu^{(1-\alpha)/2}\sigma_{13}y + O_{\mathcal{S}}(\mu),$$

$$\mathcal{F}_2 = \frac{31}{7}x + \frac{40}{7a_5}\lambda^{(0)}_{4,0,1}(\eta_{23}) + O_{\mathcal{S}}(\mu^{(1+\alpha)/2}), \quad \mathcal{F}_1 = \mathcal{F}_2 + O_{\mathcal{S}}(\mu^{(1-\alpha)/2}).$$

Obviously, for sufficiently small μ the problem (40), (41), (28) has the unique solution such that $\sigma_{i3}/\tau \to \xi_{i3}$ as $\tau \to \infty$, $i = 1, 2$. Moreover, $\sigma_{i3} - \xi_{i3}\tau$ are uniformly bounded functions which tend to constants as $\tau \to \infty$.

Finally, we integrate the Eq. (26) with $j = 1$ and obtain the function φ_{31} such that $\varphi_{31} \to$ const as $\tau \to \infty$. Now, it remains to calculate φ_{i1}, $i = 1, 2$, in accordance with (25) and conclude that the phase corrections φ_{i1} satisfy the assumptions (13). Therefore, we can summarize the main result of the paper

Theorem 1 *Under the assumption (29), the three-phase asymptotic solution (10) exists and describes a* $\bmod O_{\mathcal{D}'}(\varepsilon^2)$ *scenario of KdV type solitary waves interaction.*

Acknowledgments The research was supported by SEP-CONACYT under grant 178690.

References

1. Danilov, V., Omel'yanov, G.: Weak asymptotics method and the interaction of infinitely narrow delta-solitons. Nonlinear Anal.: Theory, Methods Appl. **54**, 773–799 (2003)
2. Omel'yanov, G.: Soliton-type asymptotics for non-integrable equations: a survey. Math. Methods Appl. Sci. **38**(10), 2062–2071 (2015)
3. Danilov, V., Shelkovich, V.: Generalized solutions of nonlinear differential equations and the Maslov algebras of distributions. Integral Transform. Special Funct. **6**, 137–146 (1997)
4. Danilov, V., Omel'yanov, G., Shelkovich, V.: Weak asymptotics method and interaction of nonlinear waves. In: Karasev, M.V. (ed.) Asymptotic methods for wave and quantum problems, AMS Trans., Ser. 2, **208**, 33–164. AMS, Providence, RI (2003)
5. Kalisch, H., Mitrovic, D.: Singular solutions of a fully nonlinear 2 × 2 system of conservation laws. Proc. Edinb. Math. Soc. II **55**, 711–729 (2012)

Explicit Solutions of the Poisson Equation in Plane Domains

H. Begehr

Dedicated to the memory of Prof. Dr. Fritz Gackstatter who has passed away in March 2016

Abstract The two basic boundary value problems for the Laplace operator are the Dirichlet and the Neumann problems. Their theories are well known and explained in all textbooks on partial differential equations. The solutions of these boundary value problems to the Poisson equation in smooth domains are given via the Green and the Neumann functions, respectively. As examples serve the unit disc or unit ball and occasionally half planes or half spaces. In the two-dimensional case where complex methods are available the conformal invariance of the Green and the Neumann functions is used to get solutions to the problems in plane domains conformally equivalent to the unit disc. But these conformal mappings are only in particular cases known in explicit form. The parqueting-reflection principle, however, serves to construct Green and Neumann functions for a large class of plane domains in explicit form [8, 9].

Keywords Poisson equation · Plane domains · Green and Neumann functions · Dirichlet and Neumann boundary value problems · Parqueting-reflection principle · Strips · Half strips · Hyperbolic half planes · Hyperbolic strips

1 The Parqueting-Reflection Principle

To explain the basic idea of this principle the particular case of a half plane, say the upper half plane $\mathbb{H}^+ = \{0 < \mathrm{Im} z\}$ is investigated. Reflecting $\mathbb{H}^+$ at its boundary,

H. Begehr (✉)
Math. Institut, FU Berlin, Arnimallee 3, 14195 Berlin, Germany
e-mail: begehrh@zedat.fu-berlin.de

V.K. Singh et al. (eds.), *Modern Mathematical Methods and High Performance Computing in Science and Technology*, Springer Proceedings in Mathematics & Statistics 171, DOI 10.1007/978-981-10-1454-3_10

the real axis $\mathbb{R} = \{\text{Im} z = 0\}$, onto the lower half plane $\mathbb{H}^- = \{\text{Im} z < 0\}$ gives a parqueting of the complex plane $\mathbb{C}$ via $\mathbb{C} = \overline{\mathbb{H}^+ \cup \mathbb{H}^-}$. Choosing an arbitrary point $z \in \mathbb{H}^+$ as a pole of an as simple as possible meromorphic function P and its reflection $\overline{z} \in \mathbb{H}^-$ at $\mathbb{R}$ as a zero of this function, leads to $P(z, \zeta) = \frac{\zeta - \overline{z}}{\zeta - z}$. The Green function for $\mathbb{H}^+$ has the form $G_1(z, \zeta) = \log |P(z, \zeta)|^2 = \log |\frac{\zeta - \overline{z}}{\zeta - z}|^2$. The Neumann function for this half plane is $N_1(z, \zeta) = -\log |(\zeta - z)(\zeta - \overline{z})|^2$. Both formulas also represent the respective Green and Neumann functions for $\mathbb{H}^-$ where now $z \in \mathbb{H}^-$.

Definition 1 A set of domains $D_j, j \in J, J$ some set of indices, in the plane $\mathbb{C}$ is called a parqueting of $\mathbb{C}$ if $D_j \cap D_k = \emptyset$ for any $j, k \in J, j \neq k$, and $\mathbb{C} = \bigcup_{j \in J} \overline{D_j}$.

Definition 2 A domain D of the complex plane $\mathbb{C}$ with piecewise smooth boundary ∂D is called admissible for the parqueting-reflection, if continued reflections at the boundary parts achieve a parqueting of $\mathbb{C}$ with possible exceptions of singular points.

Remark For reflecting, ∂D must consist of arcs from circles and straight lines.

Definition 3 For $z \in \mathbb{C}$ the point z_{re} satisfying

$$\alpha z_{re} \overline{z} + \overline{a} z_{re} + a \overline{z} + \beta = 0,$$

is called the reflection point of z at the straight line or circle

$$\Gamma = \{z \in \mathbb{C} : \alpha z \overline{z} + \overline{a} z + a \overline{z} + \beta = 0, 0 < a \overline{a} - \alpha \beta, a \in \mathbb{C}, \alpha, \beta \in \mathbb{R}\}.$$

If $z \in \Gamma$ then $z_{re} = z$!

The Principle. The original domain D is called a pole-domain. Any direct reflection of D at some parts of ∂D is called a zero-domain. A reflection of a zero-domain at some parts of its boundary is a pole-domain, a reflection of a pole-domain at some parts of its boundary is a zero-domain.

Choosing an arbitrary point $z \in D$ it will become a pole of an as simple as possible meromorphic function P. Any reflection of z into a zero-domain will become a zero of P. Any reflection of a zero (in a zero-domain) will become a pole of P in a pole-domain and vice versa.

Examples for domains admissible for the parqueting-reflection are, e.g. circles, half planes, quarter planes, certain sectors of the plane or of circles, certain convex polygons as triangles, rectangles, hexagons, strips, certain lenses, hyperbolic strips, circular rings, ring sectors, etc. [5, 6, 12–16].

Remark In case of reflection at a circle $|z - z_0| = r$ instead of the factor $\zeta - \frac{r^2}{\overline{z - z_0}}$ of P often the factor $r^2 - (\overline{z - z_0})\zeta$ is preferred.

For the quarter disc $\mathbb{Q}^{++} = \{0 < \text{Re} z, 0 < \text{Im} z, |z| < 1\}$, e.g. the point $z \in \mathbb{Q}^{++}$ is reflected at $\mathbb{R}$ onto $\overline{z}$, at $i\mathbb{R}$ onto $-\overline{z}$ and at $\partial \mathbb{D} = \{|z| = 1\}$ onto $\frac{1}{\overline{z}}$. For the parqueting of $\mathbb{C}$ one of the quarter discs $\mathbb{Q}^{+-} = \{0 < \text{Re} z, \text{Im} z < 0, |z| < 1\}$ and

$\mathbb{Q}^{-+} = \{\text{Re}z < 0,\ \ 0 < \text{Im}, |z| < 1\}$ must be reflected onto $\mathbb{Q}^{--} = \{\text{Re}z < 0, \text{Im}z < 0, |z| < 1\}$ providing $-z$ as reflection of $\overline{z}$, and of $-\overline{z}$, respectively. Moreover, the three quarter discs $\mathbb{Q}^{+-}, \mathbb{Q}^{-+}, \mathbb{Q}^{--}$ have to be reflected at $\partial\mathbb{D}$ to finalize the parqueting. This leads to the points $\frac{1}{z}, -\frac{1}{z}, -\frac{1}{\overline{z}}$.

For the meromorphic function P the points $z, -z, \frac{1}{z}, -\frac{1}{z}$ become poles while $\overline{z}, -\overline{z}, \frac{1}{\overline{z}}, -\frac{1}{\overline{z}}$ are zeros. Hence,

$$P(z,\zeta) = \frac{\zeta - \overline{z}}{\zeta - z}\frac{\zeta + \overline{z}}{\zeta + z}\frac{1 - \overline{z}\zeta}{1 - z\zeta}\frac{1 + \overline{z}\zeta}{1 + z\zeta} = \frac{\zeta^2 - \overline{z}^2}{\zeta^2 - z^2}\frac{1 - \overline{z}^2\zeta^2}{1 - z^2\zeta^2}.$$

Thus, the Green and the Neumann functions for any of the four quarter circles are $G_1(z,\zeta) = \log\left|\frac{\zeta^2 - \overline{z}^2}{\zeta^2 - z^2}\frac{1 - \overline{z}^2\zeta^2}{1 - z^2\zeta^2}\right|^2$, $N_1(z,\zeta) = -\log|(\zeta^2 - \overline{z}^2)(\zeta^2 - z^2)(1 - \overline{z}^2\zeta^2)(1 - z^2\zeta^2)|^2$.

2 Parqueting of the Plane Through Reflections of Strips

2.1 *Strip*

Let for $0 < \alpha < \pi, a \in \mathbb{R}^+$ the set S_1 be the strip

$$S_1 = \{z \in \mathbb{C} : z = e^{2i\alpha}\overline{z} + 2iate^{i\alpha}, 0 < t < 1\}.$$

The boundary parts

$$\partial^- S_1 = \{z \in \mathbb{C} : z = e^{2i\alpha}\overline{z}\}, \partial^+ S_1 = \{z \in \mathbb{C} : z = e^{2i\alpha}\overline{z} + 2iae^{i\alpha}\}$$

are parallel lines with angle α against the positive real axis, the first one passing through the origin, the second one above the first in distance a to the former.

Continued reflection of S_1 at the boundaries provides a parqueting of $\mathbb{C}$ through the strips

$$S_{k+1} = \{z \in \mathbb{C} : z = e^{2i\alpha}\overline{z} + 2ia(k+t)e^{i\alpha}, 0 < t < 1\}, k \in \mathbb{Z},$$

$$\mathbb{C} = \bigcup_{k\in\mathbb{Z}} \overline{S_k}, \quad S_k \cap S_l = \emptyset \quad for \quad k \neq l.$$

An arbitrary point $z \in S_1$ has the representation $z = e^{2i\alpha}\overline{z} + 2iate^{i\alpha}$ for some $t, 0 < t < 1$. Reflecting this point at the line $z = e^{2i\alpha}\overline{z} + 2iae^{i\alpha}$ gives the image

$$z_2 = e^{2i\alpha}\overline{z} + 2iae^{i\alpha} \in S_2.$$

Reflecting this point at the line $z = e^{2i\alpha}\overline{z} + 4iae^{i\alpha}$ leads to

$$z_3 = e^{2i\alpha}\overline{z_2} + 4iae^{i\alpha} = z + 2iae^{i\alpha} \in S_3.$$

Inductively, $z_{2k+1} = z + 2iake^{i\alpha} \in S_{2k+1}$, $z_{2k} = e^{2i\alpha}\overline{z} + 2iake^{i\alpha} \in S_{2k}$, $k \in \mathbb{Z}$, follow. Choosing the z_{2k} as zeros and the z_{2k+1} as poles leads to the meromorphic function

$$P(z,\zeta) = \frac{\zeta - e^{2i\alpha}\overline{z}}{\zeta - z}\prod_{k=1}^{\infty}\frac{\zeta - e^{2i\alpha}\overline{z} - 2iake^{i\alpha}}{\zeta - z - 2iake^{i\alpha}}\frac{\zeta - e^{2i\alpha}\overline{z} + 2iake^{i\alpha}}{\zeta - z + 2iake^{i\alpha}}$$

$$= \frac{\zeta - e^{2i\alpha}\overline{z}}{\zeta - z}\prod_{k=1}^{\infty}\frac{(\zeta - e^{2i\alpha}\overline{z})^2 - (2iake^{i\alpha})^2}{(\zeta - z)^2 - (2iake^{i\alpha})^2} = \frac{\sin\pi\frac{\zeta - e^{2i\alpha}\overline{z}}{2iae^{i\alpha}}}{\sin\pi\frac{\zeta - z}{2iae^{i\alpha}}}.$$

$G_1(z,\zeta) = \log|P(z,\zeta)|^2$ is the Green function for S_1. For $\alpha = 0$, $a = 1$ this is a classical result, see, e.g. [7]. $N_1(z,\zeta) = -\log\left|\sin\pi\frac{\zeta - e^{2i\alpha}\overline{z}}{2iae^{i\alpha}}\sin\pi\frac{\zeta - z}{2iae^{i\alpha}}\right|^2$ is the Neumann function for S_1. By the way, the symmetry of both the Green and the Neumann functions are obvious.

The outward normal vector on ∂^+S_1 is $\nu = ie^{i\alpha}$ and the outward normal derivative is $\partial_\nu = ie^{i\alpha}\partial_z - ie^{-i\alpha}\partial_{\overline{z}}$. For real functions $\partial_\nu = -2\mathrm{Im}\, e^{i\alpha}\partial_z$. For the integral representation formulas related to the Dirichlet and the Neumann boundary value problems the derivatives of the Green and Neumann functions are needed.

$$\partial_{\nu_z}G_1(z,\zeta) = -2\mathrm{Im}\, e^{i\alpha}\left(\frac{e^{-i\alpha}\pi}{2ia}\cot\pi\frac{\overline{\zeta}e^{i\alpha} - ze^{-i\alpha}}{-2ia} + \frac{\pi}{2iae^{i\alpha}}\cot\pi\frac{\zeta - z}{2iae^{i\alpha}}\right).$$

On ∂^+S_1, i.e. for $ze^{-i\alpha} = \overline{z}e^{i\alpha} + 2ia$ this becomes

$$\partial_{\nu_z}G_1(z,\zeta) = \frac{2\pi}{a}\mathrm{Re}\cot\pi\frac{\zeta - z}{2iae^{i\alpha}}.$$

As $A\cot A - 1 = o(1)$ as $A \to 0$, then on ∂^+S_1

$$\partial_{\nu_z}G_1(z,\zeta) = (1 + o(1))2ie^{-i\alpha}\frac{\zeta - \overline{\zeta}e^{2i\alpha} - 2iae^{i\alpha}}{|z - \zeta|^2}$$

for $\zeta \in S_1 \to z$.

Interchanging the roles of z and ζ gives

$$\partial_{\nu_\zeta}G_1(z,\zeta) = (1 + o(1))2ie^{-i\alpha}\frac{z - \overline{z}e^{2i\alpha} - 2iae^{i\alpha}}{|\zeta - z|^2},$$

for $z \in S_1 \to \zeta \in \partial^+ S_1$, where up to the first factor on the right-hand side the Poisson kernel for the half planes with the boundary $\partial^+ S_1$ appears. Using the reflection point $z_2 = \overline{z}e^{2i\alpha} + 2iae^{i\alpha}$, then

$$\partial_{\nu_\zeta} G_1(z,\zeta) = (1 + o(1))2ie^{-i\alpha} \frac{z - z_2}{|\zeta - z|^2}.$$

The treatment of the normal derivative of the Green function on $\partial^- S_1$ is analogue. Just the sign of the normal vector has to be changed.

The Neumann kernel for S_1 becomes on $\partial^+ S_1$ for $\zeta \in S_1$, similarly as before for the Green function,

$$\begin{aligned}\partial_{\nu_z} N_1(z,\zeta) &= 2\mathrm{Im} e^{i\alpha}\left(\frac{e^{-i\alpha}\pi}{2ia}\cot\pi\frac{\overline{\zeta}e^{i\alpha} - ze^{-i\alpha}}{-2ia} - \frac{\pi}{2iae^{i\alpha}}\cot\pi\frac{\zeta - z}{2iae^{i\alpha}}\right)\\ &= \frac{\pi}{a}\mathrm{Re} 2i\mathrm{Im}\cot\pi\frac{\zeta - z}{2iae^{i\alpha}} = 0.\end{aligned}$$

But for ζ on $\partial^+ S_1$, i.e. $\zeta = \overline{\zeta}e^{2i\alpha} + 2iae^{i\alpha}$ a singularity appears, so that

$$\begin{aligned}-2\mathrm{Im} e^{i\alpha}\partial_z N_1(z,\zeta) &= \mathrm{Im}\frac{\pi}{ia}\left(\cot\pi\frac{\overline{\zeta} - e^{-2i\alpha}z}{-2iae^{-i\alpha}} - \cot\pi\frac{\zeta - z}{2iae^{i\alpha}}\right)\\ &= -\frac{\pi}{a}\mathrm{Re}\left(\cot\pi\frac{(\zeta - z)e^{-i\alpha}}{-2ia} - \cot\pi\frac{\zeta - z}{2iae^{i\alpha}}\right)\\ &= \frac{2\pi}{a}\mathrm{Re}\cot\pi\frac{\zeta - z}{2iae^{i\alpha}}.\end{aligned}$$

Thus for $z \in S_1 \to \zeta \in \partial^+ S_1$

$$\begin{aligned}-2\mathrm{Im} e^{i\alpha}\partial_z N_1(z,\zeta) &= (1 + o(1))2ie^{-i\alpha}\frac{z - \overline{z}e^{2i\alpha} - 2iae^{i\alpha}}{|\zeta - z|^2}\\ &= (1 + o(1))2ie^{-i\alpha}\frac{z - z_2}{|\zeta - z|^2}.\end{aligned}$$

These formulas for the Neumann function also hold for $\partial^- S_1$ when changing the sign of the normal vector.

From the Green and the Neumann representation formulas for certain smooth functions w in regular domains D,

$$w(z) = -\frac{1}{4\pi}\int_{\partial D} w(\zeta)\partial_{\nu_\zeta} G_1(z,\zeta)ds_\zeta - \frac{1}{\pi}\int_D w_{\zeta\overline{\zeta}}(\zeta)G_1(z,\zeta)d\xi d\eta,$$

where s denotes the arc length parameter, $\zeta = \xi + i\eta$, and

$$w(z) = -\frac{1}{4\pi}\int_{\partial D}\{w(\zeta)\partial_{\nu_\zeta}N_1(z,\zeta) - \partial_{\nu_\zeta}w(\zeta)N_1(z,\zeta)\}ds_\zeta$$
$$-\frac{1}{\pi}\int_D w_{\zeta\bar\zeta}(\zeta)N_1(z,\zeta)d\xi d\eta,$$

see, e.g. [1, 2, 4, 8, 11], the following statements hold.

Theorem 1 *The Dirichlet problem*

$$w_{z\bar z} = f, \quad in \quad S_1, \quad w = \gamma \quad on \quad \partial S_1,$$

$$f \in L_{p,2}(S_1;\mathbb{C}), 2 < p, \ \gamma \in C(\partial S_1;\mathbb{C}),$$
$$\lim_{x\to\infty} x^{1+\epsilon}\gamma(x+iy) = 0, x+iy \in S_1, 0 < \epsilon,$$

is uniquely solvable by

$$w(z) = -\frac{1}{2a}\int\limits_{\partial^- S_1}\gamma(\zeta)Re\cot\pi\frac{\zeta - z}{2iae^{i\alpha}}ds_\zeta - \frac{1}{2a}\int\limits_{\partial^+ S_1}\gamma(\zeta)Re\cot\pi\frac{\zeta - z}{2iae^{i\alpha}}ds_\zeta$$
$$+\frac{1}{\pi}\int\limits_{S_1} f(\zeta)\log\left|\frac{\sin\pi\frac{\zeta - e^{2i\alpha}\bar z}{2iae^{i\alpha}}}{\sin\pi\frac{\zeta - z}{2iae^{i\alpha}}}\right|^2 d\xi d\eta.$$

Theorem 2 *The Neumann problem*

$$w_{z\bar z} = f, \quad in \quad S_1, \quad \partial_\nu w = \gamma \quad on \quad \partial S_1,$$

$$f \in L_{p,2}(S_1;\mathbb{C}), 2 < p, \ \gamma \in C(\partial S_1;\mathbb{C}),$$
$$\lim_{x\to\infty} x^{1+\epsilon}\gamma(x+iy) = 0, x+iy \in S_1, 0 < \epsilon,$$

is for any $c \in \mathbb{C}$ solvable by

$$w(z) = c - \frac{1}{2\pi}\int_{\partial^- S_1}\gamma(\zeta)\log\left|\sin\pi\frac{\zeta - z}{2iae^{i\alpha}}\right|^2 ds_\zeta$$
$$-\frac{1}{2\pi}\int_{\partial^+ S_1}\gamma(\zeta)\log\left|\sin\pi\frac{\zeta - z}{2iae^{i\alpha}}\right|^2 ds_\zeta$$
$$+\frac{1}{\pi}\int_{S_1} f(\zeta)\log\left|\sin\pi\frac{\zeta - e^{2i\alpha}\bar z}{2iae^{i\alpha}}\sin\pi\frac{\zeta - z}{2iae^{i\alpha}}\right|^2 d\xi d\eta.$$

Remark The Neumann problem is unconditionally solvable for the strip due to the circumstance that the normal derivative of the Neumann function vanishes at the boundary of the strip as long as the other variable lies inside the strip. The proofs of both results are straightforward by verification.

2.2 *Half Strip*

Let for $0 < \alpha < \pi$, $a \in \mathbb{R}$ the set S_1^+ be the half strip

$$S_1^+ = \{z \in \mathbb{C} : z = e^{2i\alpha}\overline{z} + 2iate^{i\alpha}, 0 < t < 1, e^{-i\alpha}z + e^{i\alpha}\overline{z} > 0\}.$$

Reflecting this half strip at the segment of its boundary part on the line $z = -e^{2i\alpha}\overline{z}$ maps it onto its complementary half strip

$$S_1^- = \{z \in \mathbb{C} : z = e^{2i\alpha}\overline{z} + 2iate^{i\alpha}, 0 < t < 1, e^{-i\alpha}z + e^{i\alpha}\overline{z} < 0\},$$

while $z \in S_1^+$ is reflected onto $\widehat{z} = -e^{2i\alpha}\overline{z}$. This latter point satisfies

$$e^{-i\alpha}\widehat{z} + e^{i\alpha}\overline{\widehat{z}} = -\{e^{-i\alpha}z + e^{i\alpha}\overline{z}\} < 0.$$

Continued reflections of the points $z, \widehat{z} \in S_1$ lead to the points $z_{2k} = e^{2i\alpha}\overline{z} + 2iake^{i\alpha}$, $\widehat{z}_{2k+1} = -e^{2i\alpha}\overline{z} + 2iake^{i\alpha}$ and to $\widehat{z}_{2k} = -z + 2iake^{i\alpha}$, $z_{2k+1} = z + 2iake^{i\alpha}$, for any $k \in \mathbb{Z}$. The first two sets of points are chosen as zeros for P the others become poles. Hence,

$$P(z, \zeta) = \frac{\zeta - e^{2i\alpha}\overline{z}}{\zeta - z}\,\frac{\zeta + e^{2i\alpha}\overline{z}}{\zeta + z}\prod_{k=1}^{\infty}\frac{(\zeta - e^{2i\alpha}\overline{z})^2 - (2iake^{i\alpha})^2}{(\zeta - z)^2 - (2iake^{i\alpha})^2}\,\frac{(\zeta + e^{2i\alpha}\overline{z})^2 - (2iake^{i\alpha})^2}{(\zeta + z)^2 - (2iake^{i\alpha})^2}$$

$$= \frac{\sin\pi\frac{\zeta - e^{2i\alpha}\overline{z}}{2iae^{i\alpha}}}{\sin\pi\frac{\zeta - z}{2iae^{i\alpha}}}\,\frac{\sin\pi\frac{\zeta + e^{2i\alpha}\overline{z}}{2iae^{i\alpha}}}{\sin\pi\frac{\zeta + z}{2iae^{i\alpha}}}.$$

The Green function for S_1^+ is thus $G_1(z, \zeta) = \log|P(z, \zeta)|^2$, its Neumann function is

$$N_1(z, \zeta) = -\log\left|\sin\pi\frac{\zeta - e^{2i\alpha}\overline{z}}{2iae^{i\alpha}}\sin\pi\frac{\zeta - z}{2iae^{i\alpha}}\sin\pi\frac{\zeta + e^{2i\alpha}\overline{z}}{2iae^{i\alpha}}\sin\pi\frac{\zeta + z}{2iae^{i\alpha}}\right|^2.$$

3 Parqueting of the Plane Through Reflections of Hyperbolic Strips

Hyperbolic geometry can in the complex plane either be studied in the unit disc or in the upper half plane. Here the unit disc $\mathbb{D} = \{|z| < 1\}$ is investigated. Straight lines in this geometry are segments of circles orthogonal to the unit circle. They are of the form $\partial D_m(r) = \{|z - m| = r\}$, where $1 < |m|, 0 < r, |m|^2 = 1 + r^2$. A hyperbolic half plane and a hyperbolic strip will be investigated. As before parquetings of the complex plane will be constructed and harmonic Green and Neumann functions attained, see [1, 3, 10].

3.1 *Hyperbolic Half Plane*

For simplicity the centre m of the circle orthogonal to $\partial\mathbb{D}$ is assumed to be real and positive, $1 < m$ and $m^2 = 1 + r^2$. The hyperplane $D = \mathbb{D} \cap D_m(r), D_m(r) = \{|z - m| < r\}$, is a convex lentil. Reflecting D at its boundary part from $\partial D_m(r)$ maps D onto $\mathbb{D}\backslash\overline{D}$ because $\partial D_m(r)$ stays pointwise fixed and as orthogonality is preserved the part of $\partial D_m(r)$ from $\partial\mathbb{D}$ is mapped on its complementary arc. Similarly, reflection at the boundary part from $\partial\mathbb{D}$ maps D onto the complement $D_m(r)\backslash\overline{D}$. Reflecting the unit disc $\mathbb{D}$ at its boundary completes the parqueting of the plane $\mathbb{C}$. The same would be reached by reflecting $D_m(r)$ at its boundary.

A point $z \in D$ is reflected at $\partial\mathbb{D}$ to $\frac{1}{\overline{z}} \in D_m(r)\backslash D$. Both these points $z, \frac{1}{\overline{z}} \in D_m(r)$ are reflected at $\partial D_m(r)$ to the points

$$m + \frac{r^2}{\overline{z} - m} = \frac{\overline{z}m - 1}{\overline{z} - m}, m + \frac{r^2}{\frac{1}{z} - m} = \frac{m - z}{1 - mz},$$

respectively. The parqueting-reflection principle leads to the rational function

$$P(z, \zeta) = \frac{1 - z\overline{\zeta}}{\zeta - z} \frac{m(\overline{\zeta} + z) - (1 + z\overline{\zeta}))}{\zeta + z - m(1 + z\zeta)}.$$

The Green and Neumann functions for D are again $G_1(z, \zeta) = \log|P(z, \zeta)|^2$, and

$$N_1(z, \zeta) = -\log|(\zeta - z)(1 - z\overline{\zeta})(\zeta + z - m(1 + z\zeta))(1 + z\overline{\zeta} - m(\overline{\zeta} + z))|^2.$$

The Poisson kernel is given for the part of ∂D on $|z| = 1$ as

$$\partial_{\nu_z} G_1(z, \zeta) = 2\text{Re}\left\{\frac{\zeta + z}{\zeta - z} + \frac{\zeta - z - m(1 - z\zeta)}{\zeta + z - m(1 + z\zeta)}\right\}.$$

Similarly, for the part on $|z - m| = r$

$$\partial_{\nu_z} G_1(z, \zeta) = 2\text{Re}\left\{\frac{\zeta + z - 2m}{\zeta - z} - \frac{1 + z\overline{\zeta} - 2m\overline{\zeta}}{1 - z\overline{\zeta}}\right\},$$

see [9]. On the boundary part of D from $\partial\mathbb{D}$

$$N_1(z,\zeta) = -2\log|(\zeta - z)(\zeta + z - m(1+z\zeta))|^2$$

and

$$\begin{aligned}\partial_{\nu_z} N_1(z,\zeta) &= 2\mathrm{Re}\{z\partial_z N(z,\zeta)\} \\ &= 2\mathrm{Re}\left\{\frac{z}{\zeta - z} + \frac{z\overline{\zeta}}{1 - z\overline{\zeta}} - \frac{z(1-m\zeta)}{\zeta + z - m(1+z\zeta)}\right. \\ &\qquad \left. - \frac{z(\overline{\zeta} - m)}{1 + z\overline{\zeta} - m(\overline{\zeta} + z)}\right\} = -4.\end{aligned}$$

Similarly, for the boundary part on $\partial D_m(r)$

$$N_1(z,\zeta) = -2\log|(\zeta - z)(1 - z\overline{\zeta})|^2 - 4\log r$$

and

$$\begin{aligned}\partial_{\nu_z} N_1(z,\zeta) &= 2\mathrm{Re}\{(z-m)\partial_z N_1(z,\zeta)\} \\ &= 2\mathrm{Re}\left\{\frac{z-m}{\zeta - z} + \frac{(z-m)\overline{\zeta}}{1 - z\overline{\zeta}} - \frac{(z-m)(1-m\zeta)}{\zeta + z - m(1+z\zeta)}\right. \\ &\qquad \left. - \frac{(z-m)(\overline{\zeta} - m)}{1 + z\overline{\zeta} - m(\overline{\zeta} + z)}\right\} = -4.\end{aligned}$$

These formulas hold for any $\zeta \in D$. For $|\zeta| = 1$ and $z \in D$ instead

$$2\mathrm{Re}\{z\partial_z N_1(z,\zeta)\} = 2\left(\frac{\zeta}{\zeta - z} + \frac{\overline{\zeta}}{\overline{\zeta - z}} - 1\right) - 4 + \mathrm{o}(1)$$

for $|z| \to 1$, and for $|\zeta - m| = r$ and $z \in D$

$$2\mathrm{Re}\{(z-m)\partial_z N_1(z,\zeta)\} = 2\left(\frac{\zeta - m}{\zeta - z} + \frac{\overline{\zeta - m}}{\overline{\zeta - z}} - 1\right) - 4 + \mathrm{o}(1)$$

for $|z - m| \to r$.

Remark From the mentioned reflections of the point z it is seen that the same expressions will appear when z belongs to any of the other domains in the parqueting of the plane. Therefore, the Green and Neumann functions attained hold in same forms for any of these four domains.

Theorem 3 *The Dirichlet problem*

$$w_{z\overline{z}} = f \quad in \quad D, f \in L_p(D;\mathbb{C}), 2 < p, w = \gamma \quad on \quad \partial D, \gamma \in C(\partial D;\mathbb{C})$$

is uniquely solvable by

$$w(z) = \frac{1}{2\pi i}\int_{\partial D\cap\partial\mathbb{D}} \gamma(\zeta) Re\left\{\frac{\zeta+z}{\zeta-z} + \frac{\zeta - z + m(1-z\zeta)}{\zeta+z-m(1+z\zeta)}\right\}\frac{d\zeta}{\zeta}$$
$$+ \frac{1}{2\pi i}\int_{\partial D\cap\partial D_m(r)} \gamma(\zeta) Re\left\{\frac{\zeta+z-2m}{\zeta-z} + \frac{1+z\overline{\zeta}-2m\overline{\zeta}}{1-z\overline{\zeta}}\right\}\frac{d\zeta}{\zeta-m}$$
$$-\frac{1}{\pi}\int_D f(\zeta)\log\left|\frac{1-z\overline{\zeta}}{\zeta-z}\frac{m(\overline{\zeta}+z)-(1+z\overline{\zeta}))}{\zeta+z-m(1+z\zeta)}\right|^2 d\xi d\eta.$$

Theorem 4 *The Neumann problem*

$$w_{z\bar{z}} = f \quad in \quad D, \quad \partial_\nu w = \gamma \quad on \quad \partial D,$$

is solvable if and only if

$$\frac{1}{4\pi}\int_{\partial D}\gamma(\zeta)ds_\zeta = \frac{1}{\pi}\int_D f(\zeta)d\xi d\eta$$

is satisfied. The solution then is given for any $c \in \mathbb{C}$ *by*

$$w(z) = c - \frac{1}{2\pi}\int_{\partial D\cap\partial\mathbb{D}} \gamma(\zeta)\log|(\zeta - z)(\zeta + z - m(1+z\zeta))|^2\frac{d\zeta}{\zeta}$$
$$-\frac{1}{2\pi}\int_{\partial D\cap\partial D_m(r)} \gamma(\zeta)\left(\log|(\zeta-z)(1-z\overline{\zeta})|^2 + \log r^2\right)\frac{d\zeta}{\zeta-m}$$
$$+\frac{1}{\pi}\int_D f(\zeta)\log|(\zeta-z)(1-z\overline{\zeta})(\zeta+z-m(1+z\zeta))$$
$$(1+z\overline{\zeta}-m(\overline{\zeta}+z))|^2 d\xi d\eta.$$

Remark Obviously, the term $\log r^2$ in the second line integral may be skipped. The proof follows by direct verification. The uniqueness—up to the constant c—follows from the general Neumann representation formula [1, 4].

3.2 Hyperbolic Strip

A strip is the set between two lines. Thus a hyperbolic strip is the complement with regard to the unit disc of two nonintersecting discs with boundaries orthogonal to the unit disc. Again for simplicity the centres of both discs are assumed to be on the real axis. Just a particular situation is investigated. Some others are discussed in [3]. For four real numbers m_1, m_2 greater than 1 and positive r_1, r_2 given such that $1 + r_1^2 = m_1^2$, $1 + r_2^2 = m_2^2$ the circles $\partial D_{-m_1}(r_1)$, $\partial D_{m_2}(r_2)$ where $D_m(r) = \{|z - m| < r\}$ for $0 < r, 1 < m, 1 + r^2 = m^2$ are orthogonal to the unit circle $\partial\mathbb{D}$. For $r_1 + r_2 < m_1 +$

m_2 the relations $-1 < r_1 - m_1 < 0 < m_2 - r_2 < 1$ hold. Both circles are disjoint and $D = \mathbb{D}\backslash\{D_{-m_1}(r_1) \cup D_{m_2}(r_2)\}$ is a hyperbolic strip, [3].

For any $z \in \mathbb{C}$ the reflected point z_{re} at $|z - m| = r$ is given by the relation $(z_{re} - m)(\overline{z - m}) = r^2$. If in particular $1 + r^2 = |m|^2$ then $z_{re} = \frac{m\overline{z}-1}{\overline{z}-m}$. Reflection at circles or lines transforms any circle or line into a circle or a line. Moreover, orthogonality is preserved.

A parqueting of the entire complex plane $\mathbb{C}$ can be achieved by successively reflecting the strip on its boundary parts as is obvious for the case of an Euclidean strip. In this way two sets on hyperbolic strips are attained matching together with the original strip D in a parqueting of the unit disc, i.e. the hyperbolic plane. Reflecting this parqueting at the unit circle $\partial\mathbb{D}$ leads to a parqueting of the complement $\mathbb{C}\ \mathbb{D}$ of the unit disc $\mathbb{D}$. Altogether they compose a parqueting of $\mathbb{C}$.

In [3] these reflections are calculated and the two sets of discs

$$D_{-m_{2k-1}}(r_{2k-1}), D_{m_{2k}}(r_{2k}), k \in \mathbb{N},$$

and the reflection points from $z \in D$

$$z_1 = -\frac{m_1\overline{z} + 1}{\overline{z} + m_1}, z_{2k+1} = -\frac{m_{2k+1}\overline{z_{2k-1}} + 1}{\overline{z_{2k-1}} + m_{2k+1}} = \frac{\alpha_{2k-1}z_{2k-3} - \beta_{2k-1}}{-\beta_{2k-1}z_{2k-3} + \alpha_{2k-1}}, k \in \mathbb{N},$$

and

$$z_2 = \frac{m_2\overline{z} - 1}{\overline{z} - m_2}, z_{2k+2} = \frac{m_{2k+2}\overline{z_{2k}} - 1}{\overline{z_{2k}} - m_{2k+2}} = \frac{\alpha_{2k}z_{2k-2} + \beta_{2k}}{\beta_{2k}z_{2k-2} + \alpha_{2k}}, k \in \mathbb{N},$$

are attained. Here $m_3^2 = 1 + r_3^2, m_4^2 = 1 + r_4^2$,

$$m_3 = \frac{2\alpha\beta - m_2(\alpha^2 + \beta^2)}{(\alpha^2 + \beta^2) - 2\alpha\beta m_2}, m_4 = \frac{(\alpha^2 + \beta^2)m_1 - 2\alpha\beta}{2\alpha\beta m_1 - \alpha^2 - \beta^2},$$

$$\alpha = m_1m_2 + 1, \beta = m_1 + m_2,$$

$$m_{2k+3}^2 = r_{2k+3}^2 + 1, \ m_{2k+3} = \frac{2\alpha_{2k-1}\beta_{2k-1} + m_{2k-1}(\alpha_{2k-1}^2 + \beta_{2k-1}^2)}{\alpha_{2k-1}^2 + \beta_{2k-1}^2 + 2\alpha_{2k-1}\beta_{2k-1}m_{2k-1}},$$

$$\alpha_{2k-1} = m_{2k-1}m_{2k+1} - 1, \ \beta_{2k-1} = m_{2k-1} - m_{2k+1}, k \in \mathbb{N},$$

and

$$r_{2k+4}^2 + 1 = m_{2k+4}^2, \ m_{2k+4} = \frac{2\alpha_{2k}\beta_{2k} + (\alpha_{2k}^2 + \beta_{2k}^2)m_{2k}}{\alpha_{2k}^2 + \beta_{2k}^2 + 2\alpha_{2k}\beta_{2k}m_{2k}},$$

$$\alpha_{2k} = m_{2k}m_{2k+2} - 1, \beta_{2k} = m_{2k} - m_{2k+2}, k \in \mathbb{N}.$$

For $k \in \mathbb{N}$ the estimates

$$1 < m_{2k+1} < m_{2k-1}, m_{4k+1}^2 - 1 \leq q^{2^k}(m_1^2 - 1), m_{4k+3}^2 - 1 \leq q^{2^k}(m_3^2 - 1),$$

$$1 < m_{2k+2} < m_{2k}, m_{4k+4}^2 - 1 < q^{2k}(m_4^2 - 1), \ m_{4k+2}^2 - 1 < q^{2k}(m_2^2 - 1),$$

$$q = \frac{m_1 + 1}{m_1 - 1}\frac{m_3 - 1}{m_3 + 1} = \frac{m_2 + 1}{m_2 - 1}\frac{m_4 - 1}{m_4 + 1} = \frac{m_1 - 1}{m_1 + 1}\frac{m_2 - 1}{m_2 + 1} < 1$$

are shown, implying

$$\lim_{k\to\infty} m_{2k+1} = 1, \ \lim_{k\to\infty} m_{2k} = 1.$$

Thus the sequence of discs $D_{-m_{2k+1}}(r_{2k+1})$ shrinks to the point -1 while the discs $D_{m_{2k}}(r_{2k})$ shrink to the point 1.

The reflections z_{2k+1}, z_{2k+2}, $(k + 1) \in \mathbb{N}$, satisfy

$$0 < 1 - |z_{4k+1}|^2 \leq 2r_{4k+1} \leq 2q^k r_1, \ 0 < 1 - |z_{4k+3}|^2 \leq 2r_{4k+3} \leq 2q^k r_2,$$

$$0 < 1 - |z_{4k+2}|^2 \leq 2r_{4k+2} \leq 2q^k r_2, \ 0 < 1 - |z_{4k+4}|^2 \leq 2r_{4k+4} \leq 2q^k r_1.$$

In [1] the simplified representations

$$m_{2k+1} = \frac{m_{2k-1}\alpha + \beta}{m_{2k-1}\beta + \alpha}, \quad m_{2k+2} = \frac{m_{2k}\alpha + \beta}{m_{2k}\beta + \alpha},$$

where $\alpha_{-1} = \alpha_0 = \alpha$, $\beta_0 = \beta_{-1} = \beta$, are proved for any $k \in \mathbb{N}$. Moreover, from here follow inductively as well

$$m_{2k+1} = \frac{m_1\gamma_k + \delta_k}{m_1\delta_k + \gamma_k}, m_{2k+2} = \frac{m_2\gamma_k + \delta_k}{m_2\delta_k + \gamma_k},$$

$$\gamma_k = \sum_{\nu=0}^{[\frac{k}{2}]} \binom{k}{2\nu} \alpha^{k-2\nu}\beta^{2\nu}, \quad \delta_k = \sum_{\nu=0}^{[\frac{k-1}{2}]} \binom{k}{2\nu + 1} \alpha^{k-2\nu-1}\beta^{2\nu+1}.$$

as

$$z_{2k+1} = \frac{\alpha z_{2k-3} - \beta}{\alpha - \beta z_{2k-3}}, \quad z_{2k+2} = \frac{\alpha z_{2k-2} + \beta}{\alpha + \beta z_{2k-2}},$$

for any $k \in \mathbb{N}$. Also

$$z_{4k-1} = \frac{\gamma_k z - \delta_k}{\gamma_k - \delta_k z}, \quad z_{4k+1} = \frac{\gamma_k z_1 - \delta_k}{\gamma_k - \delta_k z_1}, \quad z_1 = -\frac{m_1\overline{z} + 1}{\overline{z} + m_1},$$

i.e.

$$z_{4k+1} = -\frac{\widehat{\gamma_k}\overline{z} + \widehat{\delta_k}}{\widehat{\delta_k}\overline{z} + \widehat{\gamma_k}}, \quad \widehat{\gamma_k} = \gamma_k m_1 + \delta_k, \quad \widehat{\delta_k} = \delta_k m_1 + \gamma_k,$$

and

$$z_{4k} = \frac{\gamma_k z + \delta_k}{\gamma_k + \delta_k z}, \quad z_{4k+2} = \frac{\gamma_k z_2 + \delta_k}{\gamma_k + \delta_k z_2}, \quad z_2 = \frac{m_2\overline{z} - 1}{\overline{z} - m_2},$$

i.e.

$$z_{4k+2} = \frac{\widetilde{\gamma_k}\overline{z} - \widetilde{\delta_k}}{\widetilde{\delta_k}\overline{z} - \widetilde{\gamma_k}}, \quad \widetilde{\gamma_k} = \gamma_k m_2 + \delta_k, \quad \widetilde{\delta_k} = \delta_k m_2 + \gamma_k,$$

for any $k \in \mathbb{N}$ are shown.

With this point set $z \in D$, z_k, $k \in \mathbb{N}$, the harmonic Green function is constructed and the Dirichlet problem is solved for the Poisson equation in the hyperbolic strip D. For the harmonic Neumann problem besides the above parqueting two more parquetings of $\mathbb{C}$ are required. They arise from reflecting the parquetings of the discs $D_{-m_1}(r_1)$ and $D_{m_2}(r_2)$ at their boundaries rather than reflecting $\mathbb{D}$ at its boundary. These reflections produce the same parqueting of the complex plane and the same point set z_k out of the point z, but their representations differ. This turns out to be more proper for the different parts of the boundary ∂D, see [1].

The coincidence as well of the reflections

$$\widehat{z_{2k}} = -\frac{m_1\overline{z_{2k+1}} + 1}{\overline{z_{2k+1}} + m_1} = \frac{\kappa_{2k+1}z_{2k-1} + \lambda_{2k+1}}{\lambda_{2k+1}z_{2k-1} + \kappa_{2k+1}},$$
$$\kappa_{2k+1} = m_1 m_{2k+1} - 1, \lambda_{2k+1} = m_1 - m_{2k+1},$$

of

$$z_{2k+1} = -\frac{m_{2k+1}\overline{z_{2k-1}} + 1}{\overline{z_{2k-1}} + m_{2k+1}},$$

at $\partial D_{-m_1}(r_1)$ with z_{2k} as of the reflections

$$\widehat{z_{2k-1}} = \frac{m_2\overline{z_{2k+2}} - 1}{\overline{z_{2k+2}} - m_2} = \frac{\kappa_{2k+2}z_{2k} - \lambda_{2k+2}}{\kappa_{2k+2} - \lambda_{2k+2}z_{2k}},$$
$$\kappa_{2k+2} = m_2 m_{2k+2} - 1, \lambda_{2k+2} = m_2 - m_{2k+2},$$

of

$$z_{2k+2} = \frac{m_{2k+2}\overline{z_{2k}} - 1}{\overline{z_{2k}} - m_{2k+2}},$$

at $\partial D_{m_2}(r_2)$ with z_{2k-1} are shown in [1].

By the parqueting-reflection principle the harmonic Green function for D is given as $G_1(z, \zeta) = \log |P(z, \zeta|^2$ with

$$P(z,\zeta) = \frac{1-\overline{z}\zeta}{\zeta - z}\frac{\zeta - z_1}{1-\overline{z_1}\zeta}\frac{\zeta - z_2}{1-\overline{z_2}\zeta}\prod_{k=1}^{\infty}\frac{1-\overline{z_{4k-1}}\zeta}{\zeta - z_{4k-1}}\frac{1-\overline{z_{4k}}\zeta}{\zeta - z_{4k}}\frac{\zeta - z_{4k+1}}{1-\overline{z_{4k+1}}\zeta}\frac{\zeta - z_{4k+2}}{1-\overline{z_{4k+2}}\zeta}.$$

The Neumann function for D is

$$N_1(z,\zeta) = -\log|Q(z,\zeta)|^2, z \in D, \zeta \in D,$$

with

$$Q(z,\zeta) = (\zeta - z)(1-\overline{z}\zeta)\prod_{k=1}^{\infty}\frac{\zeta - z_{2k-1}}{\zeta + 1}\frac{1-\overline{z_{2k-1}}\zeta}{\overline{z_{2k-1}}(1+\zeta)}\frac{\zeta - z_{2k}}{\zeta - 1}\frac{1-\overline{z_{2k}}\zeta}{\overline{z_{2k}}(1-\zeta)}.$$

By experience it is known that Green and Neumann functions are related with one another. Choosing all the points from the parqueting-reflection construction as poles rather than as poles and zeroes of a meromorphic function provides the Neumann function. However, in case of infinitely many poles the product involved does not need to converge. If it does not, convergence providing analytic factors have to be incorporated not adding further poles and zeroes to the function.

As it turns out, it is more proper here for the case of the hyperbolic strip to alter the factors of the infinite product for the Neumann function in replacing the form $1-\overline{z_k}\zeta$ by $\frac{1}{\overline{z_k}} - \zeta$. The function $N_1(z,\zeta)$ constructed here is just some harmonic Neumann function for D, neither satisfying the often used normalization condition [5, 7] nor being symmetric in its two variables.

For the Dirichlet problem the Poisson kernel, i.e. the normal derivatives of the Green function at the boundary ∂D, for the Neumann problem the normal derivatives of the Neumann function have to be calculated. This is done in detail in [1, 3]. Here the results differ for the three different parts of the boundary ∂D.

Theorem 5

1. *For $\zeta \in \partial D \cap \partial\mathbb{D}$, the four corner points $\{-\frac{1}{2} \pm i\frac{r_1}{m_1}, \frac{1}{2} \pm i\frac{r_2}{m_2}\}$ excluded, and $z \in D$*

$$\lim_{|z|\to 1}(\zeta\partial_\zeta + \overline{\zeta}\partial_{\overline{\zeta}})G_1(z,\zeta) = -2\lim_{|z|\to 1}\left[\frac{\zeta}{\zeta - z} + \frac{\overline{\zeta}}{\overline{\zeta - z}} - 1\right],$$

$$\lim_{|z+m_1|\to r_1}(\zeta\partial_\zeta + \overline{\zeta}\partial_{\overline{\zeta}})G_1(z,\zeta) = 0,$$

$$\lim_{|z-m_2|\to r_2}(\zeta\partial_\zeta + \overline{\zeta}\partial_{\overline{\zeta}})G_1(z,\zeta) = 0.$$

2. *For $\zeta \in \partial D \cap \partial D_{-m_1}(r_1)$, the two corner points $\{-\frac{1}{2} \pm i\frac{r_1}{m_1}\}$ excluded, and $z \in D$*

$$\lim_{|z|\to 1}\partial_{\nu_\zeta}G_1(z,\zeta) = 0,$$

$$\lim_{|z+m_1|\to r_1} \partial_{\nu_\zeta} G_1(z,\zeta) = \frac{2}{r_1} \lim_{|z+m_1|\to r_1} \Big[\frac{\zeta+m_1}{\zeta-z} + \frac{\overline{\zeta}+m_1}{\overline{\zeta-z}} - 1\Big],$$

$$\lim_{|z-m_2|\to r_2} \partial_{\nu_\zeta} G_1(z,\zeta) = 0.$$

3. For $\zeta \in \partial D \cap \partial D_{m_2}(r_2)$, the two corner points $\{\frac{1}{2} \pm i\frac{r_2}{m_2}\}$ excluded, and $z \in D$

$$\lim_{|z|\to 1} \partial_{\nu_\zeta} G_1(z,\zeta) = 0,$$

$$\lim_{|z+m_1|\to r_1} \partial_{\nu_\zeta} G_1(z,\zeta) = 0,$$

$$\lim_{|z-m_2|\to r_2} \partial_{\nu_\zeta} G_1(z,\zeta) = \frac{2}{r_2} \lim_{|z-m_2|\to r_2} \Big[\frac{\zeta-m_2}{\zeta-z} + \frac{\overline{\zeta}-m_2}{\overline{\zeta-z}} - 1\Big].$$

Theorem 6 *The Dirichlet problem*

$$w_{z\overline{z}} = f \quad in \quad D, \quad f \in L_p(D;\mathbb{C}), \quad 2 < p,$$

$$w = \gamma \quad on \quad \partial D, \gamma \in C(\partial D;\mathbb{C}), \gamma(-\frac{1}{2} \pm i\frac{m_1}{r_1}) = \gamma(\frac{1}{2} \pm i\frac{m_2}{r_2}) = 0,$$

is uniquely solvable by

$$w(z) = -\frac{1}{4\pi}\int_{\partial D} \gamma(\zeta)\partial_{\nu_\zeta} G_1(z,\zeta) ds_\zeta - \frac{1}{\pi}\int_D f(\zeta) G_1(z,\zeta) d\xi d\eta.$$

The proof follows by verification on the basis of properties of the Poisson kernels and of the Pompeiu operator [4].

For any $z \in D$ the normal derivative of $N_1(z,\cdot)$ at the boundary $\partial D \backslash \{-\frac{1}{2} \pm i\frac{r_1}{m_1}, \frac{1}{2} \pm i\frac{r_2}{m_2}\}$ is

$$\partial_{\nu_\zeta} N_1(z,\zeta) = \begin{cases} 2\mathrm{Re}(\zeta\partial_\zeta)N_1(z,\zeta) = -2, \zeta \in \partial D \cap \partial\mathbb{D}, \\ -2\mathrm{Re}((\zeta+m_1)\partial_\zeta)N_1(z,\zeta) = 2 - 4\dfrac{m_1-1}{|\zeta+1|^2}, \\ \qquad\qquad z \in \partial D \cap \partial D_{-m_1}(r_1), \\ -2\mathrm{Re}((\zeta-m_2)\partial_\zeta)N_1(z,\zeta) = 2 - 4\dfrac{m_2-1}{|\zeta-1|^2}, \\ \qquad\qquad z \in \partial D \cap \partial D_{m_2}(r_2). \end{cases}$$

Obviously, $N_1(z,\zeta)$ is also harmonic in the variable z. Its normal derivative with respect to this variable is also important.

On ∂D the normal derivative of $N_1(z,\zeta)$ with respect to the variable z except at the corner points is

$$\partial_{\nu_z} N_1(z,\zeta) = \begin{cases} -2, z \in \partial D \cap \partial \mathbb{D}, \zeta \in \overline{D}\backslash \partial \mathbb{D}, \\ -\dfrac{1-|z|^2}{|z|^2}, z \in \partial D \cap \partial D_{-m_1}(r_1), \zeta \in \overline{D}\backslash \partial D_{-m_1}(r_1), \\ \qquad z \in \partial D \cap \partial D_{m_2}(r_2), \zeta \in \overline{D}\backslash \partial D_{m_2}(r_2), \\ 2\left(\dfrac{\zeta}{\zeta - z} + \dfrac{\overline{\zeta}}{\overline{\zeta - z}} - 2\right), z, \zeta \in \partial D \cap \partial \mathbb{D}, \\ -2\left(\dfrac{\zeta + m_1}{\zeta - z} + \dfrac{\overline{\zeta} + m_1}{\overline{\zeta - z}} - 1\right) - \dfrac{1-|z|^2}{|z|^2}, z, \zeta \in \partial D \cap \partial D_{-m_1}(r_1), \\ -2\left(\dfrac{\zeta - m_2}{\zeta - z} + \dfrac{\overline{\zeta} - m_2}{\overline{\zeta - z}} - 1\right) - \dfrac{1-|z|^2}{|z|^2}, z, \zeta \in \partial D \cap \partial D_{m_2}(r_2). \end{cases}$$

Any $w \in C^2(D;\mathbb{C}) \cap C^1(\overline{D};\mathbb{C})$ for regular domains D is representable as

$$w(z) = -\frac{1}{4\pi}\int_{\partial D}\{w(\zeta)\partial_{\nu_\zeta} N_1(z,\zeta) - \partial_{\nu_\zeta} w(\zeta) N_1(z,\zeta)\}ds_\zeta$$
$$-\frac{1}{\pi}\int_D w_{\zeta\overline{\zeta}}(\zeta) N_1(z,\zeta)d\xi d\eta.$$

Here $N_1(z,\zeta)$ is the harmonic Neumann function for D. This representation formula provides the solution to the Neumann problem for the Poisson equation in case it exists.

Theorem 7 *The Neumann problem for the Poisson equation*

$$w_{z\overline{z}} = f \quad in \quad D, \quad \partial_\nu w = \gamma \quad on \quad \partial D,$$

for $f \in L_p(D;\mathbb{C}), 2 < p, \gamma \in C(\partial D;\mathbb{C}), \gamma(-\frac{1}{2} \pm i\frac{m_1}{r_1}) = \gamma(\frac{1}{2} \pm i\frac{m_2}{r_2}) = 0,$ *is solvable if and only if*

$$\frac{1}{2\pi}\int_{\partial D}\gamma(\zeta)ds_\zeta = \frac{2}{\pi}\int_D f(\zeta)d\xi d\eta.$$

The solution then is with some arbitrary $c \in \mathbb{C}$

$$w(z) = c + \frac{1}{4\pi}\int_{\partial D}\gamma(\zeta)N_1(z,\zeta)ds_\zeta - \frac{1}{\pi}\int_D f(\zeta)N_1(z,\zeta)d\xi d\eta.$$

Remark The assumption on the boundary function to vanish at the corner points can in principal be abandoned for both the Dirichlet and the Neumann problems, see, e.g. [8, 15]. Obviously, the given Neumann function fails to be symmetric in its variables. If no normalization condition is required in principal some harmonic functions in just one of the variables not depending on the other one can be added. What is important here, is the fact, that the normal derivatives besides the Poisson kernels are constant in the respective variable. However, the Green function in general is symmetric. As is well known, this is a consequence from its three basic properties.

References

1. Akel, M., Begehr, H.: Neumann Function for a Hyperbolic Strip and a Class of Related Plane Domains. Preprint, FU Berlin (2015)
2. Aksoy, Ü., Celebi, A.O.: A survey on boundary value problems for complex partial differential equations. Adv. Dyn. Syst. Appl. **5**, 133–158 (2010)
3. Begehr, H.: Green function for a hyperbolic strip and a class of related plane domains. Appl. Anal. **93**, 2370–2385 (2014)
4. Begehr, H.: Complex Analytic Methods for Partial Differential Equations. An Introductory Text. World Scientific, Singapore (1994)
5. Begehr, H., Vaitekhovich, T.: Harmonic boundary value problems in half disc and half ring. Funct. Approx. Comment Math. **40**, 251–282 (2009)
6. Begehr, H., Vaitekhovich, T.: Polyharmonic Green functions for particular plane domains. In: Beznea, L. et al. (eds.) Proceedings of the 6th Congress of Roman. Mathematicians, Roman. Academic of Sciences, vol. 1, pp. 119–126. Publishing House, Bucharest (2009)
7. Begehr, H., Vaitekhovich, T.: How to find harmonic Green functions in the plane. Complex Var. Ell. Eqs. **56**, 1169–1181 (2011)
8. Begehr, H., Vaitekhovich, T.: Harmonic Dirichlet problem for some equilateral triangle. Complex Var. Ell. Eqs. **57**, 185–196 (2012)
9. Begehr, H., Vaitekhovich, T.: The parqueting-reflection principle for constructing Green function. In: Rogosin, S.V., Dubatovskaya, M.V. (eds.) Analytic Methods of Analysis and Differential Equations: AMADE-2012, pp. 11–20. Cambridge Scientific Publishers, Cottenham (2013)
10. Begehr, H., Vaitekhovich, T.: Schwarz problem in lens and lune. Complex Var. Ell. Eqs. **59**, 76–84 (2014)
11. Begehr, H., Vanegas, C.J.: Iterated Neumann problem for the higher order Poisson equation. Math. Nach. **279**, 38–57 (2006)
12. Gaertner, E.: Basic complex boundary value problems in the upper half plane. Ph.D. thesis, FU Berlin (2006). www.diss.fu-berlin.de/diss/receive/FUDISS_thesis_000000002129
13. Shupeyeva, B.: Some basic boundary value problems for complex PDEs in quarter ring and half hexagon, Ph.D. thesis, FU Berlin (2013). www.diss.fuberlin.de/diss/receive/FUDISSthesis000000094596
14. Shupeyeva, B.: Harmonic boundary value problems in a quarter ring domain. Adv. Pure Appl. Math. **3**, 393-419 (2012); **4**, 103–105 (2013)
15. Wang, Y.-F., Wang, Y.-J.: Schwarz-type problem of nonhomogeneous Cauchy-Riemann equation on a triangle. J. Math. Anal. Appl. **377**, 557–570 (2011)
16. Wang, Y.: Boundary value problems for complex partial differetial equations in fan-shaped domains. Ph.D. thesis, FU Berlin (2011). www.diss.fu-berlin.de/diss/receive/FUDISS_thesis_000000021359

A Genetically Distinguishable Competition Model

Irene Azzali, Giulia Marcaccio, Rosanna Turrisi and Ezio Venturino

Abstract A mathematical ecogenetic model of competition type is presented. The system behavior is completely assessed, either analytically when possible or through numerical simulations. No sustained oscillations are possible, while the equilibria of the system are linked via a chain of transcritical bifurcations. The most important result of the investigation shows that the principle of competitive exclusion is possibly violated in suitable situations.

Keywords Ecogenetic models · Genotypes · Competing models · Equilibria · Stability · Bifurcations

1 Introduction

Genetically distinguishable populations have been recently considered also in the context of ecology, giving rise to mathematical ecogenetic models, [1, 3]. The demographic situation envisioned in the former papers was always the interaction among predators and prey, with the possibility of having the genetically distinguishable population to be either one of the two. Other possible interactions can be considered, among which the most important one is the competition of different populations. To this purpose, in this paper we want to present and investigate a competition model in which one of the population exhibits genetic variability.

I. Azzali · G. Marcaccio · R. Turrisi · E. Venturino (✉)
Dipartimento di Matematica "Giuseppe Peano", Via Carlo Alberto 10, 10123 Torino, Italy
e-mail: ezio.venturino@unito.it

I. Azzali
e-mail: irene.azzali@studio.unibo.it

G. Marcaccio
e-mail: marcacciog@gmail.com

R. Turrisi
e-mail: Rosanna-t@hotmail.it

V.K. Singh et al. (eds.), *Modern Mathematical Methods and High Performance Computing in Science and Technology*, Springer Proceedings in Mathematics & Statistics 171, DOI 10.1007/978-981-10-1454-3_11

The system behavior is completely characterized, either analytically or when this is not possible via numerical simulations. Persistent oscillations are shown never to arise, while the equilibria of the system are related to each other via a chain of transcritical bifurcations. The major result that is achieved in this context is that the well-known principle of competitive exclusion, [4], does not necessarily hold in this context. There is a certain parameter range for which it is violated, where both populations can coexist.

The paper is organized as follows. We present the system in the next section. After easily establishing the boundedness of the trajectories, we analyze the equilibria in Sect. 3 and then their stability, Sect. 4. The overall system's behavior is examined in Sect. 5 and a final discussion on the findings concludes the paper.

2 The Model

In this article, we describe a model of competition between two populations, one of which presents two different genotypes. Let $Y(t)$ and $Z(t)$ denote the two distinct genotypes and let $X(t)$ be the variable that represents the other species. We assume that the populations live in the same environment and that $X(t)$ reproduces logistically.

We consider the following model:

$$\frac{dX}{d\tau} = r\left(1 - \frac{X}{K}\right)X - hXZ - gXY \quad \frac{dY}{d\tau} = p(aY + bZ) - cY^2 - fYZ - uYX$$
$$\frac{dZ}{d\tau} = q(aY + bZ) - nZ^2 - mYZ - vZX. \tag{1}$$

Here all the parameters are always assumed to be nonnegative. Let r denote the reproduction rate of X, let K be its carrying capacity and let h and g, respectively, be the competition rates suffered by X against the two subpopulations Z and Y. The constants c, f and n, m are the intraspecific competition coefficients of the genotypes Y and Z, with other individuals of the same genotype or the other genotype. Instead, u and v represent the interspecific competition of Y and Z against the other population X. We also assume that the two genotypically distinct subpopulations Y and Z reproduce at rates a and b, respectively; in general we take $a \neq b$ because the difference in genotype can cause different reproduction capabilities. In this situation, the key factor is here represented by the term in bracket in the last two equations of (1), that describes the fact that the two subpopulations can give rise to newborns of both genotypes. The fractions of newborns of Y and of Z, respectively, are p and q with $p + q = 1$.

The model can be nondimensionalized letting $X(\tau) := \alpha x(t)$, $Y(\tau) := \beta y(t)$, $Z(\tau) := \gamma z(t)$, $\tau := \delta t$. We then have

$$\frac{dX}{d\tau} = \alpha\frac{dx}{dt}\frac{dt}{d\tau}, \quad \frac{dY}{d\tau} = \beta\frac{dy}{dt}\frac{dt}{d\tau}, \quad \frac{dZ}{d\tau} = \gamma\frac{dz}{dt}\frac{dt}{d\tau}.$$

Substituting into the system and choosing $\alpha := K$, $\beta := \frac{b}{c}$, $\gamma := \frac{b}{f}$, $\delta := b$ and defining the new parameters $A := \frac{r}{b}$, $B := \frac{h}{f}$, $C := \frac{g}{c}$, $D := \frac{a}{b}$, $E := \frac{c}{f}$, $U := \frac{uK}{b}$, $N := \frac{n}{f}$, $M := \frac{m}{c}$, $V := \frac{vK}{b}$ we have the rescaled model

$$\frac{dx}{dt} = A(1-x)x - Bxz - Cxy \quad \frac{dy}{dt} = p(Dy + Ez) - y^2 - yz - Uyx$$
$$\frac{dz}{dt} = q\left(\frac{D}{E}y + z\right) - Nz^2 - Mzy - Vzx. \tag{2}$$

Boundedness of the system's trajectories can be easily discussed, following the steps of [1] even with some simplifications. Let $W = x + y + z$ denote the total ecosystem population. For $0 < \eta$, we have

$$\begin{aligned}\frac{dW}{dt} + \eta W &\leq (A+\eta)x - Ax^2 \\ &+ \frac{1}{E}(pDE + E\eta + qD)\,y - y^2 + (pE + q + \eta)z - Nz^2 \\ &\leq \frac{A+\eta}{4A} + \frac{pDE + E\eta + qD}{4E} + \frac{pE + q + \eta}{4N} = L\end{aligned}$$

so that ultimately we find $W(t) \leq \max\{W(0), L\eta^{-1}\}$, as desired.

3 Equilibria

The possible equilibria of the system (2) are the system total collapse, $E_0 = (0, 0, 0)$, the first-population-only point, at the carrying capacity level, $E_1 = (1, 0, 0)$, the second-population-only equilibrium $E_2 = (0, y_2, z_2)$, and possibly coexistence $E^* = (x^*, y^*, z^*)$. The first two equilibria are always feasible. We now investigate E_2. Its subpopulation levels can analytically be assessed, as the intersections of the following two conic sections:

$$y^2 + y(z - pD) - pEz = 0 \quad ENz^2 + zE(My - q) - qDy = 0. \tag{3}$$

We can rewrite these curves as

$$z(y) = y\frac{y - pD}{Ep - y}, \quad y(z) = Ez\frac{Nz - q}{qD - EMz}. \tag{4}$$

The curve $z(y)$ in the system (4) presents a vertical asymptote, $y = Ep$, and crosses the y-axis at $y = 0$ and $y = pD$. Therefore, it is necessary to distinguish the cases $Ep > pD$ and $Ep < pD$ to determine the configuration of the points belonging to the curve. Similarly, the equation for $y(z)$ presents an horizontal asymptote, $z = \frac{qD}{ME}$, and crosses the z-axis at $z = 0$ and $z = \frac{q}{N}$. Here again there are the two cases $\frac{qD}{ME} < \frac{q}{N}$ and $\frac{qD}{ME} > \frac{q}{N}$. Thus, overall there are four possible situations depicted in Fig. 1.

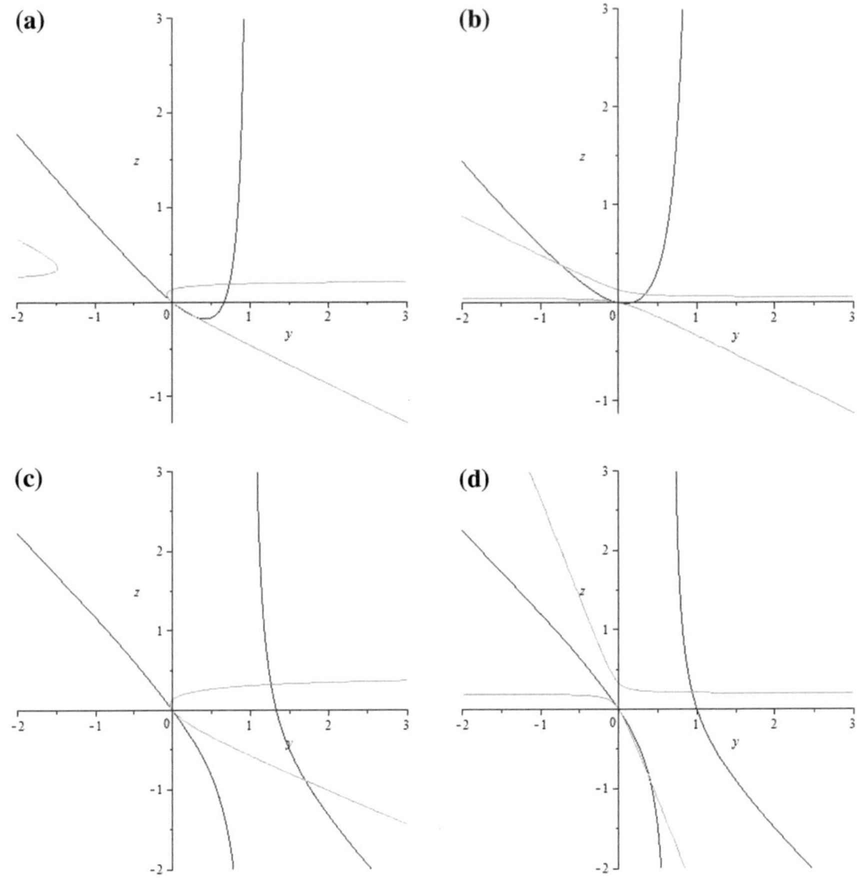

Fig. 1 The four possible positions of the conic sections **a** $\frac{ME}{N} < D < E$ **b** $D < min\left\{E_1 \frac{ME}{N}\right\}$ **c** $D > max\left\{E_1 \frac{ME}{N}\right\}$ **d** $E < D < \frac{ME}{N}$

In any case we always find an intersection in the first quadrant, which means that the equilibrium E_2 is always feasible.

We now analyze the coexistence. Solving for x from the first equation of (2),

$$x = 1 - \frac{B}{A}z - \frac{C}{A}y. \tag{5}$$

and substituting into the other two equations we obtain a system for the following two conic sections:

$$\Gamma: \quad y^2(UC - A) + y(pDA - UA) + yz(UB - A) + pAEz = 0,$$
$$\Phi: \quad z^2(EVB - NEA) + z(qAE - VAE) + yz(VCE - MAE) + qDAy = 0,$$

Since the determinants of the matrices associated with the conics are always negative, Γ and Φ are hyperbolae. To obtain a nonnegative value for x^* from (5), we need to find their intersection in the half plane $z > \frac{C}{B}y + \frac{A}{B}$.

The curve Γ has the following properties: its center is the point C_Γ, its intersections with the axes are $O = (0, 0)$, P_1; its asymptotes are $y = y_{C_\Gamma}$ and $z = m_\Gamma y + q_\Gamma$ where $q_\Gamma = z_{C_\Gamma} - m_\Gamma y_{C_\Gamma}$ and

$$C_\Gamma = \left(y_{C_\Gamma}, z_{C_\Gamma}\right) = \left(\frac{ApE}{A - UB}, \frac{2ApE(UC - A) + (ApD - AU)(A - UB)}{(A - UB)^2}\right),$$
$$P_1 = \left(\frac{AU - ApD}{UC - A}, 0\right), \quad m_\Gamma = \frac{A - UC}{UB - A}.$$

For Φ the center is C_Φ, the intersections with the axes are $O = (0, 0)$ and P_2 and the asymptotes are $z = z_{C_\Phi}$ and $y = m_\Phi z + q_\Phi$ with $q_\Phi = y_{C_\Phi} - m_\Phi z_{C_\Phi}$ and where

$$C_\Phi = \left(y_{C_\Phi}, z_{C_\Phi}\right) = \left(\frac{AqE - EVA}{MAE - VCE} + 2(EVB - NEA), \frac{AqD}{MAE - VCE}\right),$$
$$P_2 = \left(0, \frac{VAE - qAE}{EVB - NEA}\right), \quad m_\Phi = \frac{MAE - EVC}{EVB - NEA}.$$

Unfortunately, all the coefficients appearing in the above expressions are of uncertain sign, so that to study the conic sections, we need to locate their centers in every possible way in the four quadrants. In the following Table, we provide sufficient conditions for the existence of at least one intersection in the first quadrant between the hyperbolae Γ and Φ. Figure 2 contains an illustration of case (7).

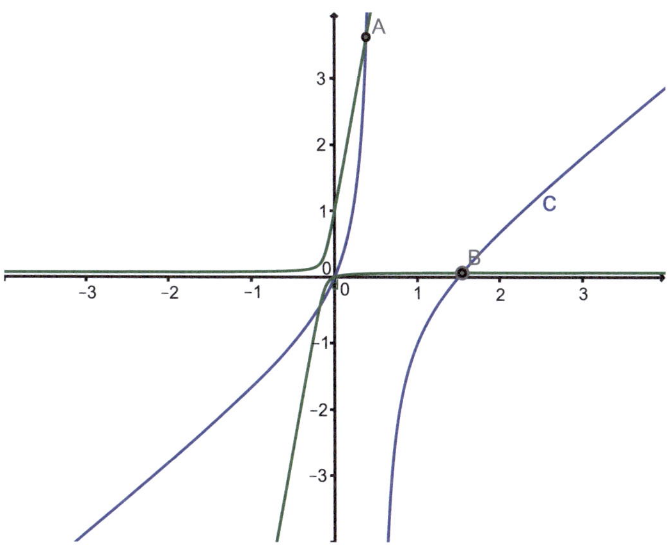

Fig. 2 Illustration of case (7)

(1) $y_{C_\Gamma}, z_{C_\Gamma}, y_{C_\Phi}, z_{C_\Phi} > 0$: $m_\Phi > m_\Gamma > 0; q_\Gamma < 0, q_\Phi > 0$	(2) $y_{C_\Gamma}, z_{C_\Gamma}, y_{C_\Phi} > 0, z_{C_\Phi} < 0$: $m_\Phi > 0 > m_\Gamma; q_\Gamma, q_\Phi > 0$
(3) $y_{C_\Gamma}, z_{C_\Gamma} > 0, y_{C_\Phi}, z_{C_\Phi} < 0$: $0 < m_\Phi < m_\Gamma; q_\Gamma > 0, q_\Phi < 0$	(4) $y_{C_\Gamma} > 0, z_{C_\Gamma}, y_{C_\Phi}, z_{C_\Phi} < 0$: $m_\Phi > 0 > m_\Gamma; q_\Gamma, q_\Phi < 0$
(5) $y_{C_\Gamma}, z_{C_\Gamma}, y_{C_\Phi}, z_{C_\Phi} < 0$: $m_\Phi > m_\Gamma > 0; q_\Gamma, q_\Phi > 0$	(6) $y_{C_\Gamma}, z_{C_\Phi} < 0, z_{C_\Gamma}, y_{C_\Phi} > 0$: $m_\Phi > m_\Gamma > 0; q_\Gamma, q_\Phi > 0$
(7) $y_{C_\Gamma}, z_{C_\Phi} > 0, z_{C_\Gamma}, y_{C_\Phi} < 0$: $m_\Phi, m_\Gamma > 0; q_\Gamma, q_\Phi < 0$	(8) $y_{C_\Gamma}, z_{C_\Gamma}, z_{C_\Phi} > 0, y_{C_\Phi} < 0$: $m_\Phi > 0, m_\Gamma < 0; q_\Gamma > 0, q_\Phi < 0$
(9) $y_{C_\Gamma}, y_{C_\Phi}, z_{C_\Phi} > 0, z_{C_\Gamma} < 0$: $m_\Phi, m_\Gamma < 0; q_\Gamma < 0, q_\Phi > 0$	(10) $z_{C_\Gamma} > 0, y_{C_\Gamma}, y_{C_\Phi}, z_{C_\Phi} < 0$: $m_\Gamma > m_\Phi > 0; q_\Gamma > 0, q_\Phi < 0$
(11) $y_{C_\Phi} > 0, y_{C_\Gamma}, z_{C_\Gamma}, z_{C_\Phi} < 0$: $m_\Phi > m_\Gamma > 0; q_\Gamma, q_\Phi > 0$	(12) $y_{C_\Gamma}, y_{C_\Phi} > 0, z_{C_\Gamma}, z_{C_\Phi} < 0$: $m_\Gamma > m_\Phi > 0; q_\Gamma < 0, q_\Phi > 0$
(13) $z_{C_\Gamma}, z_{C_\Phi} > 0, y_{C_\Gamma}, y_{C_\Phi} < 0$: $m_\Gamma > m_\Phi > 0; q_\Gamma > 0, q_\Phi < 0.$	(14) $y_{C_\Phi}, z_{C_\Phi} > 0, y_{C_\Gamma}, z_{C_\Gamma} < 0$: $m_\Phi, m_\Gamma > 0; q_\Gamma, q_\Phi < 0.$
(15) $z_{C_\Phi} > 0, y_{C_\Gamma}, z_{C_\Gamma}, y_{C_\Phi} < 0$: $m_\Gamma, m_\Phi > 0; q_\Gamma, q_\Phi < 0.$	(16) $z_{C_\Gamma}, y_{C_\Phi}, z_{C_\Phi} > 0, y_{C_\Gamma} < 0$: $m_\Gamma > m_\Phi > 0; q_\Gamma > 0, q_\Phi < 0.$

4 Stability Analysis

To assess the stability of the various equilibria of (2) we need its Jacobian:

$$\begin{pmatrix} A - 2Ax - Bz - Cy & -Cx & -Bx \\ -Uy & pD - 2y - z - Ux & pE - y \\ -Vz & q\frac{D}{E} - Mz & q - 2Nz - My - Vx \end{pmatrix} \tag{6}$$

At the origin, we find eigenvalues: $\lambda_1 = pD + q > 0$, $\lambda_2 = 0$, and $\lambda_3 = A > 0$, from which the instability of this point is immediate.

4.1 Equilibrium E_1

At E_1 one eigenvalue is immediately factorized, $\lambda_1 = -A$, while the remaining ones are the roots of the quadratic equation $P(\lambda) := \lambda^2 + (V - pD - q + U)\lambda - VpD - Uq + UV$. Its roots are

$$\lambda_{2,3} = \frac{1}{2}(pD + q - U - V) \pm \sqrt{\Delta}$$

with

$$\begin{aligned}\Delta &= (V - pD - q + U)^2 - 4(UV - Uq - VpD) \\ &= (pD)^2 + (2V - 2U + 2q)pD + q^2 + V^2 + U^2 - 2qV - 2UV + 2qU.\end{aligned}$$

We now investigate this discriminant in terms of the model parameters.

Consider at first the case $\Delta = 0$, regarding the last above expression as the following quadratic equation in terms of pD, observing that it is a convex function:

$$(pD)^2 + (2V - 2U + 2q)pD + q^2 + V^2 + U^2 - 2qV - 2UV + 2qU = 0. \quad (7)$$

Its discriminant is

$$\delta = (2V - 2U + 2q)^2 - 4(q^2 + U^2 + V^2 - 2qV - 2UV + 2qU) = 16q(V - U). \quad (8)$$

Observe that pD is the product of two parameters of the model (2) so that it has to be real and nonnegative. Hence, we must impose that the discriminant δ (8) of (7) is nonnegative, i.e., $V \geq U$. Otherwise, the two roots of (7) would not be real, since in fact (7) represents a parabola with positive concavity, it would not intersect the pD-axis. We can conclude that for $V \geq U$, for all values of the parameters we obtain $\delta > 0$, i.e., we have two different roots, which are

$$pD_{\pm} = \frac{2U - 2q - 2V \pm 4\sqrt{q(V - U)}}{2} = U - q - V \pm 2\sqrt{q(V - U)}.$$

Further we can note that if $V = U$ we have $pD_{1,2} = -q < 0$ which cannot hold, since as mentioned pD is nonnegative. Therefore, in order to have acceptable values we have to take $V > U$. Nevertheless, under this condition, the value $pD_{-} = U - q - V - 2\sqrt{q(V - U)}$ is negative and then not admissible. Similarly, we ask pD_{+} to be positive, and this amounts to have $2\sqrt{q(V - U)} > q + V - U$, which cannot be verified because from $V > U$, it would then follow $(q + U - V)^2 < 0$. Hence we are in the same situation described before: there are no possible solutions. In view of these considerations, we can conclude that $\Delta = 0$ is impossible and therefore it is of one sign. We always have $\Delta > 0$, because its expression is a convex parabola as observed above. Hence $\lambda_2, \lambda_3 \in \mathbb{R} : \lambda_2 \neq \lambda_3$.

By Descartes' rule of sign on (7) we can obtain necessary and sufficient conditions for a stable equilibrium. If there are two permanences of the sign in $P(\lambda)$ then λ_2 and λ_3 will be negative. This is verified if and only if

$$pD < V + U - q, \quad pD < \frac{U(V-q)}{V}.$$

The quantity $U(V-q)V^{-1}$ must be positive, therefore we have to impose $q < V$. Also note that

$$\min\left\{V + U - q, \frac{U(V-q)}{V}\right\} = \frac{U(V-q)}{V},$$

otherwise we would have $U + V - q \le U - UqV^{-1}$ and in turn $V^2 \le Vq - Uq$. On the other hand we took $q < V$, which implies $Vq < V^2$ so that finally we find $Vq \le Vq - Uq$ which is a contradiction since $Uq > 0$. We can therefore conclude that under the following assumptions

$$q < V, \quad pD < \frac{U(V-q)}{V} \tag{9}$$

all the eigenvalues of the characteristic polynomial are negative, thus E_1 is asymptotically stable. Note that there is a transcritical bifurcation for

$$D^{\dagger} = \frac{U}{p}\left(1 - \frac{q}{V}\right). \tag{10}$$

We give a numerical example of such situation. We change only the parameter D while maintaining all the other ones fixed, at the values: $A = 1.7$, $B = 2$, $C = 0.5$, $p = 0.67$, $E = 2.3$, $U = 2$, $q = 0.33$, $N = 2$, $M = 2.2$, $V = 3$.

Figure 3 illustrates the change of stability of E_1. As long as $0 < D < U(V - q)(Vp)^{-1}$ holds, E_1 is stable, if $D > U(V-q)(Vp)^{-1}$ it becomes unstable and the system settles first to the coexistence equilibrium and then to the equilibrium E_2.

4.2 Equilibrium E_2

In this case the Jacobian also factorizes to give one explicit eigenvalue, $A - Bz_2 - Cy_2$, while the remaining ones, using the equilibrium equations, come from the 2×2 matrix

$$\widehat{J} = \begin{pmatrix} -y_2 - pE\frac{z_2}{y_2} & pE - y_2 \\ q\frac{D}{E} - Mz_2 & -Nz_2 - q\frac{Dy_2}{Ez_2} \end{pmatrix}.$$

The Routh–Hurwitz conditions become

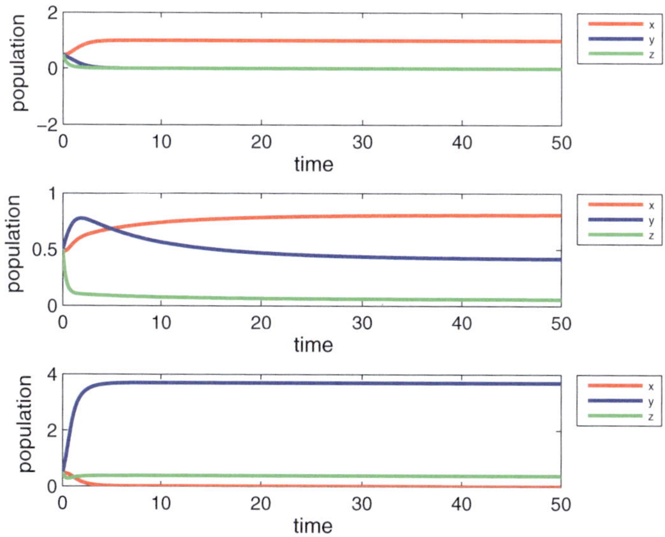

Fig. 3 Transcritical bifurcation at E_1 in terms of the parameter D. In the *top* frame we have equilibrium E_1, for low values of the bifurcation parameter, $D = 1.2$. In the central frame, we have the coexistence equilibrium, for larger values, $D = 3$. Finally in the *bottom* frame, we find equilibrium E_2 for even larger values, $D = 6$

$$-\mathrm{tr}(\widehat{J}) = y_2 + pE\frac{z_2}{y_2} + Nz_2 + q\frac{Dy_2}{Ez_2} > 0,$$
$$\det(\widehat{J}) = \left(y_2 + pE\frac{z_2}{y_2}\right)\left(Nz_2 + q\frac{Dy_2}{Ez_2}\right) - (E - y_2)\left(q\frac{D}{E} - Mz_2\right) > 0, \qquad (11)$$

the first one of which is easily seen to be satisfied.

4.3 Coexistence E^*

Using the equilibrium equations, the Jacobian evaluated at E^* becomes

$$J_{|(x^*,y^*,z^*)} = \begin{pmatrix} -Ax^* & -Cx^* & -Bx^* \\ -Uy^* & -y^* - pE\frac{z^*}{y^*} & pE - y^* \\ -Vz^* & q\frac{D}{E} - Mz^* & -Nz^* - q\frac{D}{E}\frac{y^*}{z^*} \end{pmatrix}.$$

The characteristic equation of the Jacobian evaluated at E^* is the monic cubic

$$\sum_{i=0}^{3} a_{3-i}\lambda^i = 0$$

with the coefficients

$$
\begin{aligned}
a_1 &= -\mathrm{tr}(J) = Ax^* + y^* + pE\frac{z^*}{y^*} + Nz^* + q\frac{D}{E}\frac{y^*}{z^*} \\
a_2 &= \frac{1}{2}(\mathrm{tr}(J)^2 - \mathrm{tr}(J^2)) = Ax^*\left(y^* + pE\frac{z^*}{y^*}\right) - UCx^*y^* + \left(y^* + pE\frac{z^*}{y^*}\right) \\
&\times \left(Nz^* + q\frac{D}{E}\frac{y^*}{z^*}\right) - \left(q\frac{D}{E} - Mz^*\right)(pD - y^*) \\
&+ Ax^*\left(Nz^* + q\frac{D}{E}\frac{y^*}{z^*}\right) - VBx^*z^* \\
a_3 &= -\det(J) = Ax^*\left(y^* + pE\frac{z^*}{y^*}\right)\left(Nz^* + q\frac{D}{E}\frac{y^*}{z^*}\right) + CVx^*z^*(y^* - pE) \\
&+ BUx^*y^*\left(Mz^* - q\frac{D}{E}\right).
\end{aligned}
$$

Applying the Routh–Hurwitz criterion to the cubic we have that E_2 is stable if and only if

$$a_1 > 0, \quad a_3 > 0 \quad \text{and} \quad a_1a_2 - a_3 > 0.$$

5 Bifurcations and System's Behavior

In this section, we summarize the system's behavior. In Fig. 4 we plot the bifurcation diagram of the system as function of the bifurcation parameter D. Initially, for values of D below 3, we find equilibrium E_1. Then at $D = 3$ a transcritical bifurcation occurs, for which the coexistence equilibrium arises for $3 < D < 4$. Another transcritical bifurcation is found at $D = 4$ giving equilibrium E_2 for larger values of D.

Note that Hopf bifurcations never arise in this model. In fact, they are not possible at equilibrium E_0 in view of the fact that the eigenvalues are real. At E_1 we have also shown that the characteristic equation factorizes to give a quadratic, for which the roots are once again always real. At E_2 the trace of the 2×2 reduced Jacobian is strictly negative, compare indeed the first condition in (11). Empirically, finally, we have seen that no persistent oscillations arise at coexistence, see the portion of Fig. 4 in which the coexistence equilibrium is found. Thus persistent oscillations are excluded in this context.

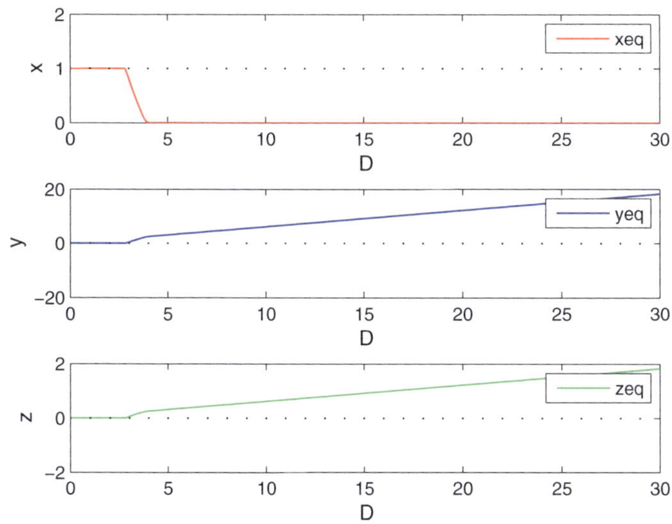

Fig. 4 System's behavior as function of the bifurcation parameter D. Starting from low values of D we encounter first equilibrium E_1, then coexistence E^*, then finally equilibrium E_2. No oscillations are found around coexistence

6 Conclusion

We have introduced and presented a competition model between two species, of which one exhibits two different genotypes. Boundedness of the system's trajectories shows that the ultimate system's behavior is captured by the equilibria analysis. We have found explicitly two equilibria: the origin E_0 corresponding to system's disappearance, which however cannot occur in view of its instability, and the equilibrium with just the population that is not genetically distinct, E_1. We have further shown the existence of the equilibrium in which only the genetically distinguishable population thrives, E_2, and of the coexistence, which has also been shown to be feasible by numerical simulations. Analytically, for the latter, we have provided sufficient conditions for its feasibility. The major result in this context however is the fact that the well-known principle of competitive exclusion, for which in classical models [4] only one of the two competitors ultimately survives, here does not necessarily hold. It is violated in fact, at coexistence, as shown explicitly in Fig. 4 for an intermediate range of the bifurcation parameter.

Comparing these results with the predator–prey models already studied, [1, 2], here we find four equilibria instead of three. The reason is that here both populations can thrive independently, while in the predator–prey situation, since the predators are assumed to be specialists, they cannot survive without prey. The essence of the equilibria in the two types of models is however the same, corresponding to system's collapse, which in all the models is not possible, a good result from the conservationist viewpoint, the one-population-only equilibrium, which is the prey-only point in [1, 2] and here is given by E_1 and E_2, and coexistence of the whole ecosystem.

The system's behavior mainly depends on the parameter $D = ab^{-1}$ representing the ratio between the growth rates of the two genetically distinguishable subpopulations y and z. Focusing on E_1, we have found that for values of D below the threshold (10) it is stable, while for values above it, it exhibits instability, and the system settles to coexistence. It is worthy to note that the same type of bifurcation appears in the model [2], relative to the equilibrium in which the distinguishable predators are extinct. In our model a further transition has been found by simulations for which from E^* we recover the second-population-only equilibrium E_2.

Also, no persistent oscillations have been discovered numerically, and in some instances also shown analytically not to arise.

Another interesting feature of the system, in line with what has already been observed in both [1, 2], is that the equilibrium in which just one of the genotypes thrives is not possible. The reason is that both genotypes generate, although possibly at different rates, individuals of the other one, so that if the latter becomes extinct, it can always be regenerated by the first subpopulation.

Acknowledgments The study was partially supported by the project "Metodi numerici nelle scienze applicate" of the Dipartimento di Matematica "Giuseppe Peano".

References

1. Venturino, E.: An ecogenetic model. Appl. Math. Lett. **25**, 1230–1233 (2012)
2. Venturino, E., Viberti, C.: A predator-prey model with genetically distinguishable predators. In: Kanarachos, A., Mastorakis N.E. (eds.) Recent Advances in Environmental Sciences, Proceedings of the 9th International Conference on Energy, Environment, Ecosystems and Sustainable Development (EEESD'13), Lemesos, Cyprus, March 21st–23rd 2013, pp. 87–92. WSEAS Press (2013). ISSN 2227-4359, ISBN 978-1-61804-167-8
3. Venturino, E., Viberti, C.: An ecosystem with HTII response and predators' genetic variability. Math. Model. Anal. **19**, 371–394 (2014). doi:10.3846/13926292.2014.925518
4. Waltman, P.: Competition Models in Population Biology, SIAM CBMS-NSF Regional Conference Series in Applied Mathematics, Philadelphia (1983)

Discrete and Phase-Only Receive Beamforming

Johannes Israel, Andreas Fischer and John Martinovic

Abstract We consider an analog receive beamforming problem for a wireless board-to-board communication scenario with a static channel and fixed positions of all transceivers. Each transceiver is equipped with an antenna array for receiving signals from the neighboring board. Every single antenna element of the array is controlled by a phase shifter and an amplifier with finite resolution only (known as discrete beamforming). Hence, maximizing the Signal-to-Interference-and-Noise Ratio (SINR) yields a difficult discrete optimization problem. As first contribution, we present an overview of recently developed inexact and exact solution methods for this discrete SINR-maximization problem. The branch-and-bound principle is a basic tool for the exact methods. In this context, upper bounds for the SINR at the nodes of the branch-and-bound tree play an important role for the efficiency of such methods. We show in particular how tight upper bounds can be obtained by means of fractional programming. Our second contribution is for the case of phase-only beamforming, i.e., if the amplitudes of the antenna elements are fixed. We show for this case how the quality of the upper bounds can be improved. Moreover, we compare the different approaches with respect to the achieved SINR and the computational expense.

Keywords Receive beamforming · Discrete antenna weights · Phase-only beamforming · Branch-and-bound

---

J. Israel · A. Fischer (✉) · J. Martinovic
Institute of Numerical Mathematics, SFB 912 - HAEC, Technische Universität Dresden, 01062 Dresden, Germany
e-mail: andreas.fischer@tu-dresden.de

J. Israel
e-mail: johannes.israel@tu-dresden.de

J. Martinovic
e-mail: john.martinovic@tu-dresden.de

V.K. Singh et al. (eds.), *Modern Mathematical Methods and High Performance Computing in Science and Technology*, Springer Proceedings in Mathematics & Statistics 171, DOI 10.1007/978-981-10-1454-3_12

1 Introduction

Analog beamforming is an important technique in wireless communications, especially for mm-wave systems [17]. In comparison to digital beamforming, the analog technique provides less functionalities but only requires a single ADC/DAC per antenna array which yields lower energy consumption and production costs [16]. Since energy efficiency is an essential criterion for intended applications [14], we consider analog beamforming in this paper. In that case, the use of finite resolution phase shifters reduces the system complexity and production costs, but implicates that the considered optimization problems are discrete.

Analog beamforming with a quantized number of phase shifts and aiming at the minimization of a mean squared error has been considered in [19]. Transmit beamforming with discrete phases and amplitudes is studied, for example in [8, 9]. In this paper, we focus on discrete receive beamforming. We present and compare different inexact and exact solution approaches for the discrete SINR-maximization problem based on results in [11–13]. Additionally, a greedy approach is discussed as an alternative heuristic method.

Moreover, we consider discrete phase-only beamforming, i.e., the beamforming weights have constant magnitudes and variable discrete phases. Phase-only beamforming yields difficult optimization problems also in the case of continuous phases. In contrast to the SINR-maximization problem with continuous beamforming weights (which has a closed-form solution), there is no direct solution to the corresponding phase-only beamforming problem [18]. Some algorithms for a local solution were introduced, for example in [1, 18]. Many other approaches for phase-only problems are based on heuristic optimization methods, e.g., [4]. Discrete phase-only beamforming in the transmit case has recently been studied in [6, 7]. For phase-only receive beamforming, the approach on discrete SINR-maximization from [12] will be improved with respect to the tightness of upper bounds needed for using the branch-and-bound principle. The results presented in this paper can also be used to obtain an approximate solution for the continuous phase-only SINR-maximization problem by replacing the continuous phases by a sufficiently fine grid of discrete phases.

The paper is organized as follows: In Sect. 2, the discrete SINR-maximization problem is introduced. In Sect. 3, we discuss inexact solution strategies: a simple rounded Capon beamformer, a greedy approach, and an approximate method based on branch-and-bound. Exact approaches for the discrete SINR-maximization problem are presented in Sect. 4. There, the existing approaches [12, 13] are briefly described and applied to the phase-only beamforming case. In Sect. 5, the different approaches are compared for a set of simulation scenarios. Section 6 concludes this paper.

2 Discrete Receive Beamforming

In the receive beamforming case, an antenna array receives the desired signal s_1 and interfering signals $s_2, \ldots, s_D$ from D different directions. It is assumed that all signals arrive as plane waves. Without loss of generality, the signal power P_1 of the desired signal is normalized to 1. The M individual antenna elements are assumed to be omnidirectional. A vector $\mathbf{w} \in \mathbb{C}^M$ is called beamformer if its k-th component w_k describes the amplification $|w_k|$ and the phase shift $\arg(w_k)$ which acts on the received signal at the k-th antenna element.

We aim at maximizing the Signal-to-Interference-and-Noise Ratio

$$\mathrm{SINR}(\mathbf{w}) := \frac{|\mathbf{w}^{\mathsf{H}}\mathbf{a}|^2}{\mathbf{w}^{\mathsf{H}}\mathbf{R}\mathbf{w}},$$

where $\mathbf{a}$ is the array steering vector for the desired signal and $\mathbf{R}$ is the interference-and-noise covariance matrix. The matrix $\mathbf{R}$ can be modeled as

$$\mathbf{R} = \sum_{d=2}^{D} P_d \mathbf{a}^d (\mathbf{a}^d)^{\mathsf{H}} + \sigma^2 \mathbf{I}.$$

Here, P_d and $\mathbf{a}^d$ are the signal power and the steering vector according to the signal s_d and σ^2 denotes the variance of the uncorrelated noise. In our static scenario with given positions of all transceivers, we can assume that the steering vector $\mathbf{a}$ and the covariance matrix $\mathbf{R}$ are known. In the case of continuous beamforming weights (phases and amplitudes) the SINR is maximized by the Capon beamformer [5]

$$\mathbf{w}_{\mathrm{cap}} := \frac{\mathbf{R}^{-1}\mathbf{a}}{\mathbf{a}^{\mathsf{H}}\mathbf{R}^{-1}\mathbf{a}} \tag{1}$$

and by all of its nonzero complex multiples.

For finite resolution phase shifters and amplifiers the discrete SINR-maximization problem is

$$\max_{\mathbf{w} \in \mathscr{D}^M} \frac{|\mathbf{w}^{\mathsf{H}}\mathbf{a}|^2}{\mathbf{w}^{\mathsf{H}}\mathbf{R}\mathbf{w}} \tag{2}$$

with $\mathscr{D}$ denoting the discrete set of all feasible phase-and-amplitude combinations. In case of discrete phase-only beamforming (only the phase of the signal can be changed at the receive antennas) we assume that the corresponding phase shifters have a resolution of m bit. Then, the discrete set $\mathscr{D}$ is given by

$$\mathscr{D} := \left\{ \mathrm{e}^{j\varphi} \,\middle|\, \varphi \in \left\{ \frac{k\pi}{2^{m-1}} \,\middle|\, k \in \{0, 1, \ldots, 2^m - 1\} \right\} \right\}. \tag{3}$$

3 Inexact Solution Strategies

We present three different approaches for an approximate solution of the discrete SINR-maximization problem (2). The easiest idea is a simple rounding strategy, where the Capon beamformer $\mathbf{w}_{\text{cap}}$ is rounded to the nearest feasible beamformer $\mathbf{w}_{\text{r}}$ in the discrete set $\mathscr{D}^M$, i.e.,

$$\mathbf{w}_{\text{r}} \in \arg\min_{\mathbf{w}\in\mathscr{D}^M} \|\mathbf{w} - \mathbf{w}_{\text{cap}}\|, \tag{4}$$

where $\|\mathbf{z}\|$ is given by $\sqrt{\mathbf{z}^{\mathsf{H}}\mathbf{z}}$ for any complex vector $\mathbf{z}$.

A more sophisticated approach for problem (2) is the use of a greedy algorithm related to [2]. We start with an arbitrary feasible beamformer $\mathbf{w}^0$, e.g., a rounded Capon beamformer $\mathbf{w}_{\text{r}}$ from (4). Then, we successively try to improve one individual component of $\mathbf{w}$ while all other components remain fixed. This is repeated with the next component of $\mathbf{w}$ and so on until no further improvement is possible. The procedure is summarized in Algorithm 1.

Algorithm 1 (Greedy coordinate search)

1: Choose $\mathbf{w}^0 \in \mathscr{D}^M$ and set $i := 0$.
2: **for** $k = 1 : M$ **do**
3: Choose $w_k^{i+1} \in \operatorname{argmax}_{d\in\mathscr{D}} \left\{\text{SINR}(w_1^{i+1}, \dots, w_{k-1}^{i+1}, d, w_{k+1}^{i}, \dots, w_M^{i})\right\}$.
4: **end for**
5: **if** $\text{SINR}(\mathbf{w}^{i+1}) > \text{SINR}(\mathbf{w}^i)$ **then**
6: $i \mapsto i + 1$ and goto line 2.
7: **else**
8: $\mathbf{w}_{\text{greedy}} := \mathbf{w}^i$ and stop.
9: **end if**

Remark 1 We emphasize that a greedy beamformer $\mathbf{w}_{\text{greedy}}$ is generally not a solution of problem (2), but it cannot be improved by only changing a single component.

Branch-and-bound is a well-known principle that can often be successfully applied to discrete optimization problems to obtain an exact solution [15]. For using this principle, it is necessary to compute (preferably tight) bounds for certain subproblems. In our case, the objective function $\text{SINR}(\cdot)$ is nonconcave, indeed it is a fraction of two convex functions—a so-called convex–convex (quadratic) fractional program. Therefore, replacing the discrete set $\mathscr{D}$ by $\mathbb{C}$ (the usual relaxation technique for obtaining bounds) leads to nonconcave subproblems. Even if these subproblems arise by simply fixing certain components of $\mathbf{w}$ (as it will be done later), the Capon beamformer is not applicable since $\mathbf{w}$ cannot be arbitrarily scaled anymore. Therefore, instead of maximizing the SINR, it was suggested in [11] to replace (2) by a minimization problem with a convex objective, namely

$$\min_{\mathbf{w}\in\mathscr{D}^M} f_{c,z}(\mathbf{w}) := \mathbf{w}^{\mathsf{H}}\mathbf{R}\mathbf{w} + c|\mathbf{w}^{\mathsf{H}}\mathbf{a} - z|^2. \tag{5}$$

The parameter $c > 0$ is a weighting parameter and the parameter $z \in \mathbb{C} \setminus \{0\}$ describes the desired antenna array output. The function $f_{c,z}$ is the weighted sum of interference and noise $\mathbf{w}^{\mathsf{H}}\mathbf{R}\mathbf{w}$ and of the squared deviation from the desired array output $|\mathbf{w}^{\mathsf{H}}\mathbf{a} - z|^2$. The parameter c enables a trade-off between the minimization of these two objectives. Obviously, $f_{c,z}$ is a convex function for any $(c, z) \in (0, \infty) \times \mathbb{C}$. The continuous relaxation of (5) ($\mathscr{D}$ is replaced by $\mathbb{C}$) has the unique solution

$$\mathbf{w}_f = c\bar{z}(\mathbf{R} + c\mathbf{a}\mathbf{a}^{\mathsf{H}})^{-1}\mathbf{a}.$$

The discrete problem (5) can now be solved by branch-and-bound. The subproblems can be generated by a successive fixing of the individual beamforming components. The continuous relaxation of the resulting subproblems is always a convex function with a closed-form solution, i.e., it can be obtained with low computational expense, see [11] for details. The solution $\mathbf{w}_{\text{approx}}$ of the discrete optimization problem (5) provides a certain approximation for the exact solution of the original discrete SINR-maximization problem (2). Simulation results with respect to the quality of the approximate solutions are given in Sect. 5.

4 Exact Solution of the Discrete SINR-Maximization Problem

For an exact solution of the discrete SINR-maximization problem (2) the branch-and-bound principle can be applied successfully as well. To this end, reasonable upper bounds for subproblems arising in the branch-and-bound tree are required. We briefly describe the corresponding approach from [13], where the SINR-maximization problem with discrete phases and amplitudes is solved exactly.

During the branch-and-bound procedure, the components of the beamforming vector $\mathbf{w}$ are fixed successively. Depending on the node in the branch-and-bound tree, let $\mathbf{w}_1 \in \mathscr{D}^k$ consist of all fixed entries of $\mathbf{w}$, whereas $\mathbf{w}_2$ contains the remaining variable components, i.e., $\mathbf{w} = (\mathbf{w}_1, \mathbf{w}_2)$. Then,

$$\mathrm{SINR}(\mathbf{w}_1, \mathbf{w}_2) = \frac{\left| \begin{pmatrix} \mathbf{w}_1 \\ \mathbf{w}_2 \end{pmatrix}^{\mathsf{H}} \begin{pmatrix} \mathbf{a}_1 \\ \mathbf{a}_2 \end{pmatrix} \right|^2}{\begin{pmatrix} \mathbf{w}_1 \\ \mathbf{w}_2 \end{pmatrix}^{\mathsf{H}} \begin{pmatrix} \mathbf{R}_{11} & \mathbf{R}_{12} \\ \mathbf{R}_{21} & \mathbf{R}_{22} \end{pmatrix} \begin{pmatrix} \mathbf{w}_1 \\ \mathbf{w}_2 \end{pmatrix}},$$

where, with $L := M - k$, the vectors $\mathbf{a}_1 \in \mathbb{C}^k$, $\mathbf{a}_2 \in \mathbb{C}^L$, and the matrices $\mathbf{R}_{11} \in \mathbb{C}^{k \times k}$, $\mathbf{R}_{12} \in \mathbb{C}^{k \times L}$, $\mathbf{R}_{21} \in \mathbb{C}^{L \times k}$, $\mathbf{R}_{22} \in \mathbb{C}^{L \times L}$ denote the appropriate components of $\mathbf{a}$ and $\mathbf{R}$, respectively. Now, the subproblems at any node of the branch-and-bound tree can be written as

$$\max_{\mathbf{w}_2 \in \mathscr{D}^L} \mathrm{SINR}(\mathbf{w}_1, \mathbf{w}_2) \tag{6}$$

with appropriate $\mathbf{w}_1 \in \mathscr{D}^k$ depending on the node. As we already mentioned in Sect. 3, an easy solution of the continuous relaxation of (6) cannot be expected. However, a weaker relaxation is given by

$$\max_{\substack{\xi \in \mathbb{C},\, \mathbf{w}_2 \in \mathbb{C}^L \\ (\xi \mathbf{w}_1, \mathbf{w}_2) \neq \mathbf{0}}} \mathrm{SINR}(\xi \mathbf{w}_1, \mathbf{w}_2). \tag{7}$$

Fortunately, this problem has the closed-form solution

$$\begin{pmatrix} \xi^\star \\ \mathbf{w}_2^\star \end{pmatrix} = \frac{\mathbf{S}^{-1}\mathbf{b}}{\mathbf{b}^{\mathsf{H}}\mathbf{S}^{-1}\mathbf{b}}, \tag{8}$$

where

$$\mathbf{S} := \begin{pmatrix} \mathbf{w}_1^{\mathsf{H}}\mathbf{R}_{11}\mathbf{w}_1 & \mathbf{w}_1^{\mathsf{H}}\mathbf{R}_{12} \\ \mathbf{R}_{21}\mathbf{w}_1 & \mathbf{R}_{22} \end{pmatrix} \quad \text{and} \quad \mathbf{b} := \begin{pmatrix} \mathbf{w}_1^{\mathsf{H}}\mathbf{a}_1 \\ \mathbf{a}_2 \end{pmatrix},$$

see [13]. Hence, the subproblems (7) can be solved with low computational burden. Moreover, it can be shown that the optimal values for (7) and for the continuous relaxation of (6) are generally equal and only differ in a very special case, see [13] for details. In other words, the introduction of the additional variable ξ simplifies the problem without worsening the optimal value. The branch-and-bound algorithm for the exact solution of the discrete SINR-maximization problem (2) is described in [13]. There, the beamforming weights are fixed successively and upper bounds for the corresponding subproblems (6) are obtained from the solution of (7).

The simulation results in [13] show that the number of required bounds for the branch-and-bound method is low compared to the cardinality of the discrete set $\mathscr{D}^M$. However, it is of interest to further reduce the number of required bounds such that the algorithm might be applied for larger numbers of antenna elements. Such a reduction can be achieved by calculating tighter bounds for the subproblems in the algorithm. Generally, the computation of improved bounds goes along with a higher computational burden. In [12] it is shown how tighter bounds can be computed efficiently by means of a fractional programming technique. There, the problem

$$\begin{aligned} \max_{\mathbf{w}_2 \in \mathbb{C}^L} \ & \mathrm{SINR}(\mathbf{w}_1, \mathbf{w}_2) \\ \text{s.t.} \ & \|\mathbf{w}_2\|^2 \leq L q_{\max} \end{aligned} \tag{9}$$

is considered as a relaxation of (6), where $q_{\max} := \max\{|d|^2 \mid d \in \mathscr{D}\}$ is the square of the maximal amplitude of a single antenna element.

For phase-only beamforming, the amplitude for each antenna element is fixed (to 1 for simplicity). Therefore, we suggest to tighten the bound that is derived from the solution of (9) by the optimal value of

$$\begin{aligned} \max_{\mathbf{w}_2 \in \mathbb{C}^L} \ & \mathrm{SINR}(\mathbf{w}_1, \mathbf{w}_2) \\ \text{s.t.} \ & \|\mathbf{w}_2\|^2 = L. \end{aligned} \tag{10}$$

Note that the optimal value of this new problem is still an upper bound for the optimal value of the subproblem (6). The concept for solving (10) is similar to the approach for (9) which is based on [3] and has been discussed in [12]. Therefore, we only briefly summarize the procedure. Instead of $\mathbf{w}_2 \in \mathbb{C}^L$, a real vector $\mathbf{x} \in \mathbb{R}^{2L}$ is used. For that purpose, let

$$\mathbf{x} := \begin{pmatrix} \mathrm{Re}(\mathbf{w}_2) \\ \mathrm{Im}(\mathbf{w}_2) \end{pmatrix}, \quad \mathbf{b}_1 := \begin{pmatrix} \mathrm{Re}(\mathbf{a}_2\mathbf{a}_1^{\mathsf{H}}\mathbf{w}_1) \\ \mathrm{Im}(\mathbf{a}_2\mathbf{a}_1^{\mathsf{H}}\mathbf{w}_1) \end{pmatrix}, \quad \mathbf{b}_2 := \begin{pmatrix} \mathrm{Re}(\mathbf{R}_{21}\mathbf{w}_1) \\ \mathrm{Im}(\mathbf{R}_{21}\mathbf{w}_1) \end{pmatrix},$$
$$\mathbf{A}_1 := \begin{pmatrix} \mathrm{Re}(\mathbf{a}_2\mathbf{a}_2^{\mathsf{H}}) & -\mathrm{Im}(\mathbf{a}_2\mathbf{a}_2^{\mathsf{H}}) \\ \mathrm{Im}(\mathbf{a}_2\mathbf{a}_2^{\mathsf{H}}) & \mathrm{Re}(\mathbf{a}_2\mathbf{a}_2^{\mathsf{H}}) \end{pmatrix}, \quad \mathbf{A}_2 := \begin{pmatrix} \mathrm{Re}(\mathbf{R}_{22}) & -\mathrm{Im}(\mathbf{R}_{22}) \\ \mathrm{Im}(\mathbf{R}_{22}) & \mathrm{Re}(\mathbf{R}_{22}) \end{pmatrix},$$
$$c_1 := \mathbf{w}^{\mathsf{H}}\mathbf{a}_1\mathbf{a}_1^{\mathsf{H}}\mathbf{w}_1, \quad c_2 := \mathbf{w}_1^{\mathsf{H}}\mathbf{R}_{11}\mathbf{w}_1$$

and the functions $f_1, f_2 : \mathbb{R}^{2L} \to \mathbb{R}$ with

$$f_k(\mathbf{x}) := \mathbf{x}^\top \mathbf{A}_k \mathbf{x} + 2\mathbf{b}_k^\top \mathbf{x} + c_k, \qquad k = 1, 2$$

be defined. Then, we have

$$\mathrm{SINR}(\mathbf{w}_1, \mathbf{w}_2) = \mathrm{SINR}(\mathbf{w}_1, \mathbf{x}) = \frac{\mathbf{x}^\top \mathbf{A}_1 \mathbf{x} + 2\mathbf{b}_1^\top \mathbf{x} + c_1}{\mathbf{x}^\top \mathbf{A}_2 \mathbf{x} + 2\mathbf{b}_2^\top \mathbf{x} + c_2} = \frac{f_1(\mathbf{x})}{f_2(\mathbf{x})}$$

and the optimization problem (10) is equivalent to

$$\max_{\mathbf{x} \in \mathscr{X}} \frac{f_1(\mathbf{x})}{f_2(\mathbf{x})}, \tag{11}$$

where

$$\mathscr{X} := \left\{ \mathbf{x} \in \mathbb{R}^{2L} \,\middle|\, \mathbf{x}^\top \mathbf{x} = L \right\}.$$

This problem is a convex–convex quadratically constrained quadratic fractional program (QQFP) and can be solved efficiently by means of the Dinkelbach method [10]. With the definition

$$F(\alpha) := \max_{\mathbf{x} \in \mathscr{X}} \left\{ f_1(\mathbf{x}) - \alpha f_2(\mathbf{x}) \right\}, \tag{12}$$

the following equivalence holds [10]:

$$\max_{\mathbf{x} \in \mathscr{X}} \frac{f_1(\mathbf{x})}{f_2(\mathbf{x})} = \alpha^\star \iff F(\alpha^\star) = 0.$$

Hence, the optimal value of (11) can be obtained by means of the unique root of $F : \mathbb{R} \to \mathbb{R}$. For the computation of $\alpha^\star$ the solution of the generally nonconcave maximization problem in (12) is needed for various values of α. However, this problem can be reformulated as a convex optimization problem whose solution can be obtained from the dual problem. Finally, we only have to determine $\eta > \lambda_1$ such that

$$\sum_{\ell=1}^{2L} \frac{h_\ell^2}{(\eta - \lambda_\ell)^2} = L \tag{13}$$

holds for known λ_ℓ and h_ℓ (see [3, 12] for details). There, η is the Lagrange multiplier according to the norm constraint for $\mathbf{x}$. In the general case with variable discrete amplitudes and the constraint $\mathbf{x}^\top \mathbf{x} \leq Lq_{\max}$ also $\eta \geq 0$ has to be satisfied which requires a case study, see [12]. In the case of phase-only beamforming, only $\eta > \lambda_1$ is needed such that the secular equation (13) always has a unique solution which can be obtained by Newton's method [3]. In summary, problem (11) can be solved by means of the following algorithm:

Algorithm 2 (Dinkelbach algorithm)

1: Choose $\varepsilon \geq 0$, $\mathbf{x}^0 \in \mathscr{X}$, set $\alpha_0 := \frac{f_1(\mathbf{x}^0)}{f_2(\mathbf{x}^0)}$ and $k := 0$.
2: **while** $F(\alpha_k) > \epsilon$ **do**
3: $\quad k \mapsto k + 1$.
4: $\quad$ Determine $\mathbf{x}^k \in \operatorname{argmax}_{\mathbf{x} \in \mathscr{X}} \{f_1(\mathbf{x}) - \alpha_{k-1} f_2(\mathbf{x})\}$.
5: $\quad$ Set $\alpha_k := \frac{f_1(\mathbf{x}^k)}{f_2(\mathbf{x}^k)}$.
6: **end while**

It can be shown that Algorithm 2 converges to the global optimum of problem (11), see [20]. Finally, the new subproblems (10) can be solved in an analog manner as proposed in [12]. For phase-only beamforming, the use of (10) instead of (9) as relaxation for the branch-and-bound subproblems yields a tighter bound, but with similar computational expense.

Although the solution of the subproblems in the Dinkelbach procedure can be obtained from the solution of a convex optimization problem and finally by solving the secular equation (13), it still requires a higher computational effort to solve (9) or (10) than the weaker relaxation (7). It is known [12] that the bound derived from (7) is equal to the bound derived from (9) if

$$\|(\xi^\star)^{-1} \cdot \mathbf{w}_2^\star\|^2 \leq Lq_{\max} \tag{14}$$

holds for a solution $(\xi^\star, \mathbf{w}_2^\star)$ of (7). Thus, to reduce the computational burden the bound derived from (9) shall only be used at nodes in a branch-and-bound algorithm if the constraint (14) is violated and if the value $\mathrm{SINR}(\xi^\star \mathbf{w}_1, \mathbf{w}_2^\star)$ is larger than the current lower bound (otherwise a tighter upper bound is not required since the node will be pruned by the algorithm anyway).

In the phase-only beamforming scenario considered here we obtain a stronger result. A tighter bound can be expected if $\|(\xi^\star)^{-1} \cdot \mathbf{w}_2^\star\|^2 \neq L$ holds for the solution $(\xi^\star, \mathbf{w}_2^\star)$ of (7). Therefore, the bound derived from problem (10) almost always improves the bound from (7). Nevertheless, to lower the computational expense, it is still beneficial to avoid too many bounds from problem (10) and, therefore, to make use of the bound derived from problem (7) which can be solved quite efficiently. This idea is described in (S3) of the following improved branch-and-bound algorithm for the exact solution of the discrete SINR-maximization problem (2) for the phase-only beamforming scenario. This algorithm generates a branch-and-bound tree whose nodes can be identified with subsets X of $\mathscr{D}^M$. By $p(X)$ the number of parent nodes of X is denoted. In particular, $p(\mathscr{D}^M)$ is set to 0.

Algorithm 3 (Branch-and-bound algorithm for phase-only SINR-maximization)

(S1) - Initialization

1: Set $\mathscr{G} := \{\mathscr{D}^M\}$ and $b(\mathscr{D}^M) := \mathrm{SINR}(\mathbf{w}_{\mathrm{cap}})$. Choose $\mathbf{x} \in \mathscr{D}^M$ and set $\hat{b} := \mathrm{SINR}(\mathbf{x})$.

(S2) - Branch:

2: Choose $X \in \mathscr{G}$ so that $p(X) = \max\{p(Y) : Y \in \mathscr{G}\}$ and $b(X) = \max\{b(Y) : Y \in \mathscr{G}$ and $p(Y) = p(X)\}$.

3: Set $k := p(X) + 1$ and $L := M - k$.

4: Partition X into $X_1, \ldots, X_{|\mathscr{D}|}$, where X_ℓ is the set of vectors in X with $w_k = d_\ell \in \mathscr{D}$.

(S3) - Bound:

5: **for** $\ell = 1 : |\mathscr{D}|$ **do**

6: Determine upper bound

$$b(X_\ell) := \max_{\substack{\xi \in \mathbb{C},\, (\mathbf{w}_1, \mathbf{w}_2) \in X_\ell \\ (\xi \mathbf{w}_1, \mathbf{w}_2) \neq \mathbf{0}}} \mathrm{SINR}(\xi \mathbf{w}_1, \mathbf{w}_2).$$

7: **if** $b(X_\ell) > \hat{b}$ **then**

8: By means of Algorithm 2 determine $\alpha^\star$ such that $F(\alpha^\star) = 0$.

9: Calculate

$$\widehat{\mathbf{w}}_2 \in \operatorname*{arg\,max}_{\mathbf{u} \in \mathbb{R}^{2L},\, \|\mathbf{u}\|^2 = L} \left\{ f_1(\mathbf{u}) - \alpha^\star f_2(\mathbf{u}) \right\}$$

10: and set

$$b(X_\ell) := \frac{f_1(\widehat{\mathbf{w}}_2)}{f_2(\widehat{\mathbf{w}}_2)}.$$

11: **end if**

12: **end for**

(S4) - Update:

13: Set $\mathscr{G} := (\mathscr{G} \backslash X) \cup \{X_1, \ldots, X_{|\mathscr{D}|}\}$.

14: **for all** $Y \in \mathscr{G}$ **do**

15: **if** $b(Y) \leq \hat{b}$ **then**

16: Set $\mathscr{G} := \mathscr{G} \backslash Y$.

17: **else if** $|Y| = 1$ $(Y = \{\mathbf{y}\})$ and $\mathrm{SINR}(\mathbf{y}) > \hat{b}$ **then**

18: Set $\hat{b} := \mathrm{SINR}(\mathbf{y})$ and $\mathbf{x} := \mathbf{y}$.

19: **end if**

20: **end for**

21: **if** $\mathscr{G} \neq \emptyset$ **then**

22: Goto line 2.

23: **else**

24: Set $\mathbf{w}^\star := \mathbf{x}$ and stop.

25: **end if**

Remark 2 A reasonable choice for the initial vector $\mathbf{x}$ in (S1) of Algorithm 3 would be a greedy beamformer $\mathbf{w}_{\text{greedy}}$ obtained from Algorithm 1 with starting point $\mathbf{w}_{\text{r}}$. The branching rule in (S2) is a depth-first search, where among the nodes with maximal number of parent nodes the node with the best bound is chosen.

The phases in the set $\mathscr{D}$ are equidistant with distance $\delta := \pi/2^{m-1}$. Thus, for any solution $\mathbf{w}^\star$ of the discrete phase-only SINR-maximization problem (2), the beamformers $\mathrm{e}^{jk\delta} \cdot \mathbf{w}^\star$ belong to $\mathscr{D}^M$ for $k = 0, 1, \ldots, 2^m - 1$ and provide the same optimal SINR as $\mathbf{w}^\star$. Therefore, the set $\mathscr{D}^M$ in Algorithm 3 is replaced by the set $\{1\} \times \mathscr{D}^{M-1}$ to reduce computational costs.

Remark 3 The additional constraint $\|\mathbf{w}\|^2 \le Lq_{\max}$ or $\|\mathbf{w}\|^2 = L$ can also be applied for the approximate approach from [11]. This would yield a similar trade-off with respect to a reduced number of required bounds and higher computational effort for the individual bounds. Further discussion on this is out of the scope of the paper.

5 Simulation Results

In this section, the beamformers previously presented are compared with respect to the achieved SINR and the computational effort within a phase-only beamforming scenario. Recall that we have introduced

- a rounded Capon beamformer $\mathbf{w}_{\text{r}}$ in (4),
- a greedy beamformer $\mathbf{w}_{\text{greedy}}$ obtained from Algorithm 1,
- an approximate beamformer $\mathbf{w}_{\text{approx}}$ as solution of (5), and
- an exact beamformer $\mathbf{w}^\star$ as solution of (2).

We created 100 random scenarios under the same conditions as in [12], assuming an AWGN channel with $\sigma^2 = 0.1$, a single desired signal, and three interfering signals. The directions of the incoming signals are chosen from the intervals

$$\theta_1 \in [-25°, -15°], \quad \theta_2 \in [-80°, -40°], \quad \theta_3 = -10°, \quad \theta_4 \in [45°, 75°],$$

such that the desired signal from direction θ_1 is always relatively close to the interfering signal from the fixed direction θ_3. The signal power of all incoming signals is equal to 1 and the receiver is a uniform linear receive antenna with $M = 8$ omnidirectional antenna elements of half-wavelength distance. We assume m-bit phase shifters and a fixed amplitude equal to 1, such that the discrete set $\mathscr{D}$ is given by (3). Figure 1 shows the (ordered) simulation results for phase shifters with resolution $m = 5$. The choices for the parameters in the approximate approach are $c = 0.2$ and $z = M$ (since, with a fixed amplitude of 1 for each antenna element, the maximal array output is $|\mathbf{w}^{\mathsf{H}}\mathbf{a}| = M$). It can be observed from Fig. 1 that the simple rounding strategy is (as expected) clearly outperformed by the alternative methods. In several cases, the SINR belonging to a greedy or an approximate beamformer is equal or close to the SINR of an exact solution. Quantitative differences can be seen in Table 1.

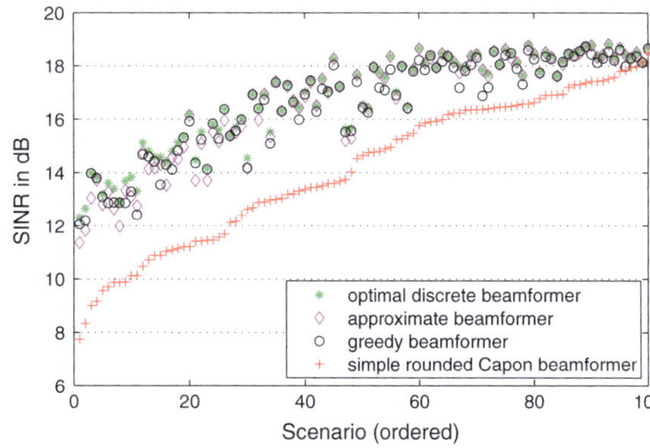

Fig. 1 Comparison of achieved SINR for different discrete beamforming approaches

Table 1 Average achieved SINR (in dB) for different beamforming strategies

	$m = 4$	$m = 5$	$m = 6$
$\mathbf{w}^{\star}$	16.40	16.80	16.91
$\mathbf{w}_{\text{approx}}$	16.06	16.62	16.73
$\mathbf{w}_{\text{greedy}}$	15.91	16.59	16.82
$\mathbf{w}_{\text{r}}$	13.52	14.15	14.21

This table provides simulation results for the same setting with $m = 4$, $m = 5$, and $m = 6$. It can be observed that the rounded Capon beamformer $\mathbf{w}_{\text{r}}$ provides bad results even for a very fine resolution. On the other hand, the greedy beamformer $\mathbf{w}_{\text{greedy}}$ yields good results which are even slightly better than the results for the approximate beamformer $\mathbf{w}_{\text{approx}}$ for $m = 6$. We want to emphasize that usually much more computational effort is required to determine $\mathbf{w}_{\text{approx}}$ compared to $\mathbf{w}_{\text{greedy}}$.

All optimization (sub)problems in Sects. 3 and 4 have at least one solution. If it is not unique, an arbitrary solution is chosen. Taking another solution might only change the SINR for the heuristic greedy beamformer. All other SINR values do not change since they are optimal in the sense described in the previous sections.

Finally, we discuss the computational effort for the exact approaches. Table 2 provides the required number of visited nodes in the branch-and-bound tree for the exact approaches, i.e., how often (S3) for the respective algorithms in [12, 13] and Algorithm 3 is executed. The simulation setting is the same as in [12]. There, a rounded Capon beamformer $\mathbf{w}_{\text{r}}$ is used as a starting point for the branch-and-bound algorithm. For the improved Algorithm 3 we chose a greedy beamformer $\mathbf{w}_{\text{greedy}}$ (see Remark 2) as initial beamformer.

If only the relaxation (7) is considered (algorithm in [13]), then much more nodes in the branch-and-bound procedure are required. For the subproblems (9) with an inequality norm constraint (algorithm in [12]) some additional bound computation in (S3) are needed but this leads to a considerable reduction of the overall number of nodes. The phase-only-specific Algorithm 3 yields a further reduction of nodes. In

Table 2 Number of computed upper bounds for standard and advanced branch-and-bound algorithm versus the number of all discrete beamformers

	$m = 4$	$m = 5$	$m = 6$	$m = 7$
# nodes for Alg. in [13]	1.2×10^4	1.2×10^5	3.0×10^6	1.4×10^8
# nodes for Alg. in [12]	3.2×10^3	1.7×10^4	2.1×10^5	5.4×10^6
# nodes for Alg. 3	2.7×10^3	1.4×10^4	1.5×10^5	3.3×10^6
# extra bounds for Alg. in [12]	3.4×10^2	1.1×10^3	9.7×10^3	1.5×10^5
# extra bounds for Alg. 3	4.0×10^2	1.2×10^3	8.8×10^3	1.2×10^5
$\lvert\mathscr{D}^M\rvert$	4.3×10^9	1.1×10^{12}	2.8×10^{14}	7.2×10^{16}

the simulations the saving is in the range of 15 % ($m = 4$) to 39 % ($m = 7$) compared to the algorithm in [12]. Clearly, the time saving for the different approaches depends on the computing system, the implementation, and also on the stopping parameter. In our implementation the time effort was approximately the same for Algorithm 3 as compared to the algorithm in [12] for $m = 4$ and $m = 5$, while a time saving of 17 and 23 % was reached in the cases $m = 6$ and $m = 7$, respectively.

6 Conclusions

We provided an overview of several approaches for the solution of the discrete and phase-only SINR-maximization problem (2). A simple rounded Capon beamformer $\mathbf{w}_r$ does not produce appropriate results but can be used as a starting point for other approaches. With an approximate beamformer $\mathbf{w}_{\text{approx}}$ better results are possible but with high computational costs. A greedy beamformer yields a good approximation of the maximal SINR and can easily be improved using multiple starting points.

For an exact solution of the discrete SINR-maximization problem, the relaxation (7) from [13] enables a reasonable application of the branch-and-bound principle. The bounds obtained by this relaxation can be improved by introducing an additional constraint in the corresponding subproblems [12]. For discrete phase-only beamforming the equality constraint $\|\mathbf{w}_2\|^2 = L$ yields even stronger bounds which can also be computed efficiently by applying the Dinkelbach method (Algorithm 2). In our simulations, this yields a significant reduction of required nodes within the branch-and-bound procedure.

Acknowledgments This work is supported by the German Research Foundation (DFG) in the Collaborative Research Center 912 "Highly Adaptive Energy-Efficient Computing".

References

1. Baird, C., Rassweiler, G.: Adaptive sidelobe nulling using digitally controlled phase-shifters. IEEE Trans. Antennas Propag. **24**, 638–649 (1976)
2. Bakr, O., Johnson, M., Mudumbai, R., Madhow, U.: Interference suppression in the presence of quantization errors. In: 47th Annual Allerton Conference on Communication, Control, and Computing (2009)
3. Beck, A., Ben-Tal, A., Teboulle, M.: Finding a global optimal solution for a quadratically constrained fractional quadratic problem with applications to the regularized total least squares. SIAM J. Matrix Anal. Appl. **28**, 425–445 (2006)
4. Boeringer, D.W., Werner, D.H.: Particle swarm optimization versus genetic algorithms for phased array synthesis. IEEE Trans. Antennas Propag. **52**, 771–779 (2004)
5. Capon, J.: High-resolution frequency-wavenumber spectrum analysis. Proc. IEEE **57**, 1408–1418 (1969)
6. Demir, O.T., Tuncer, T.E.: Optimum design of discrete transmit phase only beamformer. In: Proceedings of the 21st European Signal Processing Conference (EUSIPCO) (2013)
7. Demir, Ö.T., Tuncer, T.E.: Optimum discrete phase-only transmit beamforming with antenna selection. In: Proceedings of the 22nd European Signal Processing Conference (EUSIPCO), pp. 1282–1286 (2014)
8. Demir, O.T., Tuncer, T.E.: Optimum discrete single group multicast beamforming In: IEEE International Conference on Acoustics, Speech and Signal Processing (ICASSP), pp. 7744–7748 (2014)
9. Demir, Ö.T., Tuncer, T.E.: Optimum discrete transmit beamformer design. Digital Signal Process. **36**, 57–68 (2015)
10. Dinkelbach, W.: On nonlinear fractional programming. Manag. Sci. **13**, 492–498 (1967)
11. Israel, J., Fischer, A.: An approach to discrete receive beamforming. In: Proceedings of the 9th International ITG Conference on Systems Communications and Coding (SCC) (2013)
12. Israel, J., Fischer, A., Martinovic, J.: A branch-and-bound algorithm for discrete receive beamforming with improved bounds. In: Proceedings of the 15th IEEE International Conference on Ubiquitous Wireless Broadband (ICUWB) (2015)
13. Israel, J., Fischer, A., Martinovic, J., Jorswieck, E.A., Mesyagutov, M.: Discrete receive beamforming. IEEE Signal Process. Lett. **22**, 958–962 (2015)
14. Jenning, M., Klein, B., Hahnel, R., Plettemeier, D., Fritsche, D., Tretter, G., Carta, C., Ellinger, F., Nardmann, T., Schroter, M., Nieweglowski, K., Bock, K., Israel, J., Fischer, A., ul Hassan, N., Landau, L., Dorpinghaus, M., Fettweis, G.: Energy-efficient transceivers for ultra-highspeed computer board-to-board communication. In: Proceedings of the 15th IEEE International Conference on Ubiquitous Wireless Broadband (ICUWB) (2015)
15. Nemhauser, G.L., Wolsey, L.A.: Integer and Combinatorial Optimization. Wiley, New York (1999)
16. Ohira, T.: Analog smart antennas: an overview. In 13th IEEE International Symposium on Personal, Indoor and Mobile Radio Communications (PIMRC), vol. 4, pp. 1502–1506 (2002)
17. Poon, A.S.Y., Taghivand, M.: Supporting and enabling circuits for antenna arrays in wireless communications. Proc. IEEE **100**, 2207–2218 (2012)
18. Smith, S.T.: Optimum phase-only adaptive nulling. IEEE Trans. Signal Process. **47**, 1835–1843 (1999)
19. Venkateswaran, V., van der Veen, A.-J.: Analog beamforming in MIMO communications with phase shift networks and online channel estimation. IEEE Trans. Signal Process. **58**, 4131–4143 (2010)
20. Zappone, A., Jorswieck, E.: Energy efficiency in wireless networks via fractional programming theory. Found. Trends Commun. Inf. Theory **11**, 185–396 (2014)

On the Stability of a Variable Step Exponential Splitting Method for Solving Multidimensional Quenching-Combustion Equations

Joshua L. Padgett and Qin Sheng

Abstract This paper concerns the numerical stability of a splitting scheme for solving the three-dimensional degenerate quenching-combustion equation. The diffusion-type nonlinear equation possess highly nonlinear source terms, and is extremely important to the study of numerical combustions. Arbitrary fixed nonuniform spatial grids, which are not necessarily symmetric, are considered in our investigation. The numerical solution is advanced through a semi-adaptive exponential splitting strategy. The temporal adaptation is achieved via a suitable arc-length monitoring mechanism. Criteria for preserving the linear numerical stability of the decomposition method is proven under the spectral norm. A new stability criterion is proposed.

Keywords Combustion · Quenching singularity · Degeneracy · Nonuniform grids · Mesh adaptation · Exponential splitting · Numerical stability

1 Introduction

Let $\mathcal{E} = (0, a) \times (0, b) \times (0, c) \subset \mathbb{R}^3$, where $a, b, c > 0$, and $\partial\mathcal{E}$ be its boundary. Denote $\Omega = \mathcal{E} \times (t_0,\ T)$, $\mathcal{S} = \partial\mathcal{E} \times (t_0,\ T)$ for given $0 \le t_0 < T < \infty$. Consider the following singular reaction–diffusion problem,

$$s(x, y, z)u_t = u_{xx} + u_{yy} + u_{zz} + f(u), \quad (x, y, z, t) \in \Omega, \tag{1.1}$$

$$u(x, y, z, t) = 0, \quad (x, y, z, t) \in \mathcal{S}, \tag{1.2}$$

$$u(x, y, z, t_0) = u_0(x, y, z), \quad (x, y, z) \in \mathcal{E}, \tag{1.3}$$

J.L. Padgett · Q. Sheng (✉)
Department of Mathematics and Center for Astrophysics, Space Physics and Engineering Research, Baylor University, One Bear Place, Waco, TX 76798-7328, USA
email: qin_sheng@baylor.edu

V.K. Singh et al. (eds.), *Modern Mathematical Methods and High Performance Computing in Science and Technology*, Springer Proceedings in Mathematics & Statistics 171, DOI 10.1007/978-981-10-1454-3_13

where $s(x, y, z) = \sqrt{x^2 + y^2 + z^2}$. Further, for $0 \le u < 1$ we have

$$f(0) = f_0 > 0, \quad \lim_{u \to 1^-} f(u) = \infty.$$

In idealized solid fuel combustion applications [1, 3, 14], u represents the temperature in the combustion channel, and the x-, y-, and z-coordinates coincide with the channel walls. We require that $0 \le u_0 \ll 1$. The function $s(x, y, z)$ represents certain singularities in the temperature transportation speed within the channel wall [4, 12, 15, 19]. The solution u of (1.1)–(1.3) is said to *quench* if there exists a finite time $T > 0$ such that

$$\sup\{|u_t(x, y, z, t)| : (x, y, z) \in \mathcal{E}\} \to \infty \text{ as } t \to T^-. \tag{1.4}$$

The value T is then defined as the *quenching time* [1, 2, 11]. It has been shown that a necessary condition for quenching to occur is

$$\max\left\{|u(x, y, z, t)| : (x, y, z) \in \bar{\mathcal{E}}\right\} \to 1^- \text{ as } t \to T^-. \tag{1.5}$$

Further, such a T exists only when certain spatial references, such as the size and shape of $\mathcal{E}$, reach their critical limits. A domain $\mathcal{E}^*$ is called the *critical domain* if the solution of (1.1)–(1.3) exists for all time when $\mathcal{E} \subseteq \mathcal{E}^*$, and (1.5) occurs when $\mathcal{E}^* \subseteq \mathcal{E}$ for a finite T [11].

Numerous computational procedures, including moving mesh adaptive methods, have been constructed for solving lower dimensional quenching-combustion problems in the past decades [2, 7, 8, 15, 19]. Though in the former case, adaptations are frequently achieved via monitoring functions on the arc length of the function u; in the latter situation, adaptations are more likely to be built upon the arc length of u_t, since it is directly proportional to $f(u)$, which blows up as u stops existing [6, 11, 17].

As mentioned in many recent studies, when quenching locations can be predetermined, it is preferable to use nonuniform spatial grids throughout the computations [4, 10, 19]. In this case, key quenching characteristics such as the quenching time and critical domain, are more easily observed; Also important numerical properties of underlying algorithms, in particularly the numerical stability need to be precisely studied. To that end, this paper continues discussions on our temporally adaptive splitting scheme utilizing predetermined variable spatial grids. The numerical stability of the method will be targeted in this paper. Our discussions will be organized as follows. In the next section, the variable step exponential splitting scheme for solving (1.1)–(1.3) will be constructed and studied. Section 3 is devoted to the stability analysis of the variable step splitting scheme. The analysis will first be carried out for a fully linearized scheme, and then a more realistic stability analysis is proposed without freezing the source term. Simulations of some preliminary computational results will be provided. Finally, concluding remarks and proposed future work will be given in Sect. 4.

2 Variable Step Exponential Splitting Scheme

Utilizing the transformations $\tilde{x} = x/a,\ \tilde{y} = y/b,\ \tilde{z} = z/c$, and reusing the original variables for simplicity, we may reformulate (1.1)–(1.3) as

$$u_t = \frac{1}{a^2\phi}u_{xx} + \frac{1}{b^2\phi}u_{yy} + \frac{1}{c^2\phi}u_{zz} + g(u), \quad (x, y, z, t) \in \Omega, \tag{2.1}$$

$$u(x, y, z, t) = 0, \quad (x, y, z) \in \mathcal{S}, \tag{2.2}$$

$$u(x, y, z, t_0) = u_0, \quad (x, y, z) \in \mathcal{E}, \tag{2.3}$$

where $g(u) = f(u)/\phi,\ \phi = \phi(x, y, z) = \left(a^2x^2 + b^2y^2 + c^2z^2\right)^{q/2}$, and $\mathcal{E} = (0, 1) \times (0, 1) \times (0, 1) \subset \mathbb{R}^3$.

Let $N_1, N_2, N_3 \gg 1$. We inscribe over $\bar{\mathcal{E}}$ the following variable grid: $\mathcal{E}_h = \{(x_i, y_j, z_k) | i = 0, \ldots, N_1 + 1;\ j = 0, \ldots, N_2 + 1;\ k = 0, \ldots, N_3 + 1;\ x_0 = y_0 = z_0 = 0,\ x_{N_1+1} = y_{N_2+1} = z_{N_3+1} = 1\}$. Denote $h_{1,i} = x_{i+1} - x_i > 0,\ h_{2,j} = y_{j+1} - y_j > 0$, and $h_{3,k} = z_{k+1} - z_k > 0$ for $1 \le i \le N_1,\ 1 \le j \le N_2,\ 1 \le k \le N_3$. Let $u_{i,j,k}(t)$ be an approximation of the solution of (2.1)–(2.3) at (x_i, y_j, z_k, t) and consider the following variable step finite differences [19]:

$$\left.\frac{\partial^2 u}{\partial x^2}\right|_{i,j,k} \approx \frac{2u_{i-1,j,k}}{h_{1,i-1}(h_{1,i-1} + h_{1,i})} - \frac{2u_{i,j,k}}{h_{1,i-1}h_{1,i}} + \frac{2u_{i+1,j,k}}{h_{1,i}(h_{1,i-1} + h_{1,i})},$$

$$\left.\frac{\partial^2 u}{\partial y^2}\right|_{i,j,k} \approx \frac{2u_{i,j-1,k}}{h_{2,j-1}(h_{2,j-1} + h_{2,j})} - \frac{2u_{i,j,k}}{h_{2,j-1}h_{2,j}} + \frac{2u_{i,j+1,k}}{h_{2,j}(h_{2,j-1} + h_{2,j})},$$

$$\left.\frac{\partial^2 u}{\partial z^2}\right|_{i,j,k} \approx \frac{2u_{i,j,k-1}}{h_{3,k-1}(h_{3,k-1} + h_{3,k})} - \frac{2u_{i,j,k}}{h_{3,k-1}h_{3,k}} + \frac{2u_{i,j,k+1}}{h_{3,k}(h_{3,k-1} + h_{3,k})}.$$

Further, denote $v(t) = (u_{1,1,1}, u_{2,1,1}, \ldots, u_{N_1,1,1}, u_{1,2,1}, u_{2,2,1}, \ldots, u_{N_1,2,1}, \ldots, u_{1,N_2,1}, u_{2,N_2,1}, \ldots, u_{N_1,N_2,1}, \ldots, u_{1,N_2,N_3}, u_{2,N_2,N_3}, \ldots, u_{N_1,N_2,N_3})^{\mathsf{T}} \in \mathbb{R}^{N_1N_2N_3}$ and let $g(v)$ be a discretization of the nonhomogeneous term of (2.1). We obtain readily from (2.1)–(2.3) the following method of line system:

$$v'(t) = \sum_{\sigma=1}^{3} M_\sigma v(t) + g(v(t)), \quad t_0 < t < T, \tag{2.4}$$

$$v(t_0) = v_0, \tag{2.5}$$

where

$$M_1 = \frac{1}{a^2} B(I_{N_3} \otimes I_{N_2} \otimes T_1),$$
$$M_2 = \frac{1}{b^2} B(I_{N_3} \otimes T_2 \otimes I_{N_1}),$$
$$M_3 = \frac{1}{c^2} B(T_3 \otimes I_{N_2} \otimes I_{N_1}),$$

where $I_{N_\sigma} \in \mathbb{R}^{N_\sigma \times N_\sigma}$, $\sigma = 1, 2, 3$, are identity matrices, and

$$B = \mathrm{diag}\left(\phi_{1,1,1}^{-1}, \phi_{2,1,1}^{-1}, \ldots, \phi_{N_1,1,1}^{-1}, \phi_{1,2,1}^{-1}, \ldots, \phi_{N_1,N_2,N_3}^{-1}\right) \in \mathbb{R}^{N_1 N_2 N_3 \times N_1 N_2 N_3},$$
$$\phi_{i,j,k} = \left[a^2\left(\sum_{\ell=0}^{i-1} h_{1,\ell}\right)^2 + b^2\left(\sum_{\ell=0}^{j-1} h_{2,\ell}\right)^2 + c^2\left(\sum_{\ell=0}^{k-1} h_{3,\ell}\right)^2\right]^{q/2},$$
$$T_\sigma = \mathrm{tridiag}\left(l_{\sigma,k-2}, m_{\sigma,k-1}, n_{\sigma,k-1}\right) \in \mathbb{R}^{N_\sigma \times N_\sigma}, \quad \sigma = 1, 2, 3,$$

and for the above

$$l_{\sigma,j} = \frac{2}{h_{\sigma,j}(h_{\sigma,j} + h_{\sigma,j+1})}, \; n_{\sigma,j} = \frac{2}{h_{\sigma,j}(h_{\sigma,j-1} + h_{\sigma,j})}, \quad j = 1, \ldots, N_\sigma - 1,$$
$$m_{\sigma,j} = -\frac{2}{h_{\sigma,j-1} h_{\sigma,j}}, \quad j = 1, \ldots, N_\sigma; \; \sigma = 1, 2, 3.$$

The formal solution of (2.4) and (2.5) can thus be written as

$$v(t) = E\left(t \sum_{\sigma=1}^{3} M_\sigma\right) v_0 + \int_{t_0}^{t} E\left((t-\tau) \sum_{\sigma=1}^{3} M_\sigma\right) g(v(\tau)) d\tau, \quad t_0 < t < T. \tag{2.6}$$

Note that, based on a first-order exponential splitting [14], we have

$$E\left(\gamma \sum_{\sigma=1}^{3} M_\sigma\right) = \exp\left(\gamma \sum_{\sigma=1}^{3} M_\sigma\right) = e^{\gamma M_1} e^{\gamma M_2} e^{\gamma M_3} + \mathcal{O}\left(\gamma^2\right), \quad \gamma \to 0^+.$$

Consider [1/1] Padé approximations for the exponential matrices, we have

$$E\left(\gamma \sum_{\sigma=1}^{3} M_\sigma\right) = p(\gamma) + \mathcal{O}\left(\gamma^2\right), \quad \gamma \to 0^+,$$

where

$$p(\gamma) = \prod_{\sigma=1}^{3} \left(I - \frac{\gamma}{2} M_\sigma\right)^{-1} \left(I + \frac{\gamma}{2} M_\sigma\right).$$

Thus, a trapezoidal rule for (2.6) leads to

$$v(t) = p(t)\left[v_0 + \frac{t}{2}g(v_0)\right] + \frac{t}{2}g(v(t)) + \mathcal{O}\left((t - t_0)^2\right), \quad t \to t_0. \tag{2.7}$$

The above is a typical Local-One-Dimensional (LOD) algorithm which is exponential splitting based, rather than the Alternative-Direction-Implicit (ADI) related [16]. It provides a highly efficient way to compute numerical solutions of multidimensional problems such as (2.1)–(2.3) [13, 15]. Based on (2.7), we obtain the following first-order variable step exponential splitting scheme:

$$v_{\ell+1} = \left[\prod_{\sigma=1}^{3}\left(I - \frac{\tau_\ell}{2}M_\sigma\right)^{-1}\left(I + \frac{\tau_\ell}{2}M_\sigma\right)\right]\left(v_\ell + \frac{\tau_\ell}{2}g(v_\ell)\right) + \frac{\tau_\ell}{2}g(v_{\ell+1}), \tag{2.8}$$

where v_ℓ and $v_{\ell+1}$ are approximations of $v(t_\ell)$ and $v(t_{\ell+1})$, respectively, v_0 is the initial vector, $t_\ell = t_0 + \sum_{k=0}^{\ell-1}\tau_k,\ \ell = 0, 1, 2, \ldots$, and $\{\tau_\ell\}_{\ell\geq 0}$ is a set of variable temporal steps determined by an adaptive procedure. In order to avoid a fully implicit scheme, $g(v_{\ell+1})$ may be approximated by $g(w_\ell)$, where w_ℓ is an approximation to $v_{\ell+1}$, such as

$$w_\ell = v_\ell + \tau_\ell(Cv_\ell + g(v_\ell)), \quad 0 < \tau_\ell \ll 1, \tag{2.9}$$

in practical computations.

Due to a strong quenching singularity, the selection of proper nonuniform temporal steps τ_ℓ is vital. As an illustration, in Fig. 1, we show the projected numerical solution and its temporal derivative of a typical three-dimensional quenching-combustion initial-boundary value problem over the x-interval $[0, \pi]$. The initial function $u_0(x, y, z) = 0.001\sin(x)\sin(y)\sin(z)$, $f(u) = 1/(1 - u)$, and homogeneous Dirichlet boundary condition are employed. It is evident that v_t changes dramatically when compared with v. Recalling (1.4) and (1.5), we consider the following arc-length monitoring function on v_t,

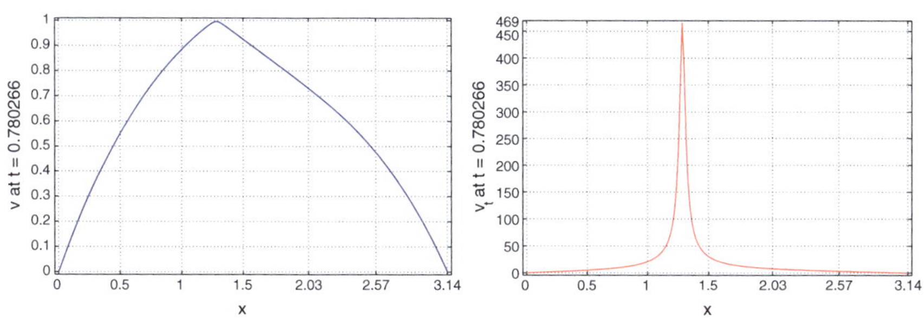

Fig. 1 Numerical solution (*left*) and its temporal derivative (*right*) immediately before quenching. It is observed that as $\max_x v(x) \to 1^-$, we have $\max_x v_t \gg 600$. The computed quenching time is $T \approx 0.780265747310047$

$$m\left(\frac{\partial v}{\partial t}, t\right) = \sqrt{1 + \left(\frac{\partial^2 v}{\partial t^2}\right)^2}, \quad t_0 < t < T.$$

Setting the two maximal arc lengths in neighboring intervals $[t_{\ell-2}, t_{\ell-1}]$ and $[t_{\ell-1}, t_\ell]$ equal [10, 18, 19], we acquire the following quadratic equations from the above,

$$\tau_\ell^2 = \tau_{\ell-1}^2 + \left(\frac{\partial v_{\ell-1}}{\partial t} - \frac{\partial v_{\ell-2}}{\partial t}\right)^2 - \left(\frac{\partial v_\ell}{\partial t} - \frac{\partial v_{\ell-1}}{\partial t}\right)^2, \quad \ell = 1, 2, 3, \ldots,$$

with τ_0 given.

3 Stability

Nonlinear stability has been an extremely difficult issue when nonlinear quenching-combustion equations are concerned [1, 4, 5, 15, 17, 19]. However, when the numerical solution varies relatively slowly, that is, before reaching a certain neighborhood of quenching, instability may be detected through a linear stability analysis of the nonlinear scheme utilized [7, 10, 20]. Although the application of such an analysis to nonlinear problems cannot be rigorously justified, it has been found to be remarkably informative in practical computations. In the following study, we will first carry out a linearized stability analysis in the von Neumann sense for (2.8) with its nonlinear source term frozen. This is equivalent to assuming that the source term is effectively accurate. The analysis will then be extended to circumstances where the nonlinear term is not frozen. In the later case, the boundedness of the Jacobian of the source term, $\|g_v(v)\|_2$, which is equivalent to assuming that we are some neighborhood away from quenching, is assumed.

In the following, let $A \in \mathbb{C}^{n\times n}$ and again denote $E(\cdot) = \exp(\cdot)$ for $n > 1$.

Definition 3.1 Let $\|\cdot\|$ be an induced matrix norm. Then the associated logarithmic norm $\mu : \mathbb{C}^{n\times n} \to \mathbb{R}$ of A is defined as

$$\mu(A) = \lim_{h\to 0^+} \frac{\|I_n + hA\| - 1}{h},$$

where $I_n \in \mathbb{C}^{n\times n}$ is the identity matrix.

For the variable step splitting method (2.8) with its nonlinear source term frozen, regularity conditions need to be imposed upon the nonuniform spatial grids for a linear stability analysis. For this purpose, let us denote $h_\sigma = \min_{j=1,\ldots,N_\sigma}\{h_{\sigma,j}\}$, $\sigma = 1, 2, 3$.

Lemma 3.1 *If*

$$\frac{1}{h_1^2\phi_{i-1,j,k}} - \frac{1}{h_{1,i-1}h_{1,i}\phi_{i,j,k}} \le \frac{K}{2}, \tag{3.1}$$

$$\frac{1}{h_2^2\phi_{i,j-1,k}} - \frac{1}{h_{2,j-1}h_{2,j}\phi_{i,j,k}} \le \frac{K}{2}, \tag{3.2}$$

$$\frac{1}{h_3^2\phi_{i,j,k-1}} - \frac{1}{h_{3,k-1}h_{3,k}\phi_{i,j,k}} \le \frac{K}{2}, \tag{3.3}$$

where the constant $K > 0$ *is independent of* $h_{\sigma,j},\ j = 1, \dots, N_\sigma,\ \sigma = 1, 2, 3$. *then*

$$\mu(M_\sigma) \le K,\ \sigma = 1, 2, 3.$$

Proof We only need to consider the case involving M_1 since the other cases are similar. Note that $\mu(M_1) = \frac{1}{2}\lambda_{\max}\left(M_1 + M_1^{\mathsf{T}}\right)$ and

$$\frac{1}{2}\left(M_1 + M_1^{\mathsf{T}}\right) = \text{diag}(X_{1,1}, \dots, X_{N_2,1}, X_{1,2}, \dots, X_{N_2,N_3}) \in \mathbb{R}^{N_1N_2N_3\times N_1N_2N_3},$$

where

$$\left(X_{j,k}\right)_{n,p} = \begin{cases} \frac{m_{1,n}}{\phi_{n,j,k}}, & \text{if } n = p, \\ \frac{n_{1,n-1}}{2\phi_{n-1,j,k}} + \frac{l_{1,n-1}}{2\phi_{n,j,k}}, & \text{if } n - p = 1, \\ \frac{n_{1,n}}{2\phi_{n,j,k}} + \frac{l_{1,n}}{2\phi_{n+1,j,k}}, & \text{if } p - n = 1, \\ 0, & \text{otherwise.} \end{cases}$$

We apply Geršchgorin's circle theorem to an arbitrary $X_{j,k}$ and note that a similar argument works for each $X_{j,k},\ j = 1, \dots, N_2,\ k = 1, \dots, N_3$. Further, notice that we only need to consider circumstances where the bandwidth of $M_1 + M_1^{\mathsf{T}}$ is three. Thus,

$$\left|\lambda_{1,i} - \frac{m_{1,i}}{\phi_{i,j,k}}\right| \le \left|\frac{n_{1,i-1}}{2\phi_{i-1,j,k}} + \frac{l_{1,i-1}}{2\phi_{i,j,k}}\right| + \left|\frac{n_{1,i}}{2\phi_{i,j,k}} + \frac{l_{1,i}}{2\phi_{i+1,j,k}}\right| \le \frac{2}{h_1^2\phi_{i-1,j,k}},$$
$$i = 2, \dots, N_1 - 1,\ j = 1, \dots, N_2,\ k = 1, \dots, N_3.$$

We then see that (3.1) follows immediately from the above and the fact that

$$\frac{2}{h_1^2\phi_{i-1,j,k}} - \frac{2}{h_{1,i-1}h_{1,i}\phi_{i,j,k}} \le K,\ i = 2, \dots, N_1 - 1,\ j = 1, \dots, N_2,\ k = 1, \dots, N_3.$$

■

Lemma 3.2 *If (3.1)–(3.3) hold then*

$$\left\| \left(I - \frac{\tau_\ell}{2} M_\sigma\right)^{-1} \left(I + \frac{\tau_\ell}{2} M_\sigma\right) \right\|_2 \le 1 + \tau_\ell K + \mathcal{O}\left(\tau_\ell^2\right), \quad \ell \ge 0, \ \sigma = 1, 2, 3, \tag{3.4}$$

for sufficiently small $\tau_\ell > 0$.

Proof Recalling the [1/1] Padé approximation utilized in Sect. 2, we have

$$\left(I - \frac{\tau_\ell}{2} M_\sigma\right)^{-1} \left(I + \frac{\tau_\ell}{2} M_\sigma\right) = E(\tau_\ell M_\sigma) + \mathcal{O}\left(\tau_\ell^3\right), \quad \sigma = 1, 2, 3.$$

Now, based on Lemma 3.1,

$$\begin{aligned} \left\| \left(I - \frac{\tau_\ell}{2} M_\sigma\right)^{-1} \left(I + \frac{\tau_\ell}{2} M_\sigma\right) \right\|_2 &\le E(\tau_\ell \mu(M_\sigma)) + \mathcal{O}\left(\tau_\ell^3\right) \\ &\le \left[1 + \tau_\ell K + \mathcal{O}\left(\tau_\ell^2\right)\right] + \mathcal{O}\left(\tau_\ell^3\right) \\ &= 1 + \tau_\ell K + \mathcal{O}\left(\tau_\ell^2\right), \end{aligned}$$

which is the desired bound. ■

Combining the above results gives the following theorem.

Theorem 3.1 *If (3.1)–(3.3) hold, then the variable step exponential splitting method (2.8) with the source term frozen is unconditionally stable in the von Neumann sense under the spectral norm, that is,*

$$\|z_{\ell+1}\|_2 \le c\|z_0\|_2, \quad \ell \ge 0,$$

where $z_0 = v_0 - \tilde{v}_0$ *is an initial error,* $z_{\ell+1} = v_{\ell+1} - \tilde{v}_{\ell+1}$ *is the* $(\ell + 1)$*th perturbed error vector, and* $c > 0$ *is a constant independent of* ℓ *and* τ_ℓ.

Proof When the nonlinear source term is frozen, $z_{\ell+1}$ takes the form of

$$z_{\ell+1} = \prod_{\sigma=1}^{3} \left(I - \frac{\tau_\ell}{2} M_\sigma\right)^{-1} \left(I + \frac{\tau_\ell}{2} M_\sigma\right) z_\ell, \quad \ell \ge 0. \tag{3.5}$$

Recall that $\sum_{k=0}^{\ell} \tau_k \le T, \ \ell > 0$. It follows by taking the norm on both sides of (3.5) that

$$
\begin{aligned}
\|z_{\ell+1}\|_2 &\le \prod_{\sigma=1}^{3} \left\| \left(I - \frac{\tau_\ell}{2}M_\sigma\right)^{-1} \left(I + \frac{\tau_\ell}{2}M_\sigma\right) \right\|_2 \|z_\ell\|_2 \\
&\le \left(1 + 3\tau_\ell K + c_2\tau_\ell^2\right) \|z_\ell\|_2 \ \le\ \prod_{k=0}^{\ell} \left(1 + 3\tau_k K + c_3\tau_k^2\right) \|z_0\|_2 \\
&\le \left(1 + 3KT + c_4 \sum_{k=0}^{\ell} \tau_k^2\right) \|z_0\|_2 \ \le\ c\|z_0\|_2,
\end{aligned}
$$

where c_1, c_2, c_3, c_4, and c are positive constants independent of $\ell,\ \tau_k,\ 0 \le k \le \ell$. Therefore, the theorem is clear. ■

We now consider the case without freezing the nonlinear source term in (2.8). In this situation, restrictions upon the Jacobian matrix $g_v(v)$ become necessary.

Theorem 3.2 *Let $\tau_k,\ 0 \le k \le \ell$, be sufficiently small and (3.1)–(3.3) hold. If there exists a constant $G < \infty$ such that*

$$
\|g_v(\xi)\|_2 \le G, \quad \xi \in \mathbb{R}^{N_1 N_2 N_3}, \tag{3.6}
$$

then the variable step exponential splitting method (2.8) is unconditionally stable in the von Neumann sense, that is,

$$
\|z_{\ell+1}\|_2 \le \tilde{c}\, \|z_0\|_2, \quad \ell > 0,
$$

where $z_0 = v_0 - \tilde{v}_0$ is an initial error, $z_{\ell+1} = v_{\ell+1} - \tilde{v}_{\ell+1}$ is the $(\ell+1)$th perturbed error vector, and $\tilde{c} > 0$ is a constant independent of ℓ and τ_ℓ.

Proof By definition we have

$$
\begin{aligned}
v_{\ell+1} &= \prod_{\sigma=1}^{3} \left(I - \frac{\tau_\ell}{2}M_\sigma\right)^{-1} \left(I + \frac{\tau_\ell}{2}M_\sigma\right) \left(v_\ell + \frac{\tau_\ell}{2}g(v_\ell)\right) + \frac{\tau_\ell}{2}g(v_{\ell+1}) \\
&= \Phi_\ell \left(v_\ell + \frac{\tau_\ell}{2}g(v_\ell)\right) + \frac{\tau_\ell}{2}g(v_{\ell+1}),
\end{aligned}
$$

where

$$
\Phi_\ell = \prod_{\sigma=1}^{3} \left(I - \frac{\tau_\ell}{2}M_\sigma\right)^{-1} \left(I + \frac{\tau_\ell}{2}M_\sigma\right).
$$

It follows that

$$
z_{\ell+1} = \Phi_\ell z_\ell + \frac{\tau_\ell}{2}\Phi_\ell g_v(\xi_\ell) z_\ell + \frac{\tau_\ell}{2} g_v(\xi_{\ell+1}) z_{\ell+1},
$$

where $\xi_k \in \mathcal{L}(v_k, \tilde{v}_k),\ k = \ell, \ell+1$. Rearranging the above equality, we have

$$\left(I - \frac{\tau_\ell}{2} g_v(\xi_{\ell+1})\right) z_{\ell+1} = \Phi_\ell \left(I + \frac{\tau_\ell}{2} g_v(\xi_\ell)\right) z_\ell.$$

Further, recall (3.6). When τ_k is sufficiently small we may claim that

$$\left(I - \frac{\tau_k}{2} g_v(\xi)\right)^{-1},\ I + \frac{\tau_k}{2} g_v(\xi) = E\left(\frac{\tau_k}{2} g_v(\xi)\right) + \mathcal{O}\left(\tau_k^2\right).$$

Thus,

$$z_{\ell+1} = \left\{\prod_{k=0}^{\ell}\left[E\left(\frac{\tau_k}{2} g_v(\xi_{k+1})\right) + \mathcal{O}\left(\tau_k^2\right)\right]\Phi_k\left[E\left(\frac{\tau_k}{2} g_v(\xi_k)\right) + \mathcal{O}\left(\tau_k^2\right)\right]\right\} z_0.$$

It follows therefore

$$\begin{aligned}\|z_{\ell+1}\|_2 &\le \left\{\prod_{k=0}^{\ell}\left\|E\left(\frac{\tau_k}{2} g_v(\xi_{k+1})\right)\right\|_2 \|\Phi_k\|_2 \left\|E\left(\frac{\tau_k}{2} g_v(\xi_k)\right)\right\|_2 + c_{1,k}\tau_k^2\right\}\|z_0\|_2 \\ &\le \left(1 + 3KT + c\sum_{k=0}^{\ell}\tau_k^2\right)\left(e^{GT} + c_1\sum_{k=0}^{\ell}\tau_k^2\right)\|z_0\|_2 \ \le\ \tilde{c}\,\|z_0\|_2,\end{aligned}$$

where $c_{1,k},\ k = 1, 2, \ldots, \ell$, are positive constants and $c, c_1, \tilde{c}$ are positive constants independent of ℓ and $\tau_\ell,\ \ell > 0$. Thus giving the desired stability. ■

The above theorem provides a precise insight as to why the standard linear analysis can reach in estimating a nonlinear stability. The extra cost paid, however, is assuming the boundedness of $\|g_v(\xi)\|_2$. Nevertheless, the approach is an improvement upon the traditional methodology of having the nonlinear source term frozen. In fact, the aforementioned bound is well observed in numerical experiments up to the situation when certain neighborhoods of quenching are reached.

To illustrate, we show our preliminary computational results in Fig. 2. The particular source function is again chosen as $f(u) = 1/(1-u)$. With $s(x, y, z) = \sqrt{x^2 + y^2 + z^2}$ and sufficiently large domain $\mathcal{E}$, the quenching can be predicted to occur about $(0.28, 0.28, 0.28)$ on the regularized domain [4, 7]. Thus we may select desirable variable spatial grids illustrated in Fig. 2. In Fig. 3, we show our preliminary numerical results immediately before quenching in a reduced three-dimensional projected space.

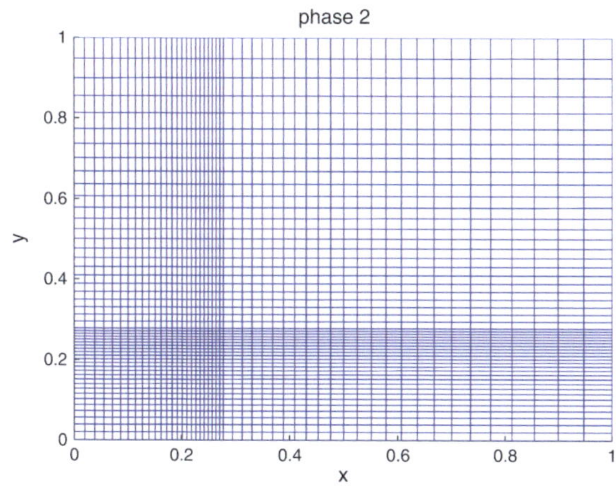

Fig. 2 An illustration of a typical variable step spatial grids in a regularized X-Y plane

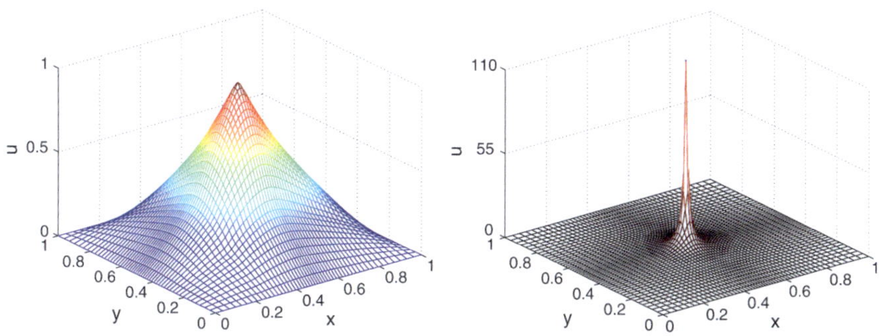

Fig. 3 Projected numerical solution (*left*) and its temporal derivative (*right*) immediately before quenching. It can be observed that as $\max_{(x,y,z)} v(x, y, z) \to 1^-$, we have $\max_{(x,y,z)} v_t(x, y, z) \gg 100$

4 Conclusions

Our variable step exponential splitting scheme is developed for solving singular reaction–diffusion equations possessing strong quenching nonlinearities in general. While a temporal adaptation is performed via an arc-length monitoring mechanism of the temporal derivative of the solution, variable spatial grids are considered for applications of variety of adaptive strategies. The novel splitting method is implicit and the impact of the degeneracy is found to be limited. Rigorous analysis is given for the stability of the numerical solution. Important criteria to guarantee the property, which depend upon the variable steps and degeneracy, are established.

Under much weaker requirements (see the latest results in [4]), the temporal step restriction for guaranteeing monotone numerical solutions of our splitting scheme has been reduced to only one-half of those in uniform spatial mesh cases [15].

Furthermore, a realistic method of targeting the realization of nonlinear stability analysis is proposed and shown to be successful. Though this new strategy needs the boundedness of $\|g_v(\xi)\|_2$, the requirement is well justified before quenching is reached. This improved methodology not only provides further insight into the stability, but also offers explanations as to why the linear stability analysis must be valid before quenching. On the other hand, simulations of real multidimensional solutions still remain as one of the most challenging tasks even with the help of parallel processors and large data storage. Possible highly accurate exponential splitting formulae, such as the Strang's formula and asymptotic formulae [16], also remain to be explored.

References

1. Acker, A., Kawohl, B.: Remarks on quenching. Nonlinear Anal. **13**, 53–61 (1989)
2. Acker, A., Walter, W.: The quenching problem for nonlinear parabolic differential equations. Ordinary and Partial Differential Equations. Lecture Notes in Mathematics, vol. 564, pp. 1–12. Springer, New York (1976)
3. Bebernes, J., Eberly, D.: Mathematical Problems from Combustion Theory. Springer, Berlin (1989)
4. Beauregard, M., Sheng, Q.: An adaptive splitting approach for the quenching solution of reaction-diffusion equations over nonuniform grids. J. Comp. Appl. Math. **241**, 30–44 (2013)
5. Cao, W., Huang, W., Russell, R.D.: A study of monitor functions for two-dimensional adaptive mesh generation. SIAM J. Sci. Comput. **20**, 1978–1994 (1999)
6. Chan, C.Y., Ke, L.: Parabolic quenching for nonsmooth convex domains. J. Math. Anal. Appl. **186**, 52–65 (1994)
7. Cheng, H., Lin, P., Sheng, Q., Tan, R.: Solving degenerate reaction-diffusion equations via variable step Peaceman-Rachford splitting. SIAM J. Sci. Comput. **25**, 1273–1292 (2003)
8. Coyle, J.M., Flaherty, J.E., Ludwig, R.: On the stability of mesh equidistribution strategies for time-dependent partial differential equations. J. Comput. Phys. **62**, 26–39 (1986)
9. Kawarada, H.: On solutions of initial-boundary value problems for $u_t = u_{xx} + 1/(1-u)$,. Publ. Res. Inst. Math. Sci. **10**, 729–736 (1975)
10. Lang, J., Walter, A.: An adaptive Rothe method for nonlinear reaction-diffusion systems. Appl. Numer. Math. **13**, 135–146 (1993)
11. Levine, H.A.: Quenching, nonquenching, and beyond quenching for solutions of some parabolic equations. Ann. Math. Pure. Appl. **4**, 243–260 (1989)
12. Ockendon, H.: Channel flow with temperature-dependent viscosity and internal viscous dissipation. J. Fluid Mech. **93**, 737–746 (1979)
13. Schatzman, M.: Stability of the Peaceman-Rachford approximation. J. Funct. Anal. **162**, 219–255 (1999)
14. Sheng, Q.: Exponential Splitting Methods for Partial Differential Equations, Ph.D. Dissertation, DAMTP, Cambridge University (1990)
15. Sheng, Q.: Adaptive decomposition finite difference methods for solving singular problems. Front. Math. China **4**, 599–626 (2009)
16. Sheng, Q.: The ADI Methods. Encyclopedia of Applied and Computational Mathematics. Springer Verlag GmbH, Heidelberg (2015)
17. Sheng, Q.: ADI, LOD and modern decomposition methods for certain multiphysics applications. J Algorithms Comput. Technol. **9**, 105–120 (2015)
18. Sheng, Q., Khaliq, A.: Linearly implicit adaptive schemes for singular reaction-diffusion equations. In: Vande Wouwer, A., Saucez, Ph., Schiesser, W.E. (eds.) Adaptive Method of Lines. Capman & Hall/CRC, London (2001)

19. Sheng, Q., Khaliq, A.: A revisit of the semi-adaptive method for singular degenerate reaction-diffusion equations. East Asia J. Appl. Math. **2**, 185–203 (2012)
20. Twizell, E.H., Wang, Y., Price, W.G.: Chaos-free numerical solutions of reaction-diffusion equations. Proc. R. Soc. London Sect. A **430**, 541–576 (1991)

Perspectives in High Performance Computing

Michael Resch

Abstract High Performance Computing is undergoing a major change in the coming years. This paper discusses the perspectives that HPC has in the coming years and how these perspectives are going to change the way we operate and use HPC systems. The paper discusses the technologies that HPC will use, discusses the current trends based on the TOP500 list and argues that a further improvement in performance will basically be driven by software rather than by hardware.

Keywords High performance computing · Trends · Hardware · Software

1 Introduction

High Performance Computing (HPC) architectures are a hotly debated issue as the designers of such systems are increasingly facing new challenges. Looking at current developments traditional approaches seem to be running out of steam. A few years ago, HPC centers were concerned about the lack of variety of architectures and suspected that a monocultural world would take hold of the HPC market. In fact, a monopoly of architectures can already be seen today with many vendors having left the market. In recent discussions on architectures at the International Supercomputing Conference at Dresden [1] it became clear that this monopoly is triggering a new development of architectures. Most of them are not yet mature enough such that it is unclear which of them will reach a level of maturity that would allow their usage in everyday production. This has made it increasingly difficult to make investment decisions when it comes to designing large-scale systems. On the other hand, HPC centers are still concerned that they might be running out of options when it comes to procuring for next-generation systems.

M. Resch (✉)
High Performance Computing Center Stuttgart (HLRS), University of Stuttgart, Nobelstrasse 19, 70569 Stuttgart, Germany
e-mail: resch@hlrs.de

V.K. Singh et al. (eds.), *Modern Mathematical Methods and High Performance Computing in Science and Technology*, Springer Proceedings in Mathematics & Statistics 171, DOI 10.1007/978-981-10-1454-3_14

The situation in the field of HPC is actually more complicated than it was 10 years ago. IBM recently decided to sell its manufacturing facilities. The step that was widely assumed to be a starting point of an exit strategy from the hardware business. At the same time, IBM made the design for the Power architecture publicly available—giving the market another visible signal for a retreat from the HPC market. These were two steps that followed the decision of IBM to sell its x86 activities to Lenovo–a Chinese vendor that already took over when IBM dropped their laptop business. Furthermore, the market receives signals that the highly successful BlueGene line of IBM may not see a follow-on product. With IBM giving mixed signals for HPC, and with the disappearance of vendors like SUN, the HPC market is left with few options—having experienced a continuous decline in number of stable vendors over the last years.

Technically HPC is facing the end of a development that used to be called Moore's law [2]. Processor clock frequencies–which carried the main load of speeding up hardware—cannot be further increased since 2004. Multi-core processors have become standard. So-called accelerators provide solutions that push the number of cores on a single chip to extremes but leave the users with adapting their codes to a new architecture and a new programming model.

In this paper, we will investigate a number of questions that come up in this context. We will explore the messages that the history of the TOP500 [3] list provides. In most recent editions, we have seen interesting developments that will be important for centers and users alike. We will furthermore look into technically new trends that may help to overcome some of the limitations that we face with massively parallel systems. Finally, we will try to explore and evaluate new technologies that might be available to the market in the near future.

2 A Little Bit of History

HPC has for a while been dominated by a development that was described by Gordon Moore in 1965 [2]. Basically, Moore figured that the number of components on a chip of the same size had doubled consistently over a certain period of time. From this, he concluded—by making an economical argument rather than diving too deep into technical details—that a similar development could be expected in the near future. Originally, he assumed a doubling of components every 12 months and later on modified this to a doubling every 18 months. As a corollary from this, it was assumed for a long time that clock frequencies of processors could be doubled every 18 months. The basic assumption was that reducing the feature size would reduce the distance for a signal to travel and hence increase the clock frequency.

For several decades, Moore's expectation proved to be right. Clock frequencies actually increased and basically followed the expected path. The development started to slow down for high-end processors in the mid-1990s. Clock frequencies were at about 0.5 GHz and higher. At the same time so-called standard processors (at the time provided by Intel and AMD) rapidly caught up with HPC systems—driven by increased clock frequencies and by a market that was eager to absorb whatever new processor became available for the consumer market.

The slow down in clock rate increase was foreseeable and parallelism was early on investigated as a concept to overcome the problem. With the introduction of parallel processing the focus shifted from single processor speed to number of cores available in a single system. Early adopters of the concept failed but provided the necessary groundwork for our current technology in HPC. By 2004—when clock speeds started to stall at a value around 2–3 GHz—parallelism took over as the leading principal in HPC [4].

3 Technology Trends

To make up for the lack of acceleration based on increased clock frequencies parallel computing was pushed to the extreme over the last decade. Parallelism is not a new paradigm. It was exploited over time in a variety of architectures. Actually, even a standard technology like pipelining is in a sense some sort of parallelism exploited in the architecture. However, at the processor level parallelism arrived relatively late and the level of parallelism employed in high performance computing systems remained relatively moderate for a while. The number of processors used was hovering around 512–1024—with the Thinking Machine approach being the notable exception.

The currently fastest system in the world is based on about 3 million compute cores bundled in a single system [3]. This development is based on the fact, that the principal idea of Gordon Moore is still true. The number of components on a chip can still be further increased and most likely will grow over the next years according to the International Technology Roadmap for Semiconductors [5].

When it comes to exploiting parallelism in hardware there are two paths the market is following.

3.1 Thin Core Concept

The concept of thin cores is mainly focused on parallelism. The basic idea is to build relatively simple cores but to build a large number of them on a single chip. Graphics processing units (GPU) heavily influenced this concept. Some solutions actually evolve originally from GPUs, which were modified in order to meet the requirements of high speed computing.

The thin core concept is based on the reasonable idea that hardware designed for high speed computing—which is typically measured in floating-point operations—should be focusing on floating-point units only. The perfect solution would be a concept in which a core is not much more than a floating-point unit.

The concept as described above is based on low clock frequencies and large numbers of cores on a chip. Very often manufacturing technology is not leading edge, as mass production for GPUs does not require high-end manufacturing technology.

Increased speed can therefore be expected for this concept from two sources mainly:

- **Higher clock frequencies:** Thin core concepts are typically based on a relatively moderate clock frequency in the range of 1 GHz or less. This allows to keep power consumption relatively low and hence squeeze more cores on a chip. Theoretically, clock frequencies still have the potential to grow by about a factor of 2–4 over the coming years. They will then have reached the level of current state-of-the-art standard processors and a further increase would lead to similar cooling and power consumption problems as with standard processors. But to increase clock frequency—even if only moderately—is an option to go for higher total performance of a chip.
- **Higher core numbers:** Using more advanced manufacturing technology, the number of components on a chip will be increased. Keeping the design of an individual core as simple as possible, the additional components can be used to increase the number of cores on a chip. This could potentially increase the number of cores on a chip by a factor of 4–8 in the coming decade.

Putting together these two trends it seems to be possible to both increase clock frequencies slightly and to further increase the number of cores on a chip. For a General Purpose GPU or for similar accelerator concepts we can expect to see a factor of 2–32 in peak performance over the coming years.

Current developments, however, seem to indicate another trend. With an increasing demand for these kinds of accelerators designers are trying to turn these cores into floating-point machines that better fit the requirements for standard simulations. Having learned that cheap GPUs can be used to speed up high-end computing systems companies increasingly see the potential in the HPC market for their product. However, the complexity of the cores has to be increased to meet the requirements of the HCP user community. Extrapolating the trend one might expect to see a stagnation in number of cores on a chip while complexity and clock frequency are increased harvesting the potential of new manufacturing technologies.

What might be expected is a moderate growth in speed only but an increase in potential for standard applications.

3.2 *Fat Core Concept*

The concept of fat cores could be described as the "classical" approach to high performance computing architectures. The increased speed required for HPC is delivered by increasing the complexity of processor architectures. A lot of complexity is added for example to overcome the limitations of slow memory subsystems. Additional available components are also used to add further functional units. By doing this we gain an additional level of parallelism directly on the chip. Performance is increased by having 4 or 8 ADD-Mult per clock cycle rather than speeding up each individual ADD-Mult.

A number of processor could be described as "fat" with each of them following a different path of development.

- **X86 architectures:** The x86 processor family is the standard architecture in HPC for several years now. The number of ores is relatively low with currently 8–16 cores for a single CPU. Each of the cores itself is a highly tuned architecture with a number of sophisticated features that—if adequately programmed—turn these processors into high-end computing engines.
- **IBM Power:** The IBM Power processors is the last surviving specialized standard processor in the arena of HPC. The first standard multi-core chip for HPC was a Power processor and since then the Power processor has proven to be always of the leading edge of processor technology.
- **Vector processors:** Vector processors seemed to be extinct when NEC dropped out of what later was to become the Japanese K Computer project. However, they reappeared in 2014 when NEC introduced its SX-ACE line. The concept follows a traditional approach with introducing vector pipes as the core to achieve performance and a relatively sophisticated memory subsystem that allows pushing sustained performance to a level hardly reachable by standard processors. However, the prize for such systems is still comparably high and hence they hardly make an appearance in the TOP500 list.

3.3 Memory

At this point it makes sense to talk about memory technology. High Performance Computing hit the memory wall about 20 years ago. Increased processor speed was not matched by memory speed—neither in terms of latency nor in terms of bandwidth. Modern architectures have become increasingly imbalanced. As a result, users can expect a sustained level of performance that varies widely. The more a code is limited by memory speed the lower its sustained performance. Experts speak of about 3–5 % of performance that can be achieved when working without cache aware programming.

Caches were seen to be the way to overcome memory speed limitations. Introducing small but fast caches on-chip, vendors were hoping to break the memory wall. Over time cache levels were introduced and as of today we expect to see three levels of caches in a high-end HPC processor. But as memory and cache systems get more complex users are facing two further problems.

- **Complexity:** With the growing complexity of cache hierarchies it gets increasingly difficult to optimize a code for a given hardware architecture. Once optimized for one architecture the changes in cache hierarchy of another architecture may cause a drop in performance by as much as a factor of 10 or more. In order to fully exploit an architecture, programmers would have to be aware of the architecture—which changes rapidly and increases its complexity continuously.

- **Imbalance:** At the same time as users struggle with cache hierarchies and their complexities architect are faced with handling the memory subsystem side effects. As a result users are increasingly facing imbalances between memory and cache hierarchies on the one hand and architectural features on the other side. The most simple problem is that for some cache hierarchies to work properly a processor needs enough registers to handle all the traffic. If this is ignored it is in the end the processor architecture that kills the memory and cache subsystem.

3.4 TOP500 Trends

The HPC community criticized the TOP500 [3] list for many years. There is at least one large-scale installation that refused to participate in the list claiming that the Linpack benchmark has no whatsoever justification to be used as a yardstick for HPC systems. Although there is some truth to this claim the TOP500 list has shown to be an interesting collection of statistical data from which at least trends can be extracted [6].

Exploring recent developments in the list allows getting a better understanding of trends and markets in HPC. Over the last years, the most striking feature is that the replacement rate of systems in the list for high-end systems is slowing down. A brief analysis of the TOP10 systems over the last years shows the following:

- In November 2011 three new systems were in the TOP 10 compared to November 2010
- In November 2012 eight new systems were in the TOP 10 compared to the previous year
- In November 2013 three new systems were in the TOP 10 compared to the previous year
- In November 2014 one system was new compared to the previous year—and even this system was not a full replacement but an upgraded version of an existing system
- In November 2015, we were back to three new systems.

When we look at the five fastest systems we see no change since 2013.

What is more interesting is the trend line that can be retrieved from the last 21 years of collecting information about the fastest systems in the world. Figure 1 shows the trend lines for the performance of

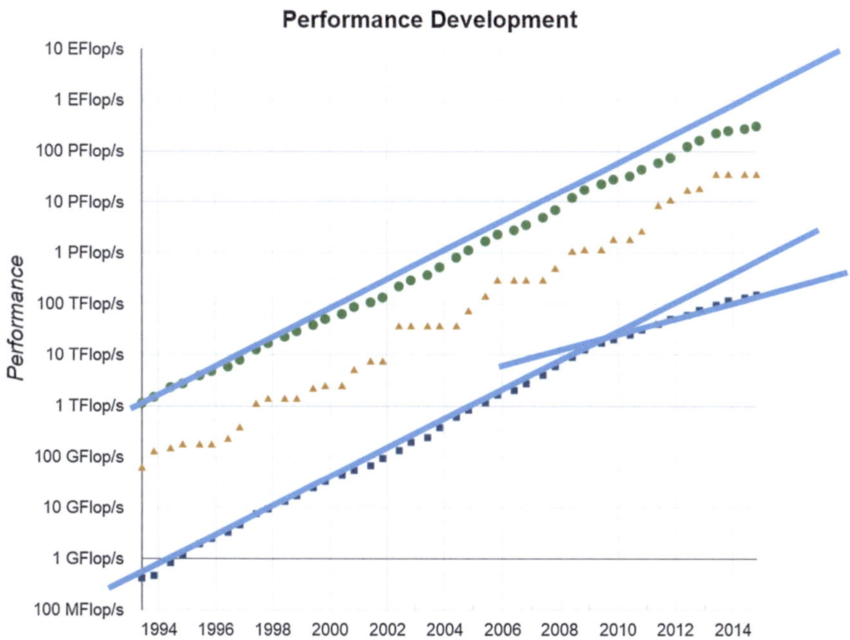

Fig. 1 Trend lines of the TOP500 list. Basic data from www.top500.org with added trend lines by the author

(a) The number 500 system (lower line),
(b) The fastest system (middle line), and
(c) The sum of all systems on the list

The figure was taken from the TOP500 webpage. The authors added trend lines.

The figure indicates that the number 500 system—the slowest system on the list—is unable to follow the general trend since about 2009/2010. A similar trend cannot yet be seen for the sum of performance of all systems. It looks though as if for the last four years the slope of the trend is smoother. It is too early to say that this is a general trend. However, it remains to be seen what is going to happen. The most recent version of the list—as published in June 2015—indicates that we may see a smoother slope for the total performance too.

There is an optimistic scenario for the trend which claims that especially the slower systems have not yet adopted the accelerator technology that allows faster systems to still keep pace with Moore's law. Following this scenario, the market should catch up over the coming 2 years and the trend of the number 500 systems should go back to what it used to be. It remains to be seen though whether the owners of smaller system are able to exploit the potential of accelerators. For those among them that work in a research environment—like universities or research labs—this should not be too difficult. However, the many industrial users of low-end HPC systems may not see an incentive in investing into a technology for which there is not yet a defined standard and for which there are not too many industrial codes that can easily run on accelerator systems. Looking back into the history of parallel computing we find that industry did catch up on parallelism but with the growing number of processors, industrial usage was increasingly decoupled from research trends.

There is also a pessimistic scenario, which claims that accelerators are a workaround for the problems of stagnating processor performance. The pessimistic scenario assumes that we will start to see a changing trend line also for the number one systems in the years ahead. The pessimistic scenario would suggest that the rule of Moore's law is over.

An interim report of the "Committee on Future Directions for NSF Advanced Computing Infrastructure to Support U.S. Science in 2017–2020" released in November 2014 states: *It is an accepted truth today that Moore's Law will end sometime in the next decade, causing significant impact to high-end systems.* And the report continues: *The transition implied by the anticipated end of Moore's Law will be even more severe—absent development of disruptive technologies; it could mean, for the first time in over three decades, the stagnation of computer performance and the end of sustained reductions in the price–performance ratio.* [7]

If we believe in the end of Moore's law we need to face the consequences and prepare for the time after.

4 What to Expect?

First, and foremost, we would have to accept that the end of Moore's law has been reached. This does not come entirely as a surprise—it was on the contrary rather surprising to see technology follow such an impressive path for more than three decades. However, it is not a reason for pessimism but rather a reason to step up the research efforts in HPC. Three main consequences follow from what we have found so far.

4.1 We Need More Investment in Better Technology

Hardware development is going to address a number of new issues beyond performance. It is reasonable to expect to see processors that are not built for floating-point performance but rather for the growing needs of data analytics. Furthermore power consumption will become a growing issue for processor architecture design. Even more than now, hardware designers will put their focus on reducing power consumption thus providing the user with lower operational costs of systems. How much this in turn will trigger a further increase in number of cores or processors remains to be seen. We may see a moderate growth into the billions of cores after a while.

We may, for example, expect to see some sort of follow-on project of the IBM BlueGene. It would be interesting to see an architecture built from billions of low-power embedded processors. As much as this would be a challenge for programmers it could yield to interesting architectural concepts.

In any case investment will not stop at hardware. There is a growing need for better programming tools. Handling millions of cores is counterintuitive for the human being. All concepts that are able to reduce this complexity—like hybrid programming models that merge MPI and OpenMP—will be extremely useful for the user. However, such concepts are in their infancy and will require a lot of effort before they can be turned into standards and supported by all necessary tools.

4.2 Convergence and Segmentation

Thin core concepts will increase clock frequencies and will—for the sake of being useful for HPC—increase complexity of each core. Hence, they will grow fatter and hit the frequency barrier. Fat core concepts will reduce complexity in order to reduce power consumption and in order to be able to increase the number of cores. Therefore, it has to be expected that the thin core concept and the fat core concept will somehow converge. What we see already today in the market is a trend to merge accelerator technology onto standard processor parts. Sometimes these are called "fused parts." We also see technologies like the Aurora concept of the Japanese vendor NEC [8]. The basic concept of Aurora is to turn the traditional vector processor concept (a typical fat core concept) into an accelerator that could be used like existing accelerators.

Given the wide variety of options in the design space for future processor architectures we can expect to see a market evolving that is similar to other mature markets. We can expect to see cheap solutions with a reasonable performance for a low price. We can also expect to see tailored high-end solutions for which niches will have to be carved out to survive. In any case we currently see a growing number of different solutions that all follow similar lines of architectural concepts but with the exception of the x86 architecture there is currently no solution available that could claim to be a standard solution.

4.3 Software Beats Hardware

Regardless of the directions that hardware design takes, software will become more important. As of today, we expect to see single digit sustained performance figures for large-scale systems. The recently initiated HPCG benchmark initiative [9] reveals that even for a standardized benchmark—which should by now be highly optimized—sustained performance numbers are embarrassingly low. In the future, optimization of codes is going to be a major issue. Given that hardware architecture development will slow down software developers will be given more time to get the most out of a given hardware concept. With hardware stagnation, it also makes sense to rethink many of the old models that are used by HPC users for decades now. In the future software will make the difference between a standard system and a high performance computing system.

5 Summary

HPC is facing the end of the basis for its success story over the last three decades. With Moore's law ending, peak performance is no longer something that just so happens. This will have implications for users, vendors, and HPC centers. Centers will have to invest much more in quality of services and will have to work intensively with software developers to be able to provide high quality services. Hardware vendors will have to focus more on improved quality of hardware rather than on speed. They will have to explore and carve out niches in which HPC as a business can create a reasonable ecosystem. This will bring industrial users much more into the focus of activities than has been the case over the last decades. HPC users will have a much harder time improving their simulations. The focus will have to move from speed to quality. This is the time when software has to be improved. This is the time when models have to be improved. This is the time when algorithms have to be improved. This is going to be a great time for HPC experts—computer scientists, mathematicians, computational scientists, and engineers.

References

1. ISC 2015, Dresden, Germany. www.supercomp.de
2. Moore, G.E.: Cramming more components onto integrated circuits. Electronics **38**(8), 114–117 (1965)
3. www.top500.org
4. Resch, M.M.: Trends in architectures and methods for high performance computing simulation. In: Topping, B.H.V., Iványi, P. (eds.) Parallel Distributed and Grid Computing for Engineering, pp 37–48. Saxe-Coburg Publications, Stirlingshire, Scotland (2009)
5. http://www.itrs.net/
6. Resch, M., Citius Altius Fortius–The TOP500 and Why Speed Matters, ISC Conference 2013, Leipzig, Germany, June 17, 2013
7. Committee on Future Directions for NSF Advanced Computing Infrastructure to Support U.S. Science and Engineering in 2017-2020, Future Directions for NSF Advanced Computing Infrastructure to Support U.S. Science and Engineering in 2017–2020: Interim Report. 2014. http://www.nap.edu/catalog/18972/future-directions-for-nsf-advanced-computing-infrastructure-to-support-us-science-and-engineering-in-2017--2020
8. IDC, NEC's HPC Vision: Bringing Vector Supercomputing to a Broader Data-Intensive User Base. http://www.marketresearch.com/IDC-v2477/NEC-HPC-Vision-Vector-Supercomputing-8573697
9. http://hpcg-benchmark.org/

Direct and Inverse Theorems for Beta-Durrmeyer Operators

Naokant Deo and Neha Bhardwaj

Abstract In this paper, we consider linear combinations of Beta-Durrmeyer operators $L_n(f, x)$ and study the direct theorem in terms of higher order modulus of continuity in simultaneous approximation and inverse theorem for these operators in ordinary approximation.

Keywords Beta-Durrmeyer operators · Ordinary approximation · Simultaneous approximation · Steklov means

2010 AMS Subject Classification 41A25 · 41A30

1 Introduction and Definitions

Gupta and Ahmad [7] defined Beta operators as

$$B_n(f, x) = \frac{1}{n} \sum_{k=0}^{\infty} p_{n,k}(x) f\left(\frac{k}{n+1}\right), \qquad x \in [0, \infty), \tag{1.1}$$

where

$$p_{n,k}(x) = \frac{(n+k)!}{k!(n-1)!} \frac{x^k}{(1+x)^{n+k+1}}$$

N. Deo
Department of Applied Mathematics, Delhi Technological University
(Formerly Delhi College of Engineering), Bawana Road, Delhi 110042, India
e-mail: dr_naokant_deo@yahoo.com

N. Bhardwaj (✉)
Department of Mathematics, Amity Institute of Applied Sciences,
Amity University, Noida 201301, India
e-mail: neha_bhr@yahoo.co.in

V.K. Singh et al. (eds.), *Modern Mathematical Methods and High Performance Computing in Science and Technology*, Springer Proceedings in Mathematics & Statistics 171, DOI 10.1007/978-981-10-1454-3_15

and the Durrmeyer variant of these operators

$$L_n(f(t);x) = \frac{1}{n}\sum_{k=0}^{\infty} p_{n,k}(x)\int_0^{\infty} p_{n,k}(t)f(t)dt = \int_0^{\infty} W_n(t,x)f(t)dt, \tag{1.2}$$

has been studied by Deo [3].

Throughout the paper, $C_\alpha[0,\infty)$ will denote the space of all real-valued continuous functions on $[0,\infty)$ satisfying growth condition $|f(t)| \le Mt^\alpha,\ M>0,\ \alpha>0$ with the norm

$$\|f\|_\alpha = \sup_{0\le t<\infty} |f(t)|t^{-\alpha}.$$

To improve the saturation order $O(n^{-1})$ for the operators (1.2), we use the technique of linear combination as described by May [12] for a sequence of positive linear operators. We consider the linear combination of the operators (1.2) as described below

The linear combinations $L_n(f,(d_0,d_1,d_2,\ldots,d_k),x)$ of $L_{d_jn}(f,x),\ j=0,1,2,\ldots,k$ are defined by

$$L_n(f,(d_0,d_1,d_2,\ldots,d_k),x) = \sum_{j=0}^{k} C(j,k)L_{d_jn}(f,x),$$

where $d_0,d_1,d_2,\ldots,d_k$ are arbitrary but fixed distinct positive integers and

$$C(j,k) = \prod_{\substack{i=0\\ i\ne j}}^{k} \frac{d_j}{d_j-d_i}\ \textit{for } k\ne 0\ \&\ C(0,0)=1.$$

Definition 1.1 The mth order modulus of continuity $\omega_m(f,\eta,a,b)$ for a function f continuous on the interval $[a,b]$ is defined by

$$\omega_m(f,\eta,a,b) = \sup\left\{\left|\Delta_h^m f(x)\right| : |h|\le\eta;\ x,x+mh\in[a,b]\right\}.$$

For $m=1,\ \omega_m(f,\eta)$ is written simply the ordinary modulus of continuity $\omega_f(\eta)$ or $\omega(f,\eta)$.

The function f is said to belong to the generalized Zygmund class Liz (α,m,a,b) if for $\eta>0$ there exists a constant M such that

$$\omega_{2m}(f,\eta,a,b) \le M\eta^{\alpha m},$$

where $\omega_{2m}(f,\eta,a,b)$ denotes the modulus of continuity of $2m$th order of $f(x)$ on the interval $[a,b]$. For the class Liz $(\alpha,1,a,b)$ is more commonly denoted by Lip*(α,a,b).

Definition 1.2 Let us assume that $0 < a < a_1 < b_1 < b < \infty$ and $f \in C_\alpha[0,\infty)$, then for $m \in N$ the Steklov mean $f_{\eta,m}$ of mth order corresponding to f, for sufficiently small values of $\eta > 0$ is defined by

$$f_{\eta,m}(x) = \eta^{-m}\left(\int_{-\eta/2}^{\eta/2}\right)^m \left\{ f(x) + (-1)^{m-1}\Delta^m_{\sum_{i=1}^m x_i} f(x)\right\} \prod_{i=1}^m dx_i, \qquad (1.3)$$

where $x \in [a,b]$ and $\Delta^m_\eta f(x)$ is the mth order forward difference with step length η. It is easily checked (see e. g., [5], [9]) that
(*i*) $f_{\eta,m} \in C[a,b]$;
(*ii*) $\|f^{(r)}_{\eta,m}\|_{C[a_1,b_1]} \leq M_1\eta^{-r}\omega_r(f,\eta,a,b),\ r = 1,2,\ldots,m$;
(*iii*) $\|f - f_{\eta,m}\|_{C[a_1,b_1]} \leq M_2\omega_m(f,\eta,a,b)$;
(*iv*) $\|f_{\eta,m}\|_{C[a_1,b_1]} \leq M_3\|f\|_{C[a,b]} \leq M_4\|f\|_{C_\alpha}$,
where $M_i, i = 1,2,3,4$ are certain constants independent of f and η.

In this paper, we obtain direct theorem in terms of higher order modulus of continuity in simultaneous approximation with the help of properties of Steklov means and in the last section of this paper, we give inverse theorem for these linear combination of the operators L_n in ordinary approximation.

2 Preliminary Results

In order to prove the Theorem, we shall require the following results:

Lemma 2.1 ([4]) *Let $m \in N^0$ (the set of nonnegative integers) and the mth moment for the operators (1.1) be defined by*

$$U_{n,m}(x) = \sum_{k=0}^{\infty}\left(\frac{k}{n+1} - x\right)^m p_{n,k}(x).$$

Then

$$(n+1)U_{n,m+1}(x) = x(1+x)\left[U'_{n,m}(x) + mU_{n,m-1}(x)\right], \quad (x \geq 0, m \geq 1).$$

Consequently

(i) $U_{n,m}(x)$ *is a polynomial in x of degree* $\leq m$;
(ii) $U_{n,m}(x) = O\left(n^{-[(m+1)/2]}\right)$ *where* $[\beta]$ *denotes the integer part of* β.

Lemma 2.2 ([3]) *Let $m \in N^0$, we define the function $T_{n,m}(x)$ as*

$$T_{n,m}(x) = \frac{1}{(n+r)}\sum_{k=0}^{\infty} p_{n+r,k}(x)\int_0^\infty p_{n-r,k+r}(t)(t-x)^m dt.$$

Then $T_{n,0}(x) = 1, \quad T_{n,1}(x) = \frac{(1+2x)(1+r)}{n-r-1}$ *and*

$$(n-m-r-1)T_{n,m+1}(x) = x(1+x)\big[T'_{n,m}(x) + 2mT_{n,m-1}(x) + (1+2x)$$
$$(r+m+1)T_{n,m}(x),\big], \quad (n > m+r+1).$$

Further, for all $x \in [0,\infty)$

$$T_{n,m}(x) = O\left(n^{-[(m+1)/2]}\right).$$

Lemma 2.3 ([3]) *If* f *is* $r-1$ *times* $(r = 1, 2, 3, \ldots)$ *differentiable on* $[0,\infty)$ *such that* $f^{(r-1)}$ *is absolutely continuous with* $f^{(r-1)}(t) = O(t^{\alpha})$ *for some* $\alpha > 0$ *as* $t \to \infty$ *and* $n > \alpha + r$*, then we have*

$$L_n^{(r)}(f,x) = \frac{(n-r-1)!(n+r-1)!}{n!(n-1)!}\sum_{k=0}^{\infty} p_{n+r,k}(x)\int_0^{\infty} p_{n-r,k+r}(t)f^{(r)}(t)dt. \tag{2.1}$$

Lemma 2.4 ([11]) *There exist polynomials* $q_{i,j,r}(x)$ *independent of* n *and* k *such that*

$$\{x(1+x)\}^r\frac{d^r}{dx^r}[p_{n,k}(x)] = \sum_{\substack{2i+j\le r\\ i,j\ge 0}}(n+1)^i\left|k-(n+1)x\right|^j q_{i,j,r}(x)p_{n,k}(x).$$

Lemma 2.5 *If* f *is a function in* $C_{\alpha}[0,\infty)$*, such that* $f^{(2k+r+2)}$ *exists at a point* $x \in (0,\infty)$*, then*

$$\lim_{n\to\infty} n^{k+1}\left\{L_n^{(r)}(f,(d_0,d_1,d_2,\ldots,d_k),x) - f^{(r)}(x)\right\} = \sum_{i=r}^{2k+r+2} Q(i,k,r,x)f^{(i)}(x),$$

where $Q(i,k,r,x)$ *are certain polynomial in* x *of degree* i*.*

The proof of Lemma 2.5 follows along the lines of [8].

Lemma 2.6 *Let* δ *and* γ *be any two positive numbers and* $[a,b] \subset [0,\infty)$*. Then, for any* $m > 0$ *there exists a constant* M_m *such that*

$$\left\|\int_{\left|t-x\right|\ge\delta} W_n(t,x)t^{\gamma}dt\right\|_{C[a,b]} \le M_m n^{-m}$$

The proof of this result follows easily by using Schwarz inequality and Lemma 2.7 from [1].

3 Direct Theorem

In this section, we study direct result in terms of higher order modulus of continuity in simultaneous approximation for the operators (1.2).

Theorem 3.1 *Let* $f^{(r)} \in C_\alpha[0,\infty)$ *and* $0 < a < a_1 < b_1 < b < \infty$. *Then for all* n *sufficiently large, we have*

$$\begin{aligned}&\left\|L_n^{(r)}(f,(d_0,d_1,d_2,\ldots,d_k),\bullet)-f^{(r)}\right\|_{C[a_1,b_1]}\\&\quad\le Max\left\{C_1\omega_{2k+2}(f^{(r)};n^{-1/2},a,b)+C_2n^{-(k+1)}\|f\|_\alpha\right\},\end{aligned}$$

where $C_1 = C_1(k,r)$ *and* $C_2 = C_2(k,r,f)$.

Proof Using linearity property

$$\begin{aligned}&\left\|L_n^{(r)}(f,(d_0,d_1,d_2,\ldots,d_k),\bullet)-f^{(r)}\right\|_{C[a_1,b_1]}\\&\quad\le\left\|L_n^{(r)}\left((f-f_{2k+2,\eta}),(d_0,d_1,d_2,\ldots,d_k),\bullet\right)\right\|_{C[a_1,b_1]}\\&\quad\quad+\left\|L_n^{(r)}\left(f_{2k+2,\eta},(d_0,d_1,d_2,\ldots,d_k),\bullet\right)-f_{2k+2,\eta}^{(r)}\right\|_{C[a_1,b_1]}\\&\quad\quad+\left\|f^{(r)}-f_{2k+2,\eta}^{(r)}\right\|_{C[a_1,b_1]}\\&\quad:=E_1+E_2+E_3.\end{aligned}$$

Since, $f_{2k+2,\eta}^{(r)}(t) = \left(f^{(r)}\right)_{2k+2,\eta}(t)$, by property (iii) of Steklov mean, we obtain

$$E_3 \le C_1\omega_{2k+2}(f^{(r)},\eta,a,b).$$

By Lemma 2.5, we get

$$E_2 \le C_2 n^{-(k+1)}\sum_{j=r}^{2k+r+2}\left\|f_{2k+2,\eta}^{(j)}\right\|_{C[a,b]}.$$

Using the interpolation property due to Goldberg and Meir [6] for each $j = r, r+1,\ldots,2k+r+2$, we get

$$\left\|f_{2k+2,\eta}^{(r)}\right\|_{C[a,b]} \le C_3\left\{\left\|f_{2k+2,\eta}\right\|_{C[a,b]}+\left\|f_{2k+2,\eta}^{(2k+r+2)}\right\|_{C[a,b]}\right\}.$$

Now using properties (ii) and (iv) of Steklov mean, we obtain

$$E_2 \le C_4 n^{-(k+1)}\left\{\|f\|_\alpha+\eta^{-(2k+2)}\omega_{2k+2}(f^{(r)},\eta)\right\}$$

To estimate E_1, choosing $a^{'}, b^{'}$ such that

$$0 < a < a^{'} < a_1 < b_1 < b^{'} < b < \infty.$$

Also let $\psi(t)$ be the characteristic function of the interval $[a^{'}, b^{'}]$, then

$$\begin{aligned} E_1 &\leq \left\| L_n^{(r)} \left(\psi(t) \left(f(t) - f_{2k+2,\eta}(t)\right) (d_0, d_1, d_2, \ldots, d_k), \bullet\right) \right\|_{C[a_1,b_1]} \\ &+ \left\| L_n^{(r)} \left((1-\psi(t)) \left(f(t) - f_{2k+2,\eta}(t)\right) (d_0, d_1, d_2, \ldots, d_k), \bullet\right) \right\|_{C[a_1,b_1]} \\ &:= E_4 + E_5. \end{aligned}$$

We note that in order to estimate E_4 and E_5, it is sufficient to consider their expressions without the linear combination. It is clear that by Lemma 2.3, we obtain

$$\begin{aligned} &L_n^{(r)} \left(\psi(t) \left(f(t) - f_{2k+2,\eta}(t)\right), x\right) \\ &= \frac{(n-r-1)!(n+r-1)!}{n!(n-1)!} \sum_{k=0}^{\infty} p_{n+r,k}(x) \int_0^{\infty} p_{n-r,k+r}(t)\psi(t) \left(f^{(r)}(t) - f_{2k+2,\eta}^{(r)}(t)\right) dt. \end{aligned}$$

Hence

$$\left\| L_n^{(r)} \left(\psi(t) \left(f(t) - f_{2k+2,\eta}(t)\right), \bullet\right) \right\|_{C[a,b]} \leq C_5 \left\| f^{(r)} - f_{2k+2,\eta}^{(r)} \right\|_{C[a^{'},b^{'}]}.$$

Now for $x \in [a_1, b_1]$ and $t \in [0,\infty) / \left[a^{'}, b^{'}\right]$ we can choose an η_1 satisfying $\left|t - x\right| \geq \eta_1$. Therefore by Lemma 2.4 and Schwarz inequality, we obtain

$$\begin{aligned} I &\equiv \left| L_n^{(r)} \left((1-\psi(t)) \left(f(t) - f_{2k+2,\eta}(t)\right), x\right) \right| \\ &\leq \frac{1}{n} \sum_{\substack{2i+j\leq r \\ i,j\geq 0}} n^i \frac{\left|q_{i,j,r}(x)\right|}{x^r} \sum_{k=0}^{\infty} p_{n,k}(x) \left|k - (n+1)x\right|^j \int_0^{\infty} p_{n,k}(t)\,(1-\psi(t)) \left|f(t) - f_{2k+2,\eta}(t)\right| dt \\ &\leq C_6 \left\|f\right\|_{\alpha} \sum_{\substack{2i+j\leq r \\ i,j\geq 0}} n^{i-1} \sum_{k=0}^{\infty} p_{n,k}(x) \left|k - (n+1)x\right|^j \int_{\left|t-x\right|\geq\eta_1} p_{n,k}(t) dt \\ &\leq C_6 \left\|f\right\|_{\alpha} \eta_1^{-2s} \sum_{\substack{2i+j\leq r \\ i,j\geq 0}} n^{i-1} \sum_{k=0}^{\infty} p_{n,k}(x) \left|k - (n+1)x\right|^j \left(\int_0^{\infty} p_{n,k}(t) dt\right)^{1/2} \left(\int_0^{\infty} p_{n,k}(t)(t-x)^{4s} dt\right)^{1/2} \\ &\leq C_6 \left\|f\right\|_{\alpha} \eta_1^{-2s} \sum_{\substack{2i+j\leq r \\ i,j\geq 0}} n^{i} \left(\frac{1}{n} \sum_{k=0}^{\infty} p_{n,k}(x) (k - (n+1)x)^{2j}\right)^{1/2} \left(\frac{1}{n} \sum_{k=0}^{\infty} p_{n,k}(x) \int_0^{\infty} p_{n,k}(t)(t-x)^{4s} dt\right)^{1/2}. \end{aligned}$$

Hence by Lemma 2.1 and Lemma 2.2, we have

$$I \leq C_7 \left\|f\right\|_{\alpha} \sum n^{(i+\frac{j}{2}-s)} \leq C_7 n^{-q} \left\|f\right\|_{\alpha},$$

where $q = (s - r/2)$. Now for $s > 0$ and $q \geq k+1$, then $I \leq C_7 n^{-(k+1)} \|f\|_\alpha$. So by property (iii) of Steklov mean, we have

$$E_1 \leq C_8 \left\| f^{(r)} - f^{(r)}_{2k+2,\eta} \right\|_{C[a',b']} + C_7 n^{-(k+1)} \|f\|_\alpha$$
$$\leq C_9 \omega_{2k+2}(f^{(r)}, \eta, a, b) + C_7 n^{-(k+1)} \|f\|_\alpha.$$

Hence with $\eta = n^{-1/2}$, the theorem follows. □

Definition 3.2 The function f is said to belong to the generalized Zygmund class $Lip(\alpha, k, a, b)$ if there exists a constant M such that

$$\omega_{2k}(f, \delta) \leq M\delta^{\alpha k}, \delta > 0$$

where $\omega_{2k}(f, \delta)$ denotes the modulus of continuity of $2k$th order on the interval $[a, b]$. The class $Lip(\alpha, k, a, b)$ is more commonly denoted by $Lip^*(\alpha, a, b)$

4 Inverse Theorem

In this section we shall prove the following inverse result.

Theorem 4.1 *If* $0 < \alpha < 2$, $0 < a_1 < a_2 < b_2 < b_1 < \infty$ *and suppose* $f \in C_\alpha[0, \infty)$, *then in the following statements are equivalent.*

(i) $\left\| L_n(f, (d_0, d_1, d_2, \ldots, d_k), \bullet) - f \right\|_{C[a_1,b_1]} = O\left(n^{-\alpha(k+1)/2}\right)$, *where* $f \in C_\alpha[a, b]$,

(ii) $f \in Lip(\alpha, k+1, a_2, b_2)$.

Proof Let us choose points a', a'', b', b'' in such a way that $a_1 < a' < a'' < a_2 < b_2 < b'' < b' < b_1$. Also suppose $g \in C_0^\infty$ with supp $g \subset (a'', b'')$ and $g(x) = 1$ on the interval $x \in [a_2, b_2]$. To prove the assertion, it is sufficient to show that

$$\left\| L_n(fg, (d_0, d_1, d_2, \ldots, d_k), \bullet) - (fg) \right\|_{C[a',b']} = O\left(n^{-\alpha(k+1)/2}\right) \Rightarrow (ii). \quad (4.1)$$

Using F in place of fg for all the values of $h > 0$, we get

$$\left\| \triangle_h^{2k+2} F \right\|_{C[a'',b'']} \leq \left\| \triangle_h^{2k+2} (F - L_n(F, (d_0, d_1, d_2, \ldots, d_k), \bullet)) \right\|_{C[a'',b'']}$$
$$+ \left\| \triangle_h^{2k+2} L_n(F, (d_0, d_1, d_2, \ldots, d_k), \bullet) \right\|_{C[a'',b'']} \quad (4.2)$$

Therefore, by definition of $\triangle_h^{2k+2}$,

$$\begin{aligned}
&\left\|\triangle_h^{2k+2} L_n\left(F,(d_0,d_1,d_2,\ldots,d_k),\bullet\right)\right\|_{C[a'',b'']}\\
&=\left\|\int_0^h\ldots\int_0^h L_n\Big(F,(d_0,d_1,d_2,\ldots,d_k),\bullet+\sum_{i=1}^{2k+2}x_i\Big)dx_1\ldots dx_{2k+2}\right\|_{C[a'',b'']}\\
&\le h^{2k+2}\left\|L_n^{(2k+2)}\left(F,(d_0,d_1,d_2,\ldots,d_k),\bullet\right)\right\|_{C[a'',b''+(2k+2)h]}\\
&\le h^{2k+2}\Big\{\left\|L_n^{(2k+2)}\left(F-F_{\eta,2k+2},(d_0,d_1,d_2,\ldots,d_k),\bullet\right)\right\|_{C[a'',b''+(2k+2)h]}\\
&\quad+\left\|L_n^{(2k+2)}\left(F_{\eta,2k+2},(d_0,d_1,d_2,\ldots,d_k),\bullet\right)\right\|_{C[a'',b''+(2k+2)h]}\Big\},
\end{aligned} \tag{4.3}$$

where $F_{\eta,2k+2}$ is the Steklov mean of $(2k+2)$th order corresponding to F. By Lemma 3 from [1], we get

$$\begin{aligned}
&\int_0^\infty\left|\frac{\partial^{2k+2}}{\partial x^{2k+2}}W_n(t,x)dt\right|\\
&\le\sum_{\substack{2i+j\le 2k+2\\ i,j\ge 0}}\frac{1}{n}\sum_{k=0}^{\infty}(n+1)^i\left|k-(n+1)x\right|^j\frac{\left|q_{i,j,2k+2}(x)\right|}{\{x(1+x)\}^{2k+2}}p_{n,k}(x)\int_0^\infty p_{n,k}(t)dt.
\end{aligned}$$

Since $\int_0^\infty p_{n,k}(t)dt=1$. By Lemma 2.1, we have

$$\sum_{k=0}^{\infty}p_{n,k}(x)\big(k-(n+1)x\big)^{2j}=(n+1)^{2j}\sum_{k=0}^{\infty}p_{n,k}(x)\left(\frac{k}{n+1}-x\right)^{2j}=O(n^j). \tag{4.4}$$

Using Schwarz inequality, we obtain

$$\left\|L_n^{(2k+2)}\left(F-F_{\eta,2k+2},(d_0,d_1,d_2,\ldots,d_k),\bullet\right)\right\|_{C[a'',b''+(2k+2)h]}\le K_1 n^{k+1}\left\|F-F_{\eta,2k+2}\right\|_{C[a'',b'']}. \tag{4.5}$$

By Lemma 2 from [1], we get

$$\int_0^\infty\left[\frac{\partial^k}{\partial x^k}W_n(t,x)\right](t-x)^i dt=0,\ for\ k>i. \tag{4.6}$$

By Taylor's expansion, we obtain

$$F_{\eta,2k+2}(t)=\sum_{i=0}^{2k+1}\frac{F_{\eta,2k+2}^{(i)}(x)}{i!}(t-x)^i+F_{\eta,2k+2}^{(2k+2)}(\xi)\frac{(t-x)^{2k+2}}{(2k+2)!}, \tag{4.7}$$

where ξ lies between t and x. By (4.6) and (4.7), we get

$$\left\| \frac{\partial^{2k+2}}{\partial x^{2k+2}} L_n \left(F_{\eta,2k+2}, (d_0, d_1, d_2, ..., d_k), \bullet\right) \right\|_{C[a'',b''+(2k+2)h]}$$
$$\leq \sum_{j=0}^{k} \frac{|C(j,k)|}{(2k+2)!} \left\| F_{\eta,2k+2}^{(2k+2)} \right\|_{C[a'',b'']} \left\| \int_0^\infty \left[\frac{\partial^{2k+2}}{\partial x^{2k+2}} W_{d_j n}(t,x) \right] (t-x)^{2k+2} dt \right\|_{C[a'',b'']}.$$

Again applying Schwarz inequality for integration and summation and Lemma 3 from [1], we obtain

$$I \equiv \int_0^\infty \left| \frac{\partial^{2k+2}}{\partial x^{2k+2}} W_n(t,x) \right| (t,x)^{2k+2} dt$$
$$\leq \frac{1}{n} \sum_{\substack{2i+j\leq 2k+2 \\ i,j\geq 0}} \sum_{k=0}^{\infty} (n+1)^i p_{n,k}(x) |k-(n+1)x|^j \frac{|q_{i,j,2k+2}(x)|}{\{x(1+x)\}^{2k+2}} \int_0^\infty p_{n,k}(t)(t-x)^{2k+2} dt$$
$$\leq \sum_{\substack{2i+j\leq 2k+2 \\ i,j\geq 0}} (n+1)^i \frac{|q_{i,j,2k+2}(x)|}{\{x(1+x)\}^{2k+2}} \left\{ \sum_{k=0}^{\infty} p_{n,k}(x)(k-(n+1)x)^{2j} \right\}^{1/2}$$
$$\times \left\{ \frac{1}{n} \sum_{k=0}^{\infty} p_{n,k}(x) \int_0^\infty p_{n,k}(t)(t-x)^{4k+4} dt \right\}^{1/2}. \tag{4.8}$$

Using Lemma 2 from [1]

$$\frac{1}{n} \sum_{k=0}^{\infty} p_{n,k}(x) \int_0^\infty p_{n,k}(t)(t-x)^{4k+4} dt = T_{n,4k+4}(x) = O\left(n^{-(2k+2)}\right). \tag{4.9}$$

Using (4.4) and (4.9) in (4.8), we obtain

$$I \leq \sum_{\substack{2i+j\leq 2k+2 \\ i,j\geq 0}} (n+1)^i \frac{|q_{i,j,2k+2}(x)|}{\{x(1+x)\}^{k+1}} O(n^{j/2}) O\left(n^{-(k+1)}\right) = O(1).$$

Hence

$$\left\| W_n^{(2k+2)} \left(F_{\eta,2k+2}, (d_0, d_1, d_2, ..., d_k), \bullet\right) \right\|_{C[a'',b''+(2k+2)h]} \leq K_2 \left\| F_{\eta,2k+2}^{(2k+2)} \right\|_{C[a'',b'']}. \tag{4.10}$$

On combining (4.2), (4.3), (4.5) and (4.10) it follows

$$\left\| \Delta_h^{2k+2} F \right\|_{C[a'',b'']} \leq \left\| \Delta_h^{2k+2} \left(F - L_n(F, (d_0, d_1, d_2, ..., d_k), \bullet)\right) \right\|_{C[a'',b'']}$$
$$+ K_3 h^{2k+2} \Big(n^{k+1} \left\| F - F_{\eta,2k+2} \right\|_{C[a'',b'']} + \left\| F_{\eta,(2k+2)}^{(2k+2)} \right\|_{C[a'',b'']} \Big).$$

For $h > 0$, the above relation holds, it follows from the properties of $F_{\eta,2k+2}$ and (4.1) that

$$\omega_{2k+2}(F, l, [a'', b'']) \leq K_4 \left\{ n^{-\alpha(k+1)/2} + l^{2k+2} \left(n^{k+1} + \eta^{-2k+2} \right)_{\omega_{2k+2}} (F, \eta, [a'', b'']) \right\}.$$

Choosing η such that $n < \eta^{-2} < 2h$ and following Berens and Lorentz [2], we obtain

$$w_{2k+2}(F, l, [a'', b'']) = O(l^{\alpha(k+1)}). \tag{4.11}$$

Since $F(x) = f(x)$ in $[a_2, b_2]$, from (4.11) we have

$$w_{2k+2}(f, l, [a_2, b_2]) = O(l^{\alpha(k+1)}), \quad i.e., \ f \in Liz(\alpha, k+1, a_2, b_2).$$

Let us assume (i). Putting $\tau = \alpha(k+1)$, we first consider the case $0 < \tau \leq 1$. For $x \in [a', b']$, we get

$$\begin{aligned} L_n(fg, (d_0, d_1, d_2, \ldots, d_k), x) - f(x)g(x) &= g(x)L_n((f(t) - f(x)), (d_0, d_1, d_2, \ldots, d_k), x) \\ &+ \sum_{j=0}^{k} C(j,k) \int_{a_1}^{b_1} W_{d_j,n}(t,x) f(x)(g(t) - g(x))\, dt \\ &+ O\left(n^{-k+1)}\right) = I_1 + I_2 + O\left(n^{-(k+1)}\right), \end{aligned} \tag{4.12}$$

where the O-term holds uniformly for $x \in [a', b']$. Since by assumption

$$\left\| L_n(f, (d_0, d_1, d_2, \ldots, d_k), \bullet) - f \right\|_{C[a_1, b_1]} = O\left(n^{-\tau/2}\right),$$

we have

$$\left\| I_1 \right\|_{C[a',b']} \leq \left\| g \right\|_{C[a',b']} \left\| L_n(f, (d_0, d_1, d_2, \ldots, d_k), \bullet) - f \right\|_{C[a',b']} \leq K_5 n^{-\tau/2}. \tag{4.13}$$

By mean value theorem, we get

$$I_2 = \sum_{j=0}^{k} C(j,k) \int_{a_1}^{b_1} W_{d_j,n}(t,x) f(t) \left\{ g'(\xi)(t-x) \right\} dt.$$

Again applying Cauchy–Schwarz inequality and Lemma 2 from [1], we get

$$\begin{aligned} \left\| I_2 \right\|_{C[a',b']} &\leq \left\| f \right\|_{C[a_1,b_1]} \left\| g' \right\|_{C[a',b']} \left(\sum_{j=0}^{k} |C(j,k)| \right) \max_{0 \leq j \leq k} \left\| \int_0^\infty W_{d_j,n}(t,x)(t-x)^2 dt \right\|_{C[a',b']}^{1/2} \\ &= O\left(n^{-\tau/2}\right). \end{aligned} \tag{4.14}$$

Combining (4.12)–(4.14), we obtain

$$\left\|L_n(fg,(d_0,d_1,d_2,\ldots,d_k),\bullet)-fg\right\|_{C[a',b']}=O\left(n^{-\tau/2}\right),\ for\ 0<\tau\leq 1.$$

Now to prove the implication for $0<\tau<2k+2$, it is sufficient to assume it for $\tau\in(m-1,m)$ and prove it for $\tau\in(m,m+1)$, $(m=1,2,3,\ldots,2k+1)$. Since the result holds for $\tau\in(m-1,m)$, we choose two points x_1,y_1 in such a way that $a_1<x_1<a'<b'<y_1<b_1$. Then in view of assumption $(i)\Rightarrow(ii)$ for the interval $(m-1,m)$ and equivalence of (ii) it follows that $f^{(m-1)}$ exists and belongs to the class $Lip(1-\delta,x_1,y_1)$ for any $0<\delta<1$. Let $g\in C_0^\infty$ be such that $g(x)=1$ on $[a'',b'']$ and $supp\ g\subset[a'',b'']$. Then with $\chi(t)$ denoting the characteristic function of the interval $[x_1,y_1]$, we have

$$\begin{aligned}&\left\|L_n(f,g,(d_0,d_1,d_2,\ldots,d_k),\bullet)-f,g\right\|_{C[a',b']}\\&\leq\left\|L_n(g(x)f(t)-f(x)),(d_0,d_1,d_2,\ldots,d_k),\bullet)\right\|_{C[a',b']}\\&\quad+\left\|L_n(f(t)(g(t)-g(x))\chi(t),(d_0,d_1,d_2,\ldots,d_k),\bullet)\right\|_{C[a',b']}+O\left(n^{-(k+1)}\right).\end{aligned}\tag{4.15}$$

Now

$$\begin{aligned}&\left\|L_n(g(x)(f(t)-f(x)),(d_0,d_1,d_2,\ldots,d_k),\bullet)\right\|_{C[a',b']}\\&\leq\left\|g\right\|_{C[a'',b'']}\left\|L_n(f,(d_0,d_1,d_2,\ldots,d_k),\bullet)-f\right\|_{C[a_1,b_1]}=O\left(n^{-\tau/2}\right).\end{aligned}\tag{4.16}$$

Applying Taylor's expansion of f, we have

$$\begin{aligned}I_3&\equiv\left\|L_n(f(t),g(t)-g(x))\chi(t),(d_0,d_1,d_2,\ldots,d_k),\bullet)\right\|_{C[a',b']}\\&=\left\|L_n\left(\left[\sum_{i=0}^{m-1}\frac{f^{(i)}(x)}{i!}(t-x)^i+\frac{\left\{f^{(m-1)}(\xi)-f^{(m-1)}(x)\right\}}{(m-1)!}\right]\right.\right.\\&\qquad\left.\left.\times(g(t)-g(x))\chi(t),(d_0,d_1,d_2,\ldots,d_k),\bullet\right)\right\|_{C[a',b']},\end{aligned}$$

where ξ lies between t and x. Since $f^{(m-1)}\in Lip(1-\delta,x_1,y_1)$,

$$\left|f^{(m-1)}(\xi)-f^{(m-1)}(x)\right|\leq K_6\left|\xi-x\right|^{1-\delta}\leq K_6\left|t-x\right|^{1-\delta},$$

where K_6 is the $Lip(1-\delta,x_1,y_1)$ constant for $f^{(m-1)}$, we have

$$
\begin{aligned}
I_3 \leq & \left\| L_n\left(\sum_{i=0}^{m-1} \frac{f^{(i)}(x)}{i!}(t-x)^i (g(t)-g(x))\chi(t), (d_0, d_1, d_2, \ldots, d_k), \bullet\right)\right\|_{C[a',b']} \\
& + \frac{K_6}{(m-1)!}\left\|g'\right\|_{C[a'',b'']}\left(\sum_{j=0}^{k}\left|C(j,k)\right|\right)\left\|L_{d_j,n}(\left|t-x\right|^{m+1-\delta}\chi(t), \bullet)\right\|_{C[a',b']} \\
= & \, I_4 + I_5 \text{ (say).}
\end{aligned} \tag{4.17}
$$

By Taylor's expansion of g and Lemma 2.5, we have

$$
I_4 = O\left(n^{-(k+1)}\right). \tag{4.18}
$$

Also, by Hölder's expansion of g and Lemma 2 from [1], we have

$$
\begin{aligned}
I_5 \leq & \frac{K_6}{(m-1)!}\left\|g'\right\|_{C[a'',b'']}\left(\sum_{j=0}^{k}\left|C(j,k)\right|\right) \\
& \max_{0\leq j\leq k}\left\|\int_{x_1}^{y_1} W_{d_j,n}(t-x)\left|t-x\right|^{m+1-\delta}dt\right\|_{C[a',b']} \\
\leq & K_7 \max_{0\leq j\leq k}\left\|\int_{x_1}^{y_1} W_{d_j,n}(t-x)(t-x)^{2(m+1)}dt\right\|_{C[a',b']}^{\frac{(m+1-\delta)}{2(m+1)}} \\
= & \, O\left(n^{-(m+1-\delta)/2}\right) = O\left(n^{\tau/2}\right),
\end{aligned} \tag{4.19}
$$

by choosing such that $0 < \delta < m+1-\delta$. Combining the estimate (4.15)–(4.19), we get

$$
\left\|L_n(fg, (d_0, d_1, d_2, \ldots, d_k), \bullet) - fg\right\|_{C[a',b']} = O\left(n^{\tau/2}\right).
$$

This completes the proof of the Theorem 4.1. □

References

1. Agrawal, P.N., Thamer, K.J.: Approximation of unbounded functions by a new sequence of linear positive operators, J. Math. Anal. Appl. **225**, 660-672 (1998)
2. Berens, H., Lorentz, G.G.: Inverse theorems for Bernstein polynomials. Indiana Univ. Math. J. **21**, 693–708 (1972)
3. Deo, N.: Direct result on the Durrmeyer variant of Beta operators. Southeast Asian Bull. Math. **32**, 283–290 (2008)
4. Deo, N.: A note on equivalence theorem for Beta operators. Mediterr. J. Math. **4**(2), 245–250 (2007)
5. Freud, G., Popov V.: On approximation by Spline functions. In: Proceedings of the Conference on Constructive Theory Functions, pp. 163-172, Budapest (1969)
6. Goldberg, S., Meir, V.: Minimum moduli of differentiable operators. Proc. London Math. Soc. **23**, 1–15 (1971)

7. Gupta, V., Ahmad, A.: Simultaneous approximation by modified Beta operators. Instanbul Uni. Fen. Fak. Mat. Der. **54**, 11–22 (1995)
8. Gupta, V., Gupta, P.: Rate of convergence by Szász-Mirakyan Baskakov type operators, Univ. Fen. Fak. Mat. **57–58**(1998–1999), 71–78
9. Hewitt, E., Stromberg, K.: Real and Abstract Analysis. McGraw-Hill, New York (1956)
10. Kasana H. S., Agrawal, P.N., Gupta, V.: Inverse and saturation theorems for linear combination of modified Baskakov operators, Approx. Theory Appl. **7**(2), 65–82 (1991)
11. Lorentz, G.G.: Bernstein Polynomials. University of Toronto Press, Toronto (1953)
12. May, C.P.: Saturation and inverse theorems for combinations of a class of exponential type operators. Canad. J. Math. **28**, 1224–1250 (1976)
13. Timan, A. F.: Theory of Approximation of Functions of Real Variable (English Translation), Pergamon Press Long Island City, New York (1963)

Big Data Gets Cloudy: Challenges and Opportunities

Pramila Joshi

Abstract Cloud computing and big data are complementary, forming a dialectical relationship. Cloud computing and the widespread use of internet application is the ultimate need of the hour. Though seen as full of promising opportunities, both the fields have their own challenges. Cloud computing is a trend in technology development, while big data is an inevitable phenomenon of the rapid development of a modern information society. Modern means like Cloud computing technologies are needed to solve big data problems. With the advent of new technologies in the field of data and computing, innumerable services are emerging on the net, generating huge volume of data. The data so generated is becoming too large and complex to be effectively processed by conventional means. How to store, manage, and create values from this huge ocean of big data has become an important research problem in today's time. Presently, users are accessing multiple data storage platforms to accomplish their operational and analytical requirements. Efficient integration of different data sources, in the merger of the two technologies, i.e., Big Data and Cloud, poses considerable challenges. Data integration here plays a very important role for both commercial and scientific domains in order to combine data from different sources and provides users with a unified view of these data. Keeping in mind the 4 V's of Big Data (volume, velocity, variety, and veracity), studying the challenges and opportunities coming in the way of efficient data integration is a key research direction for scientists. This paper will describe • How cloud and big data technologies are converging to offer a cost-effective delivery model for cloud-based big data analytics. • Big Data Challenges. • Challenges in cloud computing. • Challenges when big data moves to cloud.

Keywords Big data · Cloud · Hadoop · HDFS · MapReduce · CSP (Cloud Service Providers) · SLA (Service Level Agreements)

P. Joshi (✉)
Computer Science Department, Birla Institute of Technology Extension Centre
Noida, Noida, India
e-mail: pramila@bitmesra.ac.in

V.K. Singh et al. (eds.), *Modern Mathematical Methods and High Performance Computing in Science and Technology*, Springer Proceedings in Mathematics & Statistics 171, DOI 10.1007/978-981-10-1454-3_16

Fig. 1 Big data on cloud

1 Introduction

Two IT initiatives are currently top of mind for organizations across the globe: big data analytics and cloud computing. Whereas big data analytics offers the promise of providing valuable insights that can create competitive advantage, revolutionize the trends and more turnover by organizations, Cloud computing has the potential to enhance business agility and productivity at reduced cost while enabling greater efficiencies. Both technologies are making rapid progress and a large number of organizations are developing efficient and agile cloud solutions with cloud providers expanding their service offerings. On the other hand, IT organizations are looking up to cloud computing as the solution to support their big data projects [1] (Fig. 1).

2 Challenges in Big Data Analysis

There being no single universally accepted definition, Big Data has been defined differently by different people;

Big Data refers to datasets whose size is beyond the capability of typical database software tools to capture, store, manage, and analyze.—McKinsey.

Big Data is high volume, high velocity, and/or high variety information assets that require new forms of processing to enable enhanced decision making, insight discovery, and process optimization.—Gartner.

There are multiple phases in the Big Data analysis pipeline which pose some common challenges in the field of big data.

2.1 Heterogeneity and Incompleteness

In the current scenario, data sources are multiplying and data formats are exploding ranging from structured information to free text which is highly unstructured. Data being processed is becoming increasingly diverse in variety. Traditional data like Documents, Stock record, personal files, finances, etc., have become things of the past. A variety of data like Photographs, Pictures, Audio and Video data, 2D and 3D models, Simulations, Locations data are being stacked [2].

2.2 Scale

Whereas volume indicates more data, it is the gritty behavior of the data that makes it unique. Big Data is trending towards high volumes of low-density data which means data of unknown value, such as social media messages, twitter data feeds, sensor-enabled equipment capturing data at the speed of light, clicks on a web page, and many more. The ability of Big Data to convert low-density data into high-density data is what makes it so valuable. For some organizations, this might be in terabytes, for others it may be hundreds of petabytes [2].

2.3 Timeliness

To deliver analytical results in time keeping in mind the massive volumes and heterogeneity in formats is a great challenge. The architecture and design of a system also influences the speed of data processing. However, when one speaks of Velocity in the context of Big Data, it means how fast the data is generated and processed to meet the demands and the challenges which lie ahead in the path of growth and decision-making [2].

2.4 Privacy

Unlike other services, cloud computing involves big data and thus safety, security, and privacy of the data assumes great significance. Apart from legal issues ethical issues are also involved. Though there are stringent laws governing use of personal data, the same is not applicable to other data. These issues can best be addressed using technology and ethical training.

2.5 Human Collaboration

Whatever be the technological advancement, the human factor in any form of computing cannot be negated. The big data analytics cannot entirely be computational. The present day complexities require obtaining opinion of multiple experts who may be located in different geographical areas. The big data analytics must provide a means to collect and synthesize this data obtained from multiple experts in real time.

2.6 The Cost Problem

To begin with, let us examine the cost factor of managing centralized data storage and processing. Initially, highly expensive high-end mainframe or midrange servers with high speed, high-reliability disk arrays to guarantee data processing performance were used. The software, by virtue of the huge R&D cost involved, were also equally expensive. Requirement of trained professionals further led to cost escalation.

2.7 The Value Mining Problem

Growing volumes of details let to reduction in the value density per data unit and at the same time increased big data value. Deep data mining and analysis is essential to discern hidden patterns from massive data volumes. Big data mining, however, differs significantly from conventional data mining in that the volume of data is huge requiring distributed and parallel processing models.

3 Rethinking Data Management: The Rise of Cloud Computing and Cloud Data Stores to Handle Big Data

Cloud computing has developed greatly since 2007. The core model of Cloud computing is large-scale distributed system which provides, storage, computing, networking, and other resources which can be used as needed. Meanwhile, two parallel breakthroughs have further helped accelerate the adoption of solutions for handling Big Data:

- The availability of cloud-based solutions has dramatically lowered the cost of storage, amplified by the use of commodity hardware. Virtual file systems, either open source or vendor specific, helped transition from a managed infrastructure to a service-based approach;

- New designs for databases and efficient ways to support massively parallel processing have led to a new generation of products like the so-called noSQL databases and the Hadoop MapReduce platform [3].

4 Challenges of Cloud Computing

In Spite of the ever present challenges in the field of cloud computing, it has immense business value and companies are venturing into this field to exploit its full potential. However, like any new technology, the adoption of cloud computing has its own challenges some of which are enumerated below.

4.1 Security and Privacy

As valuable enterprise data will reside outside the corporate firewall, it will be susceptible to Hacking and various other forms of attacks, which in turn has the potential of affecting multiple clients even if only one site is attacked [4].

4.2 Service Delivery and Billing

The costs involved in providing such a service will be difficult to determine due the nature of the service unless provider has some good and comparable benchmarks to offer.

4.3 Interoperability and Portability

Clients must have the discretion to move in and out of the cloud at their will in addition to the freedom of selection among different service providers. Interoperability among different platforms will play a key role in the success or failure of the service.

4.4 Reliability and Availability

As brought earlier, uninterrupted service is a major factor as frequent outages will be detrimental to the businesses. Checks and balances to monitor the quality and reliability of the service through internal or third-party tools will be essential.

Fig. 2 Hadoop as big data solution

4.5 *Performance and Bandwidth Cost*

However, since the service is bandwidth-based, businesses may save money on hardware but will have to spend for the bandwidth. Though smaller applications may have a smaller cost, data-intensive applications will entail significant costs as delivering intensive and complex data over the network requires sufficient bandwidth.

4.6 *Moving Everything to the Cloud*

It may be premature to move everything to the cloud. It may be prudent to carefully analyse and determine the fields where cloud computing could be utilized optimally without any security challenges (Fig. 2).

5 Taking Big Data to the Cloud: Hadoop as Big Data Solution on Cloud

Hadoop, as a cloud computing solution service, enables access to speedy and accurate processing of medium and large-scale data at reduced costs. Insignificant operational challenges of running Hadoop enable emphasis on other more relevant business activities.

New Opportunities—Why Hadoop in the Cloud

Since cloud computing offers unlimited scaling and on-demand access to compute and storage capacity, it is the perfect match for big data processing [5].

5.1 On-Demand Elastic Cluster: Scale and Performance

One big advantage of taking big data to Hadoop cluster is the ease with which extra nodes can be added or removed from clusters automatically depending on data size to improve performance [5].

5.2 Integrated Big Data Software

Hadoop platform is comprised of two main components, HDFS and MapReduce. HDFS is a reliable fully distributed file system which includes full integration with the Hadoop MapReduce, Hive, Pig, Oozie, Sqoop, Spark, and Presto.

5.3 Simplified Cluster Management

Organisations using Hadoop need not worry about devoting extra time and resources to manage nodes, set up clusters and infrastructure scaling as everything is handled very efficiently by Qubole Data Service which offers a fully managed Hadoop-based cluster.

5.4 Lower Costs

No advance expenditure is required for on-site hardware or IT support in Hadoop Cloud. Costs greatly come down by 90 % because of spot instant pricing as compared to on-demand instances.

5.5 Cost Efficiency

Built-in software scalability, elasticity, flexibility, and availability features make HDFS highly reliable on industry-standard hardware. Organizations can optimize their hardware expenditures working in tandem with Hadoop's open source economics resulting in reduced costs as compared to their traditional architecture costs (Figs. 3 and 4).

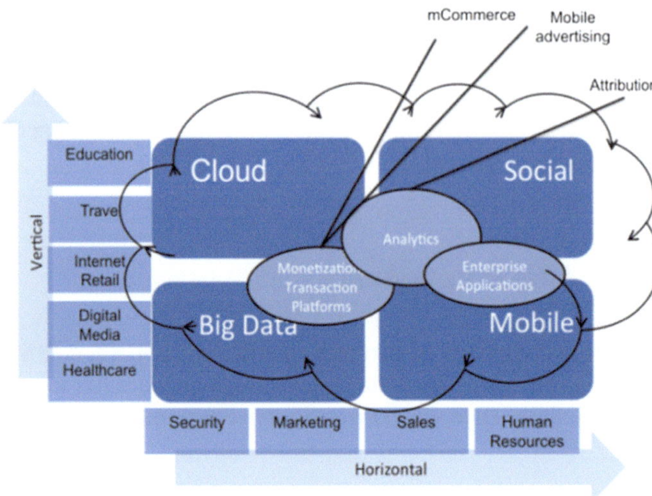

Fig. 3 Big data in cloud growing

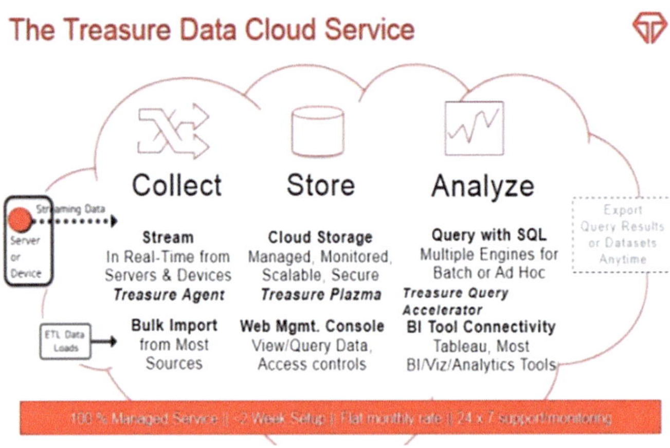

Fig. 4 How it works

5.6 *Integrated Computing and Storage*

HDFS and the Hadoop frameworks are tightly integrated and physically collocated within the same server in a cluster to provide the shortest path between data and computing, which brings accessibility and throughput to any workload with any data within the system.

5.7 *Unified and Flexible Storage*

The design of HDFS lets organizations capture data once into efficient, vernacular, and open formats that permits shared simultaneous access to that data by all current and future Hadoop processing and analytic frameworks.

5.8 *Durability and Security*

Durability and Security assurance by HDFS facilitates uninterrupted business continuity through built-in software high-availability, snapshots, and data replication facilities.

6 Challenges Coming on the Way: Big Data on Cloud

After the data management structure is established and operating well, one is ready to take on the new frontier of data management in the cloud.

6.1 *Institutional Data Management*

When a firm decides to move its data on the cloud, what happens to data management? The data to be stored, managed, and analyzed by multiple providers can be a big security threat. Also what happens to firm's data management plan in this new environment? [6].

6.2 *Data Dictionary*

One particularly significant challenge might be the management of data dictionary. Cloud services may or may not be different from the organisation's in-house architecture [6].

6.3 *How to Handle Basic Change Management*

The challenges of moving big data to cloud also involve basic change management: new options including data feeds will be available at a greater pace as compared to when the system was hosted locally [6, 7].

6.4 Access and Security Issues

Access and security issues due to physical location of the servers away from their businesses make clients hesitate to adopt cloud services [6].

6.5 Data Life Cycle and Retention

Another threatening challenge in a cloud environment is data life cycle and retention even though the cloud companies promise to copy and protect your data, often with exorbitant possession claims [6].

6.6 Data Governance

The data governance group itself is a challenge. While the group primarily focuses on internally controllable aspects of data management, as more and more data moves to cloud it might need to add a senior contracts administrator or an attorney [6, 7].

6.7 Moving Large Data Sets to the Cloud

Cloud adoption by businesses has been limited because of the problem of moving their data into and out of the cloud. Data migration in large volumes to and from the cloud may be cost prohibitive [7].

6.8 Data Location as a Security Challenge

Cloud computing technology allows cloud servers to reside anywhere, thus the organisation may not know the physical location of the server being used to store and process their data and applications.

6.9 Commingled Data as a Security Threat

Application sharing and multi-tenancy of data is one of the characteristics associated with cloud computing. Although many Cloud Service Providers have secure multi-tenant applications which are also scalable and customizable, still security and privacy issues often come up among enterprises.

6.10 Cloud Security Policy/Procedures Transparency

Some CSPs may have less transparency than others about their information security policy. The rationalization for such difference is the policies may be proprietary. As a result, it may create conflict with the enterprise's information compliance requirement.

6.11 Cloud Data Ownership

Cloud data ownership can be a big challenge. It may be the case that the Cloud Provider owns the data stored in the cloud computing environment according to the contract agreements [8].

6.12 Lock-in with CSPs Proprietary Application Programming Interfaces (API)

Currently many Cloud service Providers adopt proprietary APIs and then implement their application. As a result, it has become extremely difficult and time-consuming for the enterprise to make transition from one CSP to another CSP if it wishes [8].

6.13 Compliance Requirements

The enterprise taking cloud services for its data does not actually know where does the data reside and also is unaware if it is fully compliant with laws and regulations? The enterprise is still responsible for its data.

6.14 Disaster Recovery

When an enterprise decides to move its data on the cloud, deciding about how resilient the services of the CSP is a challenging task since data may be stored at geographically far apart locations around multiple servers. Also the data can be scattered and

Fig. 5 Big data security in cloud: major issue

commingled. In conventional hosting platforms, the enterprise knows exactly where their data is located to be able to be retrieved rapidly in the event of disaster [8] (Fig. 5).

7 Conclusions and Future Directions

In an increasingly competitive and complex business environment, organizations are exploring new ways and means to improve their competitive edge. Cloud computing provides them with one such tool. However, this service is not without its pitfalls. Business must carry out a deliberate evaluation of their security concerns taking into account the existing architecture and legacy systems prior to switching to a potentially complex private or hybrid cloud deployment. Issues regarding the process which need to be changed and the legacy processes which can be accommodated should be deliberated upon at length. Placing data pertaining to core competencies on cloud may not be prudent.

However, the fact that cloud computing and big data is here to stay cannot be negated. Ways and means will have to be designed to address and overcome the various concerns associated with cloud computing in order to exploit this technology optimally.

References

1. The cloud as an enabler for big data analytic, Intel IT Centre, Big data in the cloud (April 2015)
2. https://www.oracle.com/big-data/index.html
3. NESSI: Big data white paper (2012)
4. www.cloudtweaks.com/2012/08/top-five-challenges-of-cloud-computing/
5. http://www.qubole.com/hadoop-as-a-service (2015)
6. Waggener, S.: Cloud computing: managing data in the cloud. EDUCAUSE Q. **33**(3) (2010)

7. https://inews.berkeley.edu/articles/Oct-Nov2010/cloud-computing-EQ3
8. http://www.moorestephens.com/cloud_computing_benefits_challenges.aspx
9. Labrinidis, A., Jagadish, H.V.: Challenges and opportunities with big data. Proc VLDB Endow. **5**(12), 2032–2033 (2012)
10. Sarkar, D., Nath, A.: Big data – a pilot study on scope and challenges. IJARCSMS **2**(12) (2014). www.ijarcsms.com
11. http://www.qubole.com/resources/articles/big-data-cloud-database-computing
12. http://mahout.apache.org/
13. Big data science: myth and reality (2015)
14. White paper: hadoop and HDFS: next generation data management
15. http://www.webopedia.com/TERM/B/big_data.html
16. http://www.jeffince.co.uk/big-data--analytics.html
17. https://www.linkedin.com/pulse/20140306073407-64875646-big-data-the-5-vs-everyone-must-know
18. Zheng, Z., Zhu, J., Lyu, M.R.: Service-generated big data and big data-as-a-service: an overview (2013)
19. Ali Ahmed, E.S., Saeed, R.A.: A survey of big data cloud computing security (2014). www.Academia.edu
20. http://www.thbs.com/knowledge-zone/cloud-computing-overview
21. Evans, M., Huynh, T., Evans, A., Huynh, T., Le, K., Singh, M.: Cloud storage (2011)
22. Rexha, B., Likaj, B., Lajqi, H.: Assuring security in private clouds using ownCloud (2012) www.ijacit.com
23. Meenaskhi, A.C.: An overview on cloud computing technology. Int. J. Adv. Comput. Inf. Technol. (2012)
24. Hurwitz, J., Bloor, R., Kaufman, M., Halper, F.: Comparing public, private, and hybrid cloud computing options. Cloud computing for dummies (2009)
25. Vineetha, V.: Performance monitoring in cloud (2012). http://www.infosys.com/engineering-services/features-pinions/Documents/cloud-performance-monitoring.pdf
26. http://www.brainypro.com/cloudComputing.html (2013)
27. Shivi, G., Narayanan, T.: A review on matching public, private, and hybrid cloud computing options. Int. J. Comput. Sci. Inf. Technol. Res. **2**(2) (2014)
28. Hemlatha, S.M., Ganesh, S.: A brief survey on encryption schemes on cloud environments. Int. J. Comput. Org. Trends **3**(9) (2013)
29. http://searchcloudcomputing.techtarget.com/definition/hybrid-cloud (2015)
30. http://www.computerweekly.com/feature/Big-data-storage-Hadoop-storage-basics
31. www.cloudera.com/content/cloudera/en/.../hdfs-and-mapreduce.html (2013)
32. Patil, A., Bagban, T.I.: Improved utilization of infrastructure of clouds by using upgraded functionalities. Int. J. Innov. Res. Adv. Eng. **1**(7) (2014)
33. www.qubole.com/resources/articles/what-is-hadoop
34. www.qubole.com/resources/articles/big-data-cloud-database-computing
35. Sharma, T.: Modelling cloud services for big data using hadoop. Int. J. Comput. Sci. Inf. Technol. **6**(2) (2015)
36. www.hadoop.apache.org > Hadoop > Apache Hadoop Project Dist POM
37. Ye, X., Huang, M., Zhu, D., Xu, P.: A novel blocks placement strategy for hadoop. In: Conference IEEE/ACIS 11th International Conference on Computer and Information Science (2012)
38. Sharir, R.: Cloud database service: the difference between dbaas, daas and cloud storage - what's the difference (2011). http://xeround.com/blog/2011/02/dbaas-vs-daas-vs-cloud-storage-difference
39. Lenzerini, M.: Data integration: a theoretical perspective. In: Proceedings of the 21st ACMSIGMOD-SIGACT-SIGART Symposium on Principles of database systems. ACM (2002)
40. Slack, E.: Storage infrastructures for big data workflows. Technical Report, Storage Switchland, LLC (2012)

41. Zheng, Z., Zhu, J., Lyu, M.R.: Service-generated big data and big data-as-a-service: an overview (2013)
42. http://en.wikipedia.org/wiki/MapReduce
43. http://www.cloudera.com/content/cloudera/en/products-and-services/cdh/hdfs-and-mapreduce.html
44. http://www.qubole.com/resources/articles/what-is-hadoop/#sthash.Cnsov1wL.dpuf
45. http://www.qubole.com/resources/articles/big-data-cloud-database-computing/#sthash.p8s4FGVu.dpuf
46. An enterprise architect's guide to big data reference architecture overview oracle enterprise architecture white paper (2015)
47. https://hadoop.apache.org/docs/r1.2.1/hdfs_design.html
48. http://hadoop.apache.org/hdfs
49. http://www.qubole.com/resources/articles/big-data-cloud-database-computing/#sthash.p8s4FGVu.dpuf
50. Optimized cloud resource management and scheduling. Elsevier Inc (2015) http://dx.doi.org/10.1016/B978-0-12-801476-9.00002-1
51. Singh, D., Reddy, C.K.: A survey of platforms of big data analytics. J. Big Data (2014)
52. www.searchtelecom.techtarget.com > Cloud networks

A Moored Ship Motion Analysis in Realistic Pohang New Harbor and Modified PNH

Prashant Kumar, Gulshan Batra and Kwang Ik Kim

Abstract In recent decades, loading and unloading of moored ship is difficult task during the extreme wave oscillation in a harbor during the seasonal weather condition. In this paper, we have analyzed the six different modes of ship motion such as surge, sway, heave, roll, pitch, and yaw components using six degree of freedom for Pohang New Harbor (PNH) and modified PNH. A general mathematical formulation is designed based on Helmholtz and Laplace equation, which is solved byusing 3-D Boundary Element Method (BEM). Hydrodynamic forces acting on the mooring ropes and fenders also considered to determine added mass and damping coefficient in PNH. Based on simulation results, some tactics such as adding breakwater at entrance is implemented in modified PNH. In result, the added mass and damping coefficient of six modes of moored ship motion is reduced in modified PNH. Thus, the present numerical model can be implemented to any other realistic harbor with complex geometry to analyze the moored ship motion.

Keywords Equation of motion · Boundary element method · Moored ship motion · Pohang New Harbor

1 Introduction

During the loading and unloading activities, the ship is severely affected by moored ship motion (surge, sway, heave, roll, pitch, and yaw) that is induced by long waves with small amplitude, shallow water waves, harbor oscillation, mooring systems, and wind velocity. Harbor resonance causes several problems such as breaking mooring

P. Kumar (✉) · G. Batra
Department of Applied Sciences (Mathematics), National Institute of Technology, Delhi 110040, India
e-mail: prashantkumar@nitdelhi.ac.in

K.I. Kim
Department of Mathematics, Pohang University of Science and Technology, Pohang, South Korea
e-mail: kimki@postech.ac.kr

V.K. Singh et al. (eds.), *Modern Mathematical Methods and High Performance Computing in Science and Technology*, Springer Proceedings in Mathematics & Statistics 171, DOI 10.1007/978-981-10-1454-3_17

ropes, fenders, and breaking coastal structure. In order to ensure safe cargo handling, we required to predict the wave field, resonance frequencies near moored ship, establishing the effective countermeasures are also required to improve the moored ship motion, and increase the effective harbor working day.

Several researchers have analyzed the moored ship motion in a harbor using various numerical schemes [1–5] and predict the wave field under the resonance conditions. However, these numerical approaches are based on harbor model but they are very useful to predict the wave field in an irregular geometry. The Boundary Integral Equation Method (BIEM) is used by Sawaragi and Kubo [2], in which 3-D Green's functions was applied on a rectangular floating body in a rectangular harbor. Takagi and Naito [6] investigated mild slope equation model along with the variable bathymetry, which solved by Finite Element Method (FEM). Further, a combined method is formulated with the combination of 3-D BEM and 2-D FEM, which was applied on moored ship motion in a harbor [3, 7]. A hybrid Boussinesq panel method is utilized to predict the various modes of moored ship motion in restricted water depth [8–12].

The moored ship motion in a harbor under the resonance conditions are analyzed in realistic Pohang New Harbor (PNH). After adding breakwater at entrance, geometry of original PNH is modified, so we say it modified PNH. Then we compute the six components of the moored ship motion such as surge, sway, heave, roll, pitch and yaw for original and modified PNH. In comparison, harbor resonance is reduced for modified PNH as compared to original PNH. In this study, investigation of moored ship oscillation helped to find the cause and countermeasure of harbor oscillation generated by typhoon and applied some tactics to reduce the oscillation.

2 Mathematical Formulation

The geometry of the mathematical model of PNH is described in Fig. 1. The model geometry is divided into three domains, i.e., bounded, unbounded and the ship domain. The boundary of the ship region is denoted by S_0, the depth is uniformly both bounded and unbounded region as h. The origin is located at the entrance, x-axis and y-axis is directed along the shoreline and towards the open sea, respectively, and z-axis is directed vertically upwards from the sea surface. The ship region is small enveloped area in the bounded region, which encircled the moored ship S_M. The Incident wave directed toward the entrance is shown in Fig. 1 at various directions and the exterior and interior boundary is given at entrance of PNH (Fig. 1).

The Helmholtz equation $(\nabla^2 + k^2)f = 0$ is derived from the continuity equation both bounded and unbounded region and wave function f_b in the bounded region and f_u in the unbounded region, which is determined in term of unknown normal derivatives at the entrance [13]. Hence the matching boundary condition at the entrance is given by $f_b = f_u$ and $\partial f_b/\partial n = -\partial f_u/\partial n$, where, flow into the harbor is the opposite direction to outward the normal vector, so negative sign is taken in the boundary condition at the entrance.

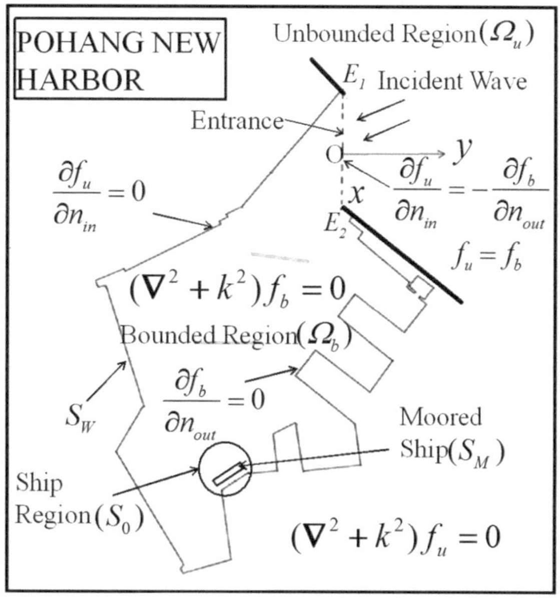

Fig. 1 Model sketch of PNH along with governing equations and boundary conditions. The ship region S_0 includes moored ship S_M

In order to determine the velocity potential completely, we have to determine the bounded wave function by using the following integral equation

$$f_b(x, y) = -\frac{i}{4} \int_{\Omega_b} \left[f(x_0, y_0) \frac{\partial}{\partial n} (H_0^{(1)}(kr)) - H_0^{(1)}(kr) \frac{\partial}{\partial n} (f(x_0, y_0)) \right] ds, \quad (2.1)$$

where $f_b(x, y)$ the wave function inside the harbor at any point (x, y), (x_0, y_0) is defined as the integration variable on the boundary, $r = \sqrt{(x - x_0)^2 + (y - y_0)^2}$ is distance between interior point to the boundary points, $H_0^{(1)}(kr)$ the Hankel function of zeroth order of first kind and k is the wave number can be defined as dispersion relation $\omega^2 = gk \tanh kh$. In ship region, the velocity potential is defined as the sum of diffraction potential and radiation potential

$$\phi^{(s)}(x, y, z, t) = Re \left[\left\{ A_0 \phi_0(x, y, z) + \sum_{j=1}^{6} \xi_j \phi_j(x, y, z) \right\} e^{-i\omega t} \right], \quad (2.2)$$

where A_0 is the incident wave amplitude, ϕ_0 is diffraction potential, and ϕ_j is radiation potential for various mode of ship motion $j = 1, 2, \ldots, 6$ and ξ_j is the incident wave amplitude for jth mode, and ω is the radian frequency. The diffraction potential is described such that $\phi_0 = \phi_{IR} + \phi_s = \phi_I + \phi_R + \phi_s$, where, ϕ_I is the incident wave is potential, ϕ_R is the reflection wave potential, and ϕ_s is the scattering wave potential representing the disturbance of the incident and reflected waves by the moored ship. The potential function ϕ_j satisfies the Laplace equation

$$\nabla^2 \phi_j = 0 \quad \text{for} \quad j = 0, 1, 2, \ldots, 6. \tag{2.3}$$

Thus the boundary condition acting on the moored ship S_M can be define as the kinematic boundary condition for the free water surface $\partial\phi_i/\partial z - (\omega^2/g)\phi_i = 0$ and boundary condition at bottom of sea floor is given as $\partial\phi_i/\partial z = 0$ for $i = 0, 1, 2, \ldots, 6$. Further the boundary conditions applicable to moored ship S_M is given

$$\frac{\partial\phi_i}{\partial n} = -i\omega n_i \quad \text{on} \quad S_M \quad \text{for } i = 1, 2, 3, \tag{2.4}$$

$$\frac{\partial\phi_i}{\partial n} = -i\omega(r_s \times n)_{i-3} \quad \text{on} \quad S_M \quad \text{for } i = 4, 5, 6. \tag{2.5}$$

The normal vectors $n_1, n_2, \ldots, n_6$ is defined the as the generalized direction cosine acting on moored ship. The Green's theorem can be applied to solve the Eq. (2.3) in fluid domain Ω_s, where we employed a 3D-BEM model and the velocity potential in ship region is determined by following integral

$$\phi_j^{(s)}(\vec{x}) = -\frac{1}{c}\left\{\int_{S_0}\left\{\phi_j(\vec{x}_0)\frac{\partial G}{\partial n} - G\,\frac{\partial\phi_j(\vec{x}_0)}{\partial n}\right\}ds + \int_{S_M \cup S_B}\left\{\phi_j(\vec{x}_0)\frac{\partial G}{\partial n}ds - \int_{S_M} n_j G ds\right\}\right\}, \tag{2.6}$$

where G denotes the Green's function, i.e., $G = 1/4\pi r$, where $r = |\vec{x} - \vec{x}_0|$. and the coefficient c is $1/2\pi$ depending on the interior point $(x, y) \in S_0$ and $1/4\pi$ for $(x, y) \notin S_0$.

The hydrodynamic forces and moments acting on the body surface are calculated by integration on the surface. The hydrodynamic forces can be described as

$$X_k = -\rho \int\int_{S_0} \phi_0 \frac{\partial\phi_k}{\partial n} dS, k = 1, 2, 3, \ldots, 6 \tag{2.7}$$

where X_k is the complex amplitude of the forces or moments which is acting toward on the body kth direction for the incident wave with unit amplitude. The added mass coefficients, damping coefficients and wave exciting forces are calculated for the rectangular ship which moves with six degree of freedom. The equation of motion with six degree of freedom considering the mooring ropes and fenders is written as follows:

$$\sum_{j=1}^{6}\left[-\omega^2(M_{kj} + a_{kj}) + i\omega b_{kj} + C_{kj}\right]\zeta_j = X_k + \sum_{i=1}^{N_l} L_{kj} + \sum_{i=1}^{N_f} F_{kj}, \quad k = 1, 2, 3, \ldots, 6. \tag{2.8}$$

Here, M_{kj} is the moment of Inertia matrix, b_{kj} is a damping coefficient, and a_{kj} is the added mass coefficient, C_{kj} is a matrix of buoyancy(hydrostatic resorting force) coefficients, X_k is the wave exciting force with respect to time [14] and N_l and N_f is the total no of mooring ropes and fenders, respectively. L_{kj} denotes the resulting force acting of kth mooring ropes acting about the center of gravity of ship motion and F_{kj} denotes the fender force in kth fender.

3 Convergence of the Numerical Method

The convergence of numerical scheme in PNH domain, the boundary of the PNH is discretized into N1 = 251 segment divisions, N2 = 502 segment divisions and N3 = 1004 segment divisions and entrance is divided into P1 = 17 segments, P2 = 34 segments and P3 = 68 segments, respectively.

In Fig. 2, the amplification factor for N1 = 251, N2 = 502 and N3 = 1004 segment divisions are represented by red, green and blue asterisk line, respectively, for the boundary of PNH. The numerical scheme has high accuracy as the segment divisions increased across the boundary PNH. An analytical approximation of the numerical scheme obtained optimum solution for the discretization N = 1004. It is difficult to get good convergence on sharp corners besides that we have reasonably good convergence at corner also because of the discretization near the corner have been taken into account very precisely.

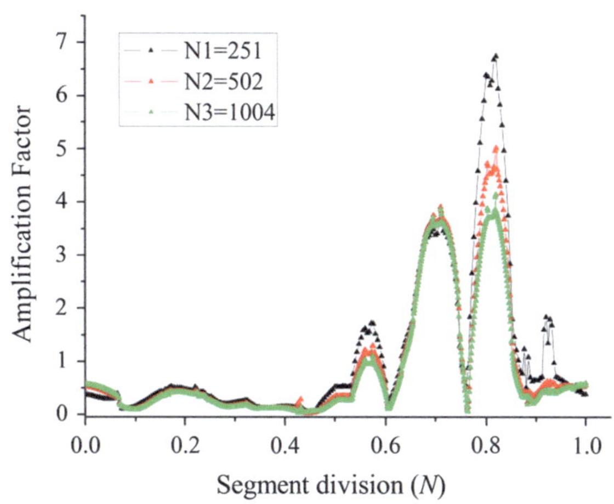

Fig. 2 Convergence graph for different number of discrete boundary points, N = 251 N2 = 502 and N3 = 1004 boundary segment divisions

4 Numerical Simulation Results

The numerical simulation has been carried in realistic PNH domain for the analysis of moored ship motion. The six different modes of ship motion such as surge, sway, heave, roll, pitch, and yaw motion have been analyzed in PNH. The current numerical scheme applied to original PNH (see Fig. 1) and modified PNH (additional breakwater on entrance E_1E_2) to analyze the various modes of ship motions as surge added mass (A11/M), sway added mass (A22/M), heave added mass (A33/M), roll added mass (*A44/MLs*^{2}), pitch added mass (*A55/MLs*^{2}) and yaw added mass (*A66/MLs*^{2}) coefficient (see Fig. 3, upper part). Further, damping coefficient of different modes as the surge ($B11/\omega M$), sway ($B22/\omega M$), heave ($B33/\omega M$), roll ($B44/\omega M L_s^2$), pitch ($B55/\omega M L_s^2$) and yaw damping coefficient ($B66/\omega M L_s^2$) is also analyzed for original and modified PNH (see Fig. 3). The added mass and damping coefficients for various modes of moored ship motion has been reduced in modified PNH as compare to original PNH. In modified PNH, we have restricted the incident waves

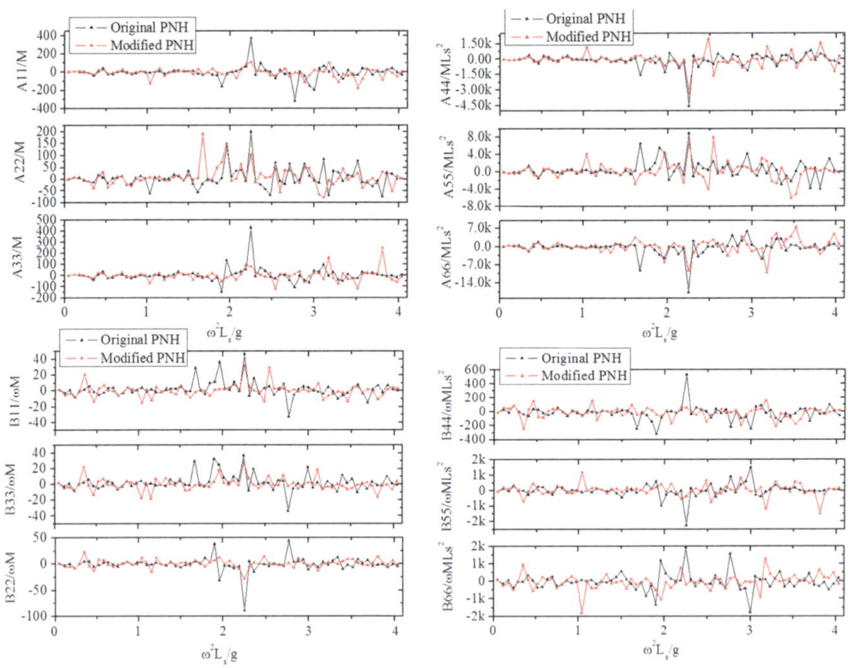

Fig. 3 Surge added mass ($A11/M$) sway added mass ($A22/M$) heave added mass ($A33/M$), roll added mass ($A44/MLs^2$), pitch added mass ($A55/MLs^2$) and yaw added mass ($A66/MLs^2$) for original and modified PNH is given in upper part of figure.Surge damping coefficient ($B11/\omega M$), sway damping coefficient ($B22/\omega M$), heave damping coefficient ($B33/\omega M$), roll damping coefficient ($B44/\omega M$), pitch damping coefficient ($B55/\omega M L_s^2$) and yaw damping coefficient ($B66/\omega M L_s^2$) is given for the same region. On the x-axis nondimensional frequency $\omega^2 L_s/g$ is taken, where ω angular frequency is, L_S is the length of model ship, and g the gravitational constant

entering directly towards entrance. The resonant frequencies for moored ship motion can be predicted for various modes of moored ship.

Once we implant the breakwater on the harbor's entrance to restrict such incident waves, the various modes of moored ship motion is subjugated significantly for modified PNH compared to original PNH. The impact of each component of moored ship motion can be identified. Therefore moored ship motion including surge, sway, heave, roll, pitch and yaw motion at specific location near boundary of harbor can be predicted under the resonance conditions. The numerical scheme can be implemented to any complex geometry harbors in the world.

5 Conclusion

We have modeled the realistic PNH domain to predict the linear and nonlinear response of a moored ship motion in the restricted water. In simulation results, the added masses and damping coefficients for original PNH and modified PNH are compared. Further, the various modes of moored ship motion in modified PNH are reduced as compared to original PNH. Thus, the direction of the incident wave dramatically affects the wave oscillation in a harbor. Small tactics such as implant breakwaters in a harbor can significantly reduce the oscillation in a harbor. The resonance modes of moored ship motion are reduced in modified PNH as compared to original PNH (see Fig. 3).

Acknowledgments This research work is currently supported by Department of Applied Science (Mathematics), National Institute of Technology, Delhi.

References

1. Oortmerssen, G.V.: The motions of a ship in shallow water. Ocean Eng. **3**, 221–255 (1976)
2. Sawaragi, T., Kubo, M.: The motion of a ship in a harbor basin. In: Proceedings of 18th conference on Coastal Engineering, pp. 2743–2762. ASCE, Cape Town, South Africa (1982)
3. Ohyama,T., Tsuchida, M.: Development of a partially three dimensional model for ship motion in a harbor with arbitrary bathymetry. In: Proceedings of 24th Conference Coastal Engineering, pp. 871–885. ASCE, Kobe, Japan (1994)
4. Sakakibara, S., Kubo, M.: Characteristic of low frequency motion of ships moored inside ports and harbors on the basis of field observations. Marine Struct. **21**, 196–223 (2008)
5. Dusseljee, D., Klopman, G., Vledder, G.V., Riezebos, H.J.: Impact of harbor navigation channel on waves: A numerical Modeling guideline.In: Proceeding of 34th Coastal Engineering, pp. 1–12. ASCE, Seoul, South Korea (2014)
6. Takagi, K., Naito, S.: Hydrodynamic forces acting on a floating body in a harbor of arbitrary geometry. Int. J. Offshore Polar Eng. **4**(1), 97–104 (1994)
7. Bingham, H.B.: A hybrid Boussinesq panel method for predicting the motion of a moored ship. Coast. Eng. **40**, 21–38 (2000)
8. Molen, W., Wenneker, I.: Time-domain calculation of moored ship motions in nonlinear waves. Coast. Eng. **55**, 409–422 (2008)

9. Guerrini, M., Bellotti, G., Fan, Y., Franco, L.: Numerical modeling of long waves amplification at Marina di Carrara harbor. Appl. Ocean Res. **48**, 322–330 (2014)
10. Lee, H.S., Kim, S.D., K-H., Wang, S, Eom: Boundary element modeling of multidirectional random wave in a harbor with arectangular navigation channel. Ocean Eng. **36**, 1287–1294 (2009)
11. Ohyama, T., Tsuchida, M.: Expanded mild-slope equations for the analysis of wave-induced ship motions in a harbor. Coast. Eng. **30**, 77–103 (1997)
12. Lee, H.S., Kim, S.D., Wang, K.-H., Eom, S.: Boundary element modeling of multidirectional random wave in a harbor with arectangular navigation channel. Ocean Eng. **36**, 1287–1294 (2009)
13. Mei, C.C., Stiassnie, M., Yue, D.K.-P.: Theory and Applications of Ocean Surface Waves, Part-I: Linear Aspects. World Scientific Press, Singapore (2005)
14. Newman, J.N.: Marine Hydrodynamics, p. 361. The MIT Press, Cambridge (1977)

The Legacy of ADI and LOD Methods and an Operator Splitting Algorithm for Solving Highly Oscillatory Wave Problems

Qin Sheng

Abstract Different splitting methods have been playing an important role in computations of numerical solutions of partial differential equations. Modern numerical strategies including mesh adaptations, linear and nonlinear transformations are also utilized together with splitting algorithms in applications. This survey concerns two cornerstones of the splitting methods, that is, the Alternating Direction Implicit (ADI) and Local One-Dimensional (LOD) methods, as well as their applications together with an eikonal mapping for solving highly oscillatory paraxial Helmholtz equations in slowly varying envelope approximations of active laser beams. The resulted finite difference scheme is not only oscillation-free, but also asymptotically stable. This ensures the high efficiency and applicability in optical wave applications.

Keywords Splitting methods · Decompositions · Eikonal transformation · Wave equations · Oscillations · Numerical stability

AMS Subject Claasifications: 65N06 · 65N12 · 65Y05 · 35M06

1 Introduction

It has been known that many natural, human or biological, chemical, mechanical, economical or financial systems and processes can be described at a macroscopic level by a set of partial differential equations governing averaged quantities such as density, temperature, concentration, velocity, etc. [6]. In some sense, partial differential equations are the basis of all physical theorems. As the computer and computational technologies develop, numerical partial differential equations has become an extremely important branch of numerical analysis that studies the numerical solution of partial differential equations for the real world.

Q. Sheng (✉)
Department of Mathematics and Center for Astrophysics, Space Physics and Engineering Research, Baylor University, Waco, TX 76798-7328, USA
e-mail: qin_sheng@baylor.edu

V.K. Singh et al. (eds.), *Modern Mathematical Methods and High Performance Computing in Science and Technology*, Springer Proceedings in Mathematics & Statistics 171, DOI 10.1007/978-981-10-1454-3_18

Finite difference methods have been extremely important to the numerical solution of partial differential equations. Splitting methods have been playing a key role in finite difference strategies. Associated with finite differences, finite elements, hybrid multi-scale settings or adaptations, various kinds of splitting algorithms are widely used and proven to be very effective and efficient for solving all major classes of different differential equations in various applications.

While the mathematical foundation of splitting methods can be traced back to the pioneering work of Hausdorff, Trotter et al. [7, 22], the philosophic inspirations of the modern computational strategy came actually from René Descartes in 1637 [2]: *...The first rule was never to accept anything as true unless I recognized it to be certainly and evidently such.... The second was to divide each of the difficulties which I encountered into as many parts as possible, and as might be required for an easier solution.*

This brief survey on splitting methods consists of the following four consecutive parts:

1. Alternating Direction Implicit (ADI) Methods.
2. Local One-Dimensional (LOD) Methods.
3. Connections and Further Developments.
4. Applications in Highly Oscillatory Wave Computations.
5. Typical Oscillatory Waves Computed.

2 Alternating Direction Implicit Methods

ADI methods have been a family of classical splitting methods with extraordinary features in structure simplicity, computational efficiency and flexibility in applications.

The original ADI idea was due to D.W. Peaceman and H.H. Rachford, Jr. in 1955 [5, 14]. Later, J. Douglas, Jr. and H.H. Rachford, Jr. were able to implement the algorithm by splitting the time-step procedure into two fractional steps. The strategy of the ADI approach can be readily explained in a contemporary way of modern numerical analysis.

To see the underlying strategy of ADI, we may let $\mathcal{E}$ be an n-dimensional domain, and consider the following evolution equation:

$$\frac{\partial u}{\partial t}(x,t) = \mathcal{F}u(x,t) \quad x \in \mathcal{E}, \quad t > t_0, \tag{2.1}$$

where $\mathcal{F} = \mathcal{F}_1 + \mathcal{F}_2 + \cdots + \mathcal{F}_m$, $m \geq 2$, is a differential operator.

For the simplicity of discussion, we may set $n = m = 2$. Assume that a semidiscretization of (2.1) together with suitable boundary conditions yields the following ordinary differential system:

$$v' = Av + Bv, \quad t > t_0, \tag{2.2}$$

where $A,\ B \in \mathbb{C}^{n\times n}$, $AB \neq BA$ in general, and $v \in \mathbb{C}^n$.

If $v(t_0) = v_0$ is an initial vector, then for $\tau > 0$, the solution of (2.2) can be expressed as

$$v(t+\tau) = e^{\tau A}v(t) + \int_0^\tau e^{(\tau-\xi)A}Bv(t+\xi)d\xi, \quad t \ge t_0.$$

An application of the left-point rule and [0/1] Padé approximant yields

$$w(t+\tau) = (I - \tau A)^{-1}(I + \tau B)w(t), \quad t \ge t_0, \tag{2.3}$$

where w approximates v. Proceed for one more step, we have

$$w(t+2\tau) = (I - \tau B)^{-1}(I + \tau A)w(t+\tau), \quad t + \tau \ge t_0. \tag{2.4}$$

Both (2.3) and (2.4) are first order approximations.

Combining (2.3) and (2.4) we acquire immediately that

$$w(t+2\tau) = (I - \tau B)^{-1}(I + \tau A)(I - \tau A)^{-1}(I + \tau B)w(t), \quad t \ge t_0. \tag{2.5}$$

The above is the standard *ADI formula.* It is frequently called a *Peaceman–Rachford splitting.* It is a splitting algorithm not based on any *exponential splitting* [15]. It is of second order in accuracy for solving (2.2).

The formula (2.5) occasionally shows us in a slightly different form. To see it, we denote $\Delta t = 2\tau,\ w^\ell = w(t),\ w^{\ell+1} = w(t + \Delta t u)$. Thus, (2.5) becomes

$$\left(I - \frac{\Delta t}{2}B\right)w^{\ell+1} = \left(I + \frac{\Delta t}{2}A\right)\left(I - \frac{\Delta t}{2}A\right)^{-1}\left(I + \frac{\Delta t}{2}B\right)w^\ell. \tag{2.6}$$

Example Consider the two-dimensional advection-diffusion equation

$$\frac{\partial u}{\partial t} = \nabla(a(x,y)\nabla u), \quad a \le x, y \le b,$$

where a is sufficiently smooth, together with homogeneous Dirichlet boundary conditions. A straightforward semidiscretization leads to

$$\begin{aligned} v'_{k,j} = &\frac{1}{h^2}\left[a_{k-1/2,j}v_{k-1,j} - \left(a_{k-1/2,j} + a_{k+1/2,j}\right)v_{k,j} + a_{k+1/2,j}v_{k+1,j}\right] \\ &+\frac{1}{h^2}\left[a_{k,j-1/2}v_{k,j-1} - \left(a_{k.j-1/2} + a_{k,j+1/2}\right)v_{k,j} + a_{k,j+1/2}v_{k,j+1}\right] \\ &\quad k, j = 1, 2, \ldots, n,\ 0 < h \ll 1. \end{aligned}$$

This yields the system (2.2) where block tridiagonal A, B contain the contribution of the differentiation in the x- and y- variables, respectively. An ADI procedure (2.5) or (2.6) can be applied immediately.

Remark 1 ADI methods can be used for solving linear nonhomogeneous equations with nonhomogeneous boundary conditions too.

Remark 2 ADI methods can be extended for the numerical solution of certain nonlinear, or even singular, partial differential equations.

Remark 3 The ADI strategy can be utilized for solving partial differential equations consisting of multiple components, such as Schrödinger equations.

Remark 4 ADI methods can be used together with other highly effective numerical strategies, such as temporal and spacial adaptations, and compact finite difference schemes.

Remark 5 ADI methods can be modified for solving other types of equations including integro-differential equations.

Remark 6 We note that A, B in (2.2) are not necessary matrices. They can be more general linear or nonlinear operators. This leads to an exciting research field of operator splitting, in which important mathematical tools, such as semigroups, Hopf algebra, graph theory, and symplectic integrations, can be applied.

Remark 7 Key ideas of ADI methods have been extended well beyond the territory of computational mathematics to areas such as fractional PDEs, hybrid modeling and realizations.

3 Local One-Dimensional Methods

Needless to say, the introduction and original analysis of this type of splitting methods are due to E.G. D'Yakonov, G.I. Marchuk, A.A. Samarskii and N.N. Yanenko [4, 11, 23].

To see the general LOD splitting strategy, we recall the first order exponential splitting [9, 12, 13, 15],

$$e^{\tau(A+B)} = e^{\tau A}e^{\tau B} + O(\tau^2), \quad \tau \to 0^+. \tag{3.1}$$

Thus, the solution of the semidiscretized system (2.2) can be approximated through

$$v(t+2\tau) \approx e^{2\tau A}e^{2\tau B}v(t), \quad t \geq t_0.$$

Now, an application of the [1/1] Padé approximant leads immediately to the Local One-Dimensional configuration:

$$w(t+2\tau) = (I-\tau A)^{-1}(I+\tau A)(I-\tau B)^{-1}(I+\tau B)w(t), \quad t \ge t_0. \tag{3.2}$$

This new formula may look like another Peaceman–Rachford splitting by a first glance. However, (2.5) and (3.2) are fundamentally different.

To see this, we may re-list both the formulas as follows:

- LOD splitting formula:

$$w(t+2\tau) = (I-\tau A)^{-1}(I+\tau A)(I-\tau B)^{-1}(I+\tau B)w(t).$$

- ADI splitting formula:

$$w(t+2\tau) = (I-\tau B)^{-1}(I+\tau A)(I-\tau A)^{-1}(I+\tau B)w(t), .$$

Apparently, LOD methods are *exponential splitting* oriented [15]. It can be viewed as a consecutive application of two one-dimensional *Crank–Nicolson methods.* It can be readily proven that the LOD method is unconditionally stable if all eigenvalues of A, B lie in the left half of the complex plane. This also leads to the convergence. Further, LOD methods can be conveniently extended for approximating solutions of multidimensional problems. However, (3.2) is first order in accuracy.

But, can LOD methods be more accurate?

The answer is YES. This leads to the study of exponential splitting. Some of the basic exponential splitting formulas include

$$\begin{aligned}
e^{2\tau(A+B)} &= e^{2\tau A}e^{2\tau B} + O(\tau^2),\\
e^{2\tau(A+B)} &= \frac{1}{2}\left(e^{2\tau A}e^{2\tau B} + e^{2\tau B}e^{2\tau A}\right) + O(\tau^3),\\
e^{2\tau(A+B)} &= e^{\tau A}e^{2\tau B}e^{\tau A} + O(\tau^3).
\end{aligned}$$

They are considered as the discretization parameter $\tau \to 0^+$.

The next question may be whether exponential splitting formulas can be as accurate as we wish?

The answer is NO. This is because of the following Sheng–Suzuki Theorem:

Theorem 3.1 *The order of accuracy of a exponential splitting based method cannot be more than two, if a diffusion type stability needs to be observed.*

For more detailed discussions, the reader is referred to [1, 15, 17, 21].

4 Connections and Further Developments

To see possible internal connections between ADI and LOD formulations, let us start from the ADI formula:

$$\begin{aligned}w(t+4\tau) &= \left[(I-\tau B)^{-1}(I+\tau A)(I-\tau A)^{-1}(I+\tau B)\right]^2 w(t)\\ &= (I-\tau B)^{-1}(I+\tau A)(I-\tau A)^{-1}(I+\tau B)(I-\tau B)^{-1}\\ &\quad \times(I+\tau A)(I-\tau A)^{-1}(I+\tau B)w(t), \quad t\ge t_0.\end{aligned}$$

Thus, by denoting

$$w_0(\xi) = (I+\tau B)w(\xi)$$

and drop all truncation errors incurred, we have

$$w_0(t+4\tau) = e^{2\tau B}e^{2\tau A}e^{2\tau B}e^{2\tau A}w_0(t) = \left(e^{2\tau B}e^{2\tau A}\right)^2 w_0(t), \quad t\ge t_0. \tag{4.1}$$

The above implies repeated applications of the same LOD method!

The research over generalized ADI, LOD and exponential splitting methods has been very active. Interested readers may search recent publications of the following researchers:

- R. McLachlan, R. Quispel, S. Descombes
- S. Blanes, F. Casas, J. M. Sanz-Serna
- L. Einkemmer, A. Iserles, E. Hansen, Alex Ostermann
- V. K. Singh, G. Strang, S. Chin, Q. Sheng...

Intensive research information and the latest meeting and workshops can also be found in the *Splitting in Action!* website.

Beyond the traditional local traction error estimates, the study of global error estimates for exponential splitting has been in excellent progresses. To see some of the recent results, we may let $A,\ B\in\mathbb{C}^{n\times n},\ t\ge 0$ and

$$\begin{aligned}E_1(t) &= e^{tA}e^{tB} - e^{t(A+B)}\\ E_2(t) &= e^{\frac{t}{2}A}e^{tB}e^{\frac{t}{2}A} - e^{t(A+B)},\\ E_3(t) &= \frac{1}{2}\left(e^{tA}e^{tB}+e^{tB}e^{tA}\right) - e^{t(A+B)}.\end{aligned}$$

It can be shown [13, 16] that

$$\begin{aligned}\|E_1(t)\| &\le \frac{t^2}{2}\|[A,B]\|\max\left\{e^{t\mu(A+B)},\, e^{t(\mu(A)+\mu(B))}\right\},\\ \|E_2(t)\| &\le \frac{t^3}{6}\left\|\frac{1}{2}A+B\right\|\,\|[A,B]\|\max\left\{e^{t\left(\frac{1}{2}\mu(A)+\mu(\frac{1}{2}A+B)\right)},\, e^{t\mu(A+B)}\right\}\\ &\quad\times\max\left\{e^{\theta\left(\frac{1}{2}\mu(A)+\mu(B)\right)},\, e^{\theta\mu\left(\frac{1}{2}A+B\right)}\right\},\\ \|E_3(t)\| &\le \frac{t^3}{6}\|A-B\|\,\|[A,B]\|\max\left\{e^{t\mu(A+B)},\, e^{t(\mu(A)+\mu(B))}\right\},\end{aligned}$$

where $\mu(M)$ is the logarithmic norm of M [12].

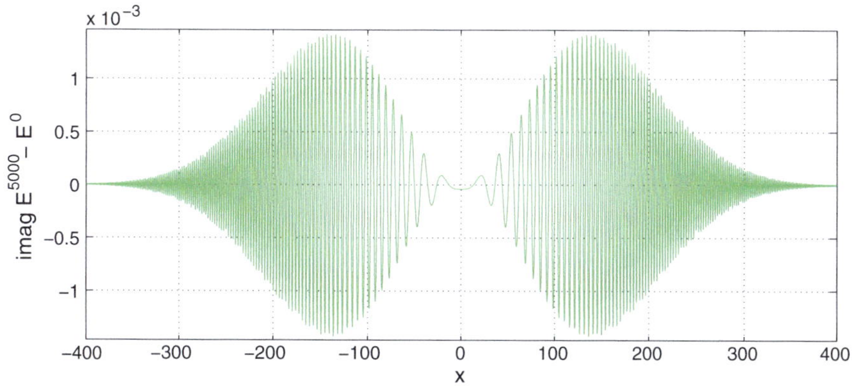

Fig. 1 The imaginary part of a typical numerical error of computed highly oscillatory waves via LOD methods [18, 19]

However, the study of global error estimates when multiple operators are involved is still in its infancy. Since the Pandora's box has been opened, there are just more questions than answers in split computations nowadays [3, 10]. For example, in multiple physics applications, we often need to evaluate (Fig. 1)

1. global errors of multiple component splitting such as

$$E_{1,n}(t) = e^{tA_1}e^{tA_2}\ldots e^{tA_{n-1}}e^{tA_n} - e^{t(A_1+A_2+\cdots A_{n-1}+A_n)},$$
$$E^*_{1,n}(t) = e^{tA_1}e^{tA_2}\ldots e^{tA_{n-1}}e^{tA_n} - e^{tA_n}e^{tA_2}\ldots e^{tA_{n-1}}e^{tA_1};$$

2. complex time exponential splitting such as

$$S = \prod_{k=1}^{n} e^{b_k hA}e^{a_k hB}, \quad a_k, b_k \in \mathbb{C};$$

3. asymptotically perturbed exponential splitting;
4. compact splitting, domain splitting and physical preservations.

5 Applications in Highly Oscillatory Wave Computations

We consider a highly oscillatory wave problem in which a slowly varying envelope approximation of the laser beam is considered. In the case, the *paraxial Helmholtz equation*,

$$2i\kappa u_z = u_{xx} + u_{yy} + f(u), \quad 0 \le x, y \le \ell, \ z \ge z_0, \tag{5.1}$$

plays an important role. When f is linear, then (5.1) can be simplified to

$$2i\kappa u_z = u_{xx} + u_{yy}, \quad 0 \le x, y \le \ell, \ z \ge z_0. \tag{5.2}$$

There have been many modern numerical methods for solving similar wave problems, including

- Different coordinate transforms for overcoming the numerical inefficiency
- Stratton-Chu scattering diffraction integral configurations
- Other integral methods possessing different advantages under particular circumstances
- Filon and Levin type collocations
- Spectral methods, finite element methods.

However, since the wave number κ is extremely large in optical applications, consequently the energy function u is highly oscillatory. Therefore mesh steps often need to be extremely small in computations. Therefore Eq. (5.1) is costly to solve via any existing conventional methods. Challenges remain in balancing the algorithmic simplicity, accuracy and efficiency [8, 19].

Recall the nonlinear *ray transformation* in geometrical optics,

$$u(x, y, z) = \phi(x, y, z)\exp\{i\kappa\psi(x, y, z)\}, \tag{5.3}$$

where $\phi, \ \psi$ are nonoscillatory real functions and $\phi \ne 0$ is for wavefronts of the disturbance.

Utilizing (5.3), we acquire from (5.1) the following coupled eikonal system

$$\phi_z = \alpha\left(\psi_{xx} + \psi_{yy}\right) + f_1, \tag{5.4}$$

$$\psi_z = \beta\left(\phi_{xx} + \phi_{yy}\right) + f_2, \tag{5.5}$$

where

$$\alpha = \frac{\phi}{2}, \ \beta = -\frac{1}{2\kappa^2\phi}, \ f_1 = \phi_x\psi_x + \phi_y\psi_y, \ f_2 = \frac{\psi_x\psi_x + \psi_y\psi_y}{2}.$$

Note that solutions $\phi, \ \psi$ are not oscillatory! Further, (5.4) and (5.5) can be written as

$$w_z = Mw_{xx} + Mw_{yy} + f$$

together with

$$w(x, y, z_0) = g_0(x, y).$$

Denote

$$w = \begin{pmatrix} \phi \\ \psi \end{pmatrix}, \quad f = \begin{pmatrix} f_1 \\ f_2 \end{pmatrix}, \quad M = \begin{pmatrix} 0 & \alpha \\ \beta & 0 \end{pmatrix}.$$

Lemma 5.1 *M is similar to a skew symmetric matrix and thus its eigenvalues are pure imaginary. Further, for any $\phi \neq 0$, the matrix has a pair of constant eigenvalues $\lambda_M = \pm\, i/(2\kappa)$ and thus a spectral radius $\rho(M) = 1/(2\kappa)$. Hence, $cond_2(M) = 1$, and $\|M\|_2 = \max\{|\alpha|, |\beta|\}$ is small when κ is large.*

Consider

$$w(0, y, z) = w(\ell, y, z) = 0, \;\; w(x, 0, z) = w(x, \ell, z) = 0, \tag{5.6}$$

and then

$$w_x(0, y, z) = w_x(\ell, y, z) = 0, \;\; w_y(x, 0, z) = w_y(x, \ell, z) = 0. \tag{5.7}$$

The above boundary settings are typical. They imply that laser beams are concentrated symmetrically around its geometric center and the light intensity is negligible near the boundaries in both x and y directions. Certainly more general conditions, such as the reflective or absorbing boundary conditions, can be employed.

Note that our eikonal equation can be reconfigured to

$$w_z = (A + B)w + \phi,$$

where $A,\ B$ are spatial differential operators involved. Recall the discussion of (2.2). We have the formal solution

$$w(z + h) = e^{h(A+B)} w(z) + \int_0^h e^{(h-\xi)(A+B)} \phi(z + \xi) d\xi.$$

An exponential splitting leads to

$$\begin{aligned} w(z + h) &= e^{hB} e^{hA} \left(w(z) + \frac{h}{2}\phi(z) \right) + \frac{h}{2}\phi(z + h) \\ &= e^{hB} v(z) + \frac{h}{2}\phi(z + h). \end{aligned}$$

Subsequently,

$$\begin{aligned} v(z) &= e^{hA} \left(w(z) + \frac{h}{2}\phi(z) \right), \\ w(z + h) &= e^{hB} v(z) + \frac{h}{2}\phi(z + h). \end{aligned}$$

This leads to the complete LOD scheme

$$\left(I - \frac{h}{2}A\right) v(z) = \left(I + \frac{h}{2}A\right) w(z) + \frac{h}{2}\left(I + \frac{h}{2}A\right)\phi(z),$$
$$\left(I - \frac{h}{2}B\right) w(z+h) = \left(I + \frac{h}{2}B\right) v(z) + \frac{h}{2}\left(I - \frac{h}{2}B\right)\phi(z+h)$$

via an [1/1] Padé approximation.

Now, denote

$$\left(I + \frac{h}{2}A\right) f(z) = f_1(z), \quad \left(I - \frac{h}{2}B\right) f(z) = f_2(z),$$

$$w^r = w(z_r), \;\; w^{r+1/2} = v(z_r), \;\; w^{r+1} = w(z_r + h)$$

on a z-mesh.

We obtain the following semidiscretized PDE system,

$$w^{r+1/2} - w^r = \frac{h}{2}\left(Mw_{xx}^{r+1/2} + Mw_{xx}^r\right) + \frac{h}{2}f_1^r,$$
$$w^{r+1} - w^{r+1/2} = \frac{h}{2}\left(Mw_{yy}^{r+1/2} + Mw_{yy}^{r+1}\right) + \frac{h}{2}f_2^{r+1/2}.$$

Since each of the above equations is implicit in one spacial dimension, we may employ standard central difference approximations in space:

$$\delta_x^2 v_{s,t}^r = \left(v_{s+1,t}^r - 2v_{s,t}^r + v_{s-1,t}^r\right) h_x^{-2},$$
$$\delta_y^2 v_{s,t}^r = \left(v_{s,t+1}^r - 2v_{s,t}^r + v_{s,t-1}^r\right) h_y^{-2}.$$

The above leads to the following fully discretized coupled difference equations

$$w_{s,t}^{r+1/2} - w_{s,t}^r = \frac{h}{2}M_{s,t}^r\left(\delta_x^2 w_{s,t}^{r+1/2} + \delta_x^2 w_{s,t}^r\right) + \frac{h}{2}f_{1,s,t}^r,$$
$$w_{s,t}^{r+1} - w_{s,t}^{r+1/2} = \frac{h}{2}M_{s,t}^{r+1/2}\left(\delta_y^2 w_{s,t}^{r+1/2} + \delta_y^2 w_{s,t}^{r+1}\right) + \frac{h}{2}f_{2,s,t}^{r+1/2},$$

where (x_s, y_t, z_r) is any internal mesh point with steps h_x, h_y, h. For the simplicity in discussions, we may let $h_x = h_y = \ell/(n+1)$.

Under the first boundary condition (5.6), we obtain a block tridiagonal system from the above:

$$\left(I - \mu M^r\right) w^{r+1/2} = \left(I + \mu M^r\right) w^r + \frac{h}{2}f^r, \tag{5.8}$$

$$\left(I - \eta M^{r+1/2}\right) v^{r+1} = \left(I + \eta M^{r+1/2}\right) v^{r+1/2} + \frac{h}{2}g^{r+1/2}, \tag{5.9}$$

where

- $I \in \mathbb{R}^{2n^2 \times 2n^2}$ is an identity matrix,
- $\mu = h/(2h_x^2),\ \eta = h/(2h_y^2)$ are dimensional Courant numbers [9],
- $v^\sigma = Pw^\sigma,\ f^\sigma = f_1^\sigma,\ g^\sigma = Pf_2^\sigma$, P is a permutation matrix,
- $\sigma = 0, 1/2, 1, 3/2, 2, \ldots, r, r + 1/2, r + 1, \ldots$

Further, M^σ is block-diagonal. Let $N^\sigma = P^{-1}M^\sigma P$. Then (5.8) and (5.9) can be conveniently reformulated as

$$\left(I - \mu M^r\right) w^{r+1/2} = \left(I + \mu M^r\right) w^r + \frac{h}{2} f^r,$$

$$\left(I - \eta N^{r+1/2}\right) w^{r+1} = \left(I + \eta N^{r+1/2}\right) w^{r+1/2} + \frac{h}{2} g^{r+1/2}.$$

Lemma 5.2 *The eigenvalues of matrices $M^\sigma,\ N^\sigma$ are pure imaginary.*

Lemma 5.3 *We have $\rho(S^\sigma) = \mathcal{O}(1/\kappa)$, where κ is the wave number.*

Theorem 5.4 *Suppose that the homogeneous Dirichlet boundary condition* (5.6) *is used and $\rho(M^\sigma) = \mathcal{O}\left(\kappa^{-c}\right),\ c > 1/2$. Then the semidiscretization based splitting method* (5.8) *and* (5.9) *is unconditionally asymptotically stable. Further, the asymptotical stability index of the scheme is c.*

Proofs of the lemmas and theorem are straightforward. A key to notice is that the perturbation matrix $(I - \mu M^r)^{-1}(I + \mu M^r)$ is actually an [1/1] Padé approximant of the matrix exponential $e^{2\mu M^r}$.

Now, what may happen if the Neumann boundary condition (5.7) is adopted? In the case we may acquire the linear system

$$\left(I - \mu Q^r\right) w^{r+1/2} = \left(I + \mu Q^r\right) w^r + \frac{h}{2} f^r, \tag{5.10}$$

$$\left(I - \eta Q^{r+1/2}\right) v^{r+1} = \left(I + \eta Q^{r+1/2}\right) v^{r+1/2} + \frac{h}{2} g^{r+1/2}. \tag{5.11}$$

Again, there exists a permutation matrix P such that $v^\sigma = Pw^\sigma$. We denote $R^\sigma = P^{-1}Q^\sigma P$.

Lemma 5.5 *The eigenvalues of the matrices $Q^\sigma,\ R^\sigma$ are pure imaginary.*

Theorem 5.6 *Suppose that the homogeneous Neumann boundary condition* (5.7) *is used and $\rho(Q^\sigma) = O\left(\kappa^{-c}\right),\ c > 1/2$. Then the semidiscretization based splitting method* (5.10) *and* (5.11) *is unconditionally asymptotically stable. Further, the asymptotical stability index of the scheme is c.*

Their proofs are similar to that of the last results for the Helmholtz equation under Dirichlet boundary conditions.

How to compute such eikonal systems? In fact, either the system (5.8), (5.9) or (5.10), (5.11) can be comprised to yield

$$A_r\psi^{r+1/2} = \mu D_{\beta^r} S g_1^r + g_2^r, \tag{5.12}$$
$$\phi^{r+1/2} = \mu D_{\alpha^r} S\psi^{r+1/2} + g_1^r, \tag{5.13}$$
$$B_{r+1/2}\tilde{\psi}^{r+1} = \eta D_{\beta^{r+1/2}} S g_3^{r+1/2} + g_4^{r+1/2}, \tag{5.14}$$
$$\tilde{\phi}^{r+1} = \eta D_{\alpha^{r+1/2}} S\tilde{\psi}^{r+1} + g_3^{r+1/2}, \tag{5.15}$$
$$r = 0, 1, 2, \ldots,$$

where $A_r = I - \mu^2 D_{\beta^r} S D_{\alpha^r} S,\ B_{r+1/2} = I - \eta^2 D_{\beta^{r+1/2}} S D_{\alpha^{r+1/2}} S$ are quintic diagonal.

The ordered systems can further be compressed into an imbedded form

$$\phi^{r+1/2} = \mu D_{\alpha^r} S A_r^{-1}\left(\mu D_{\beta^r} S g_1^r + g_2^r\right) + g_1^r,$$
$$\tilde{\phi}^{r+1} = \eta D_{\alpha^{r+1/2}} S B_{r+1/2}^{-1}\left(\eta D_{\beta^{r+1/2}} S g_3^{r+1/2} + g_4^{r+1/2}\right) + g_3^{r+1/2},$$
$$r = 0, 1, 2, \ldots$$

Needless to say, above procedures are considerably simple, effective, efficient and user-friendly. They can be conveniently parallelized. These are desirable characters for engineering and industrial software package implementations.

Since $D_{\beta^r} S$ and $D_{\alpha^r} S$ are both tridiagonal, $D_{\beta^r} S D_{\alpha^r} S$ is quintic diagonal. We have

Theorem 5.7 *There exist reasonable values of μ, η such that A_r and $B_{r+1/2}$ are nonsingular. Therefore the solution of system* (5.12)–(5.15) *exists and is unique.*

The above study indicates that splitting methods are highly successful for solving highly oscillatory wave problems, given that proper auxiliary tools, such as the eikonal transformation, can be equipped [20]. However, are there any unanswered questions or out cries from the physical or engineering applications? YES, there are many. Such as the splitting methods, in particular ADI and LOD schemes, when nonlinear stabilities are involved. Further, difficult research associate with splitting methods include:

1. Physical and mathematical preservations,
2. High dimensional decomposition and analysis,
3. Parallel computation realizations...

Back to the nonlinear waves. How to use splitting methods for similar singular wave applications? Are there higher order splitting algorithms for multidimensional Schrödinger equations, Korteweg–de Vries equations, quenching-combustion equations, sine-Gordon equation and stochastic Black–Scholes equations and inequalities? These are apparently open challenges to everyone here for the M3HPCST Conference in India.

6 Typical Oscillatory Wave Examples

In this section, we show a few typical highly oscillatory wave projections computed via our LOD based eikonal splitting schemes. Boundary conditions (5.6) is utilized. Parallel MapLab software packages and cluster computer platforms are used.

6.1 Highly Oscillatory Gaussian Beam (Laser Optical Waves)

See Figs. 2, 3 and 4.

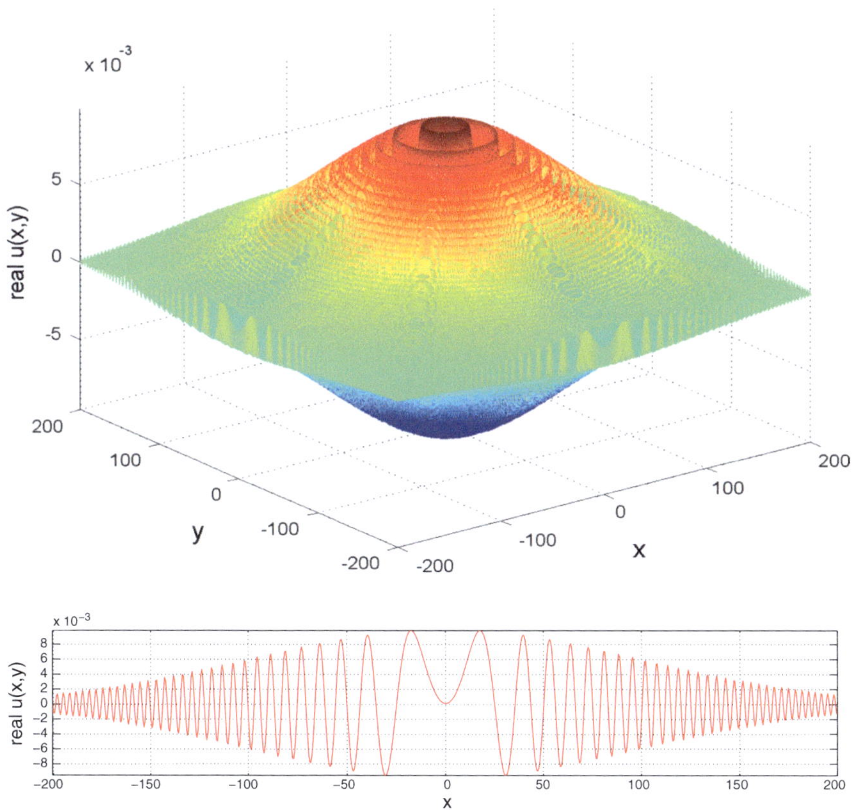

Fig. 2 3D Gaussian beam plot: real part of a computed highly oscillatory optical wave via LOD methods

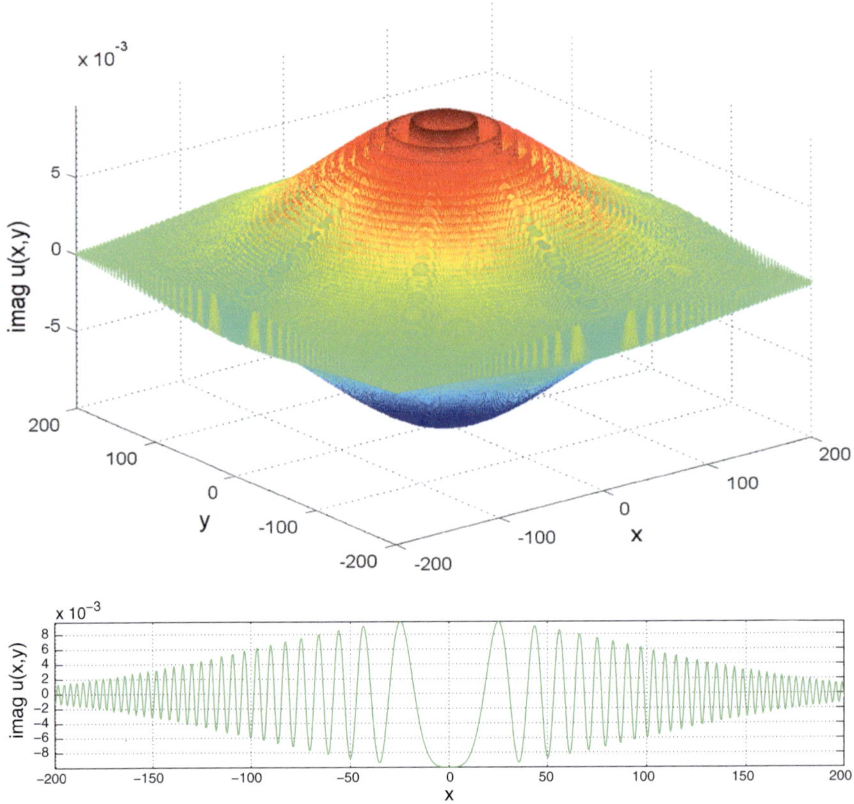

Fig. 3 3D Gaussian beam plot: imaginary part of a computed highly oscillatory optical wave via LOD methods

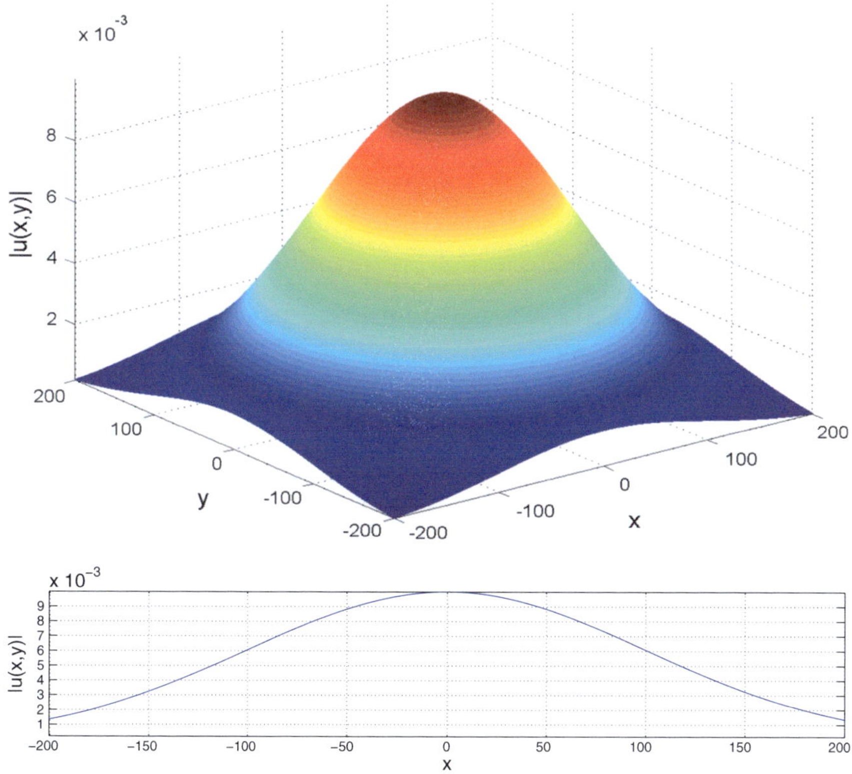

Fig. 4 3D Gaussian beam plot: modules of a computed highly oscillatory optical wave via LOD methods

References

1. Chin, S.A.: A fundamental theorem on the structure of symplectic integrators. Phys. Lett. A **354**, 373–376 (2006)
2. Descartes, R.: Discourse on the Méthod (trans: Lafleur, L.J., 1637). The Liberal Arts Press, New York (1960)
3. Descombes, S., Thalhammer, M.: An exact local error representation of exponential operator splitting methods for evolutionary problems and applications to linear Schrödinger equations in the semi-classical regime. BIT **50**, 729–749 (2010)
4. D'Yakonov, E.G.: Difference schemes with splitting operator for multi-dimensional nonstationary problems. Zh. Vychisl. Mat. i Mat. Fiz. **2**, 549–568 (1962)
5. Douglas Jr., J., Rachford Jr., H.H.: On the numerical solution of heat conduction problems in two and three space variables. Trans. Am. Math. Soc. **82**, 421–439 (1956)
6. Frey, P., George, P.-L.: Mesh Generation, 2nd edn. Wiley-ISTE, New York (2008)
7. Hausdorff, F.: Die symbolische Exponentialformel in der Gruppentheorie. Ber Verh Saechs Akad Wiss Leipzig **58**, 19–48 (1906)
8. Hundsdorfer, W.H., Verwer, J.G.: Stability and convergence of the Peaceman-Rachford ADI method for initial-boundary value problems. Math. Comput. **53**, 81–101 (1989)

9. Iserles, A.: A First Course in the Numerical Analysis of Differential Equations, 2nd edn. Cambridge University Press, London (2011)
10. Jahnke, T., Lubich, C.: Error bounds for exponential operator splitting. BIT **40**, 735–744 (2000)
11. Marchuk, G.I.: Some applicatons of splitting-up methods to the solution of problems in mathematical physics. Aplikace Matematiky **1**, 103–132 (1968)
12. McLachlan, R.I., Quispel, G.R.W.: Splitting methods. Acta Numerica **11**, 341–434 (2002)
13. Moler, C., Van Loan, C.: Nineteen dubious ways to compute the exponential of a matrix, twenty-five years later. SIAM Rev. **45**, 3–46 (2003)
14. Peaceman, D.W., Rachford Jr., H.H.: The numerical solution of parabolic and elliptic differential equations. J. Soc. Ind. Appl. Math. **3**, 28–41 (1955)
15. Sheng, Q.: Solving linear partial differential equations by exponential splitting. IMA J. Numer. Anal. **9**, 199–212 (1989)
16. Sheng, Q.: Global error estimate for exponential splitting. IMA J. Numer. Anal. **14**, 27–56 (1993)
17. Sheng, Q.: The ADI Methods. Encyclopedia of Applied and Computational Mathematics. Springer Verlag GmbH, Heidelberg (2015)
18. Sheng, Q.: ADI, LOD and modern decomposition methods for certain multiphysics applications. J. Algorithms Comput. Technol. **9**, 105–120 (2015)
19. Sheng, Q., Sun, H.: On the stability of an oscillation-free ADI method for highly oscillatory wave equations. Commun. Comput. Phys. **12**, 1275–1292 (2012)
20. Sheng, Q., Sun, H.: Exponential splitting for n-dimensional paraxial Helmholtz equation with high wavenumbers. Comput. Math. Appl. **68**, 1341–1354 (2014)
21. Suzuki, M.: General theory of fractal path integrals with applications to manybody theories and statistical physics. J. Math. Phys. **32**, 400–407 (1991)
22. Trotter, H.F.: On the product of semi-groups of operators. Proc. Am. Math. Soc. **10**, 545–551 (1959)
23. Yanenko, N.N.: The Method of Fractional Steps; the Solution of Problems of Mathematical Physics in Several Variables. Springer, Berlin (1971)

Generalized Absolute Convergence of Trigonometric Fourier Series

R.G. Vyas

Abstract Recently, Moricz and Veres generalized the classical results of Bernstein, Szasz, Zygmund and others related to the absolute convergence of single and multiple Fourier series. In this paper, we have extended this result for single Fourier series of functions of the classes $\Lambda BV(\overline{\mathbb{T}})$ and $\Lambda BV^{(p)}(\overline{\mathbb{T}})$.

Keywords Generalized $\beta-$absolute convergence · Fourier series · $\Lambda BV^{(p)}(\overline{\mathbb{T}})$

2010 AMS Mathematics Subject Classification: Primary: 42A20 · 42B99. Secondary: 26A16 · 26B30

1 Introduction

The classical result of Zygmund, for the absolute convergence of Fourier series if a function of bounded variation on $\overline{T}$, where $\mathbb{T} = [-\pi, \pi)$ is the torus, is generalized in many ways and many interesting results are obtained for different generalized absolute convergence of Fourier of functions of different generalized classes (see [1, 4]). In 2006, Gogoladze and Meskhia [1] obtained sufficient conditions for the generalized absolute convergence of a single Fourier series. Moricz and Veres [2] obtained sufficient conditions for the generalized absolute convergence of single and multiple Fourier series of functions of the classes $BV^{(p)}(\overline{\mathbb{T}})$ and $BV^{(p)}(\overline{\mathbb{T}}^N)$, respectively (also see [5]). In this paper, generalizing such results for single Fourier series, we have obtained sufficient conditions for the generalized absolute convergence of single Fourier series of functions of the classes $\Lambda BV(\overline{\mathbb{T}})$ and $\Lambda BV^{(p)}(\overline{\mathbb{T}})$.

In the sequel, $\mathbb{L}$ is the class of non-decreasing sequence $\Lambda = \{\lambda_i\}$ $(i = 1, 2, \ldots)$ of positive numbers such that $\sum_i \frac{1}{\lambda_i}$ diverges, a real number $p \geq 1$ and C represents a constant vary time to time.

R.G. Vyas (✉)
Faculty of Science, Department of Mathematics, The Maharaja Sayajirao University of Baroda, Vadodara, Gujarat, India
e-mail: drrgvyas@yahoo.com

V.K. Singh et al. (eds.), *Modern Mathematical Methods and High Performance Computing in Science and Technology*, Springer Proceedings in Mathematics & Statistics 171, DOI 10.1007/978-981-10-1454-3_19

2 Notations and Definitions

For a complex valued, 2π-periodic, function $f \in L^1(\overline{\mathbb{T}})$, its Fourier series is defined as

$$f(x) \sim \sum_{m\in\mathbb{Z}} \hat{f}(m)e^{imx}, \quad x \in \overline{\mathbb{T}},$$

where

$$\hat{f}(m) = \left(\frac{1}{2\pi}\right)\int_{\overline{\mathbb{T}}} f(x)e^{-imx}\,dx$$

denotes the mth Fourier coefficient of f.

For $p \geq 1$, the p-integral modulus of continuity of f over $\overline{\mathbb{T}}$ is define as

$$\omega^{(p)}(f;\delta) := \sup_{0<h\leq\delta} \| T_h f - f \|_p,$$

where $T_h f(x) = f(x+h)$ for all x and $\| (\cdot) \|_p$ denotes the L^p-norm over $\overline{\mathbb{T}}$. $p = \infty$ gives the modulus of continuity $\omega(f;\delta)$ of f.

Following the definition in [1], a sequence $\gamma = \{\gamma_m : m \in \mathbb{N}\}$ of nonnegative numbers is said to belongs to the class $\mathcal{A}_\alpha$ for some $\alpha \geq 1$ if

$$\left(\sum_{m\in\mathcal{D}_\mu} \gamma_m^\alpha\right)^{1/\alpha} \leq \kappa 2^{\mu(1-\alpha)/\alpha} \sum_{m\in\mathcal{D}_{\mu-1}} \gamma_m, \quad \mu \in \mathbb{N}, \tag{2.1}$$

where

$$\mathcal{D}_0 := \{1\}; \quad \mathcal{D}_\mu := \{2^{\mu-1}+1, 2^{\mu-1}+2, \ldots, 2^\mu\}, \quad \mu \in \mathbb{N}; \tag{2.2}$$

and the constant κ does not dependent on μ. Without the loss of generality, we assume that $\kappa \geq 1$.

Note that,

$$\mathcal{A}_{\alpha_2} \subset \mathcal{A}_{\alpha_1}, \quad where \quad 1 \leq \alpha_1 < \alpha_2 < \infty. \tag{2.3}$$

If a sequences γ is such that

$$\max\{\gamma_m : m \in \mathcal{D}_\mu\} \leq \kappa \, \min\{\gamma_m : m \in \mathcal{D}_{\mu-1}\}, \quad \mu \in \mathbb{N}, \tag{2.4}$$

then $\gamma \in \mathcal{A}_\alpha$ for every $\alpha \geq 1$. This inequality was introduced by Ul'yanov [3]. Moreover, Moricz and Veres [2] observed that, if a sequence $\gamma = \{\gamma_m\}$ is of the form

$$\gamma_m = m^\tau w(m), \quad m \in \mathbb{N},$$

where $\tau \in \mathbb{R}$ and $w : \mathbb{R}_+ \to \mathbb{R}_+$ is a slowly varying function, that is,

$$\lim_{x \to \infty} \frac{w(\lambda x)}{w(x)} = 1, \quad \textit{for every } 0 < \lambda < \infty, \tag{2.5}$$

then $\gamma \in \mathcal{A}_\alpha$ for every $\alpha \geq 1$.

For convenience in writing, put

$$\gamma_{-m} := \gamma_m, \quad m \in \mathbb{N}. \tag{2.6}$$

Definition 2.1 Given $\Lambda = \{\lambda_n\} \in \mathbb{L}$. A complex valued function f defined on an interval $I := [a, b]$ is said to be of $p - \Lambda$-bounded variation (that is, $f \in \Lambda BV^{(p)}(I)$) if

$$V_{\Lambda_p}(f, I) = \sup_{\{I_k\}} \left(\sum_k \frac{|f(I_k)|^p}{\lambda_k} \right)^{1/p} < \infty,$$

where $\{I_k\}$ is a finite collections of non-overlapping subintervals $I_k = [a_k, b_k] \subset [a, b]$ and $f(I_k) = f(b_k) - f(a_k)$.

Note that, for $p = 1$ and $\Lambda = \{1\}$ (that is, $\lambda_n = 1$, for all n,) the class $\Lambda BV^{(p)}(I)$ reduces to the class $BV(I)$ (the class of functions of bounded variation). For $p = 1$ the class $\Lambda BV^{(p)}(I)$ reduces to the class $\Lambda BV(I)$; and for $\Lambda = \{1\}$ the class $\Lambda BV^{(p)}(I)$ reduces to the class $BV^{(p)}(I)$ (the class of functions of p-bounded variation).

3 Results for Functions of Single Variable

Theorem 3.1 *If $f \in \Lambda BV(\overline{\mathbb{T}})$ and $\gamma = \{\gamma_m\} \in \mathcal{A}_{2/(2-\beta)}$ for some $\beta \in (0, 2)$ then*

$$\sum(\gamma; f)_\beta = \sum_{|m|\geq 1} \gamma_m |\hat{f}(m)|^\beta \leq \kappa C \sum_{\mu=0}^{\infty} 2^{-\mu\beta/2} \Gamma_{\mu-1} \left(\frac{(\omega(f; \frac{\pi}{2^\mu}))}{\sum_{i=1}^{2^\mu} \frac{1}{\lambda_i}} \right)^{\beta/2},$$

where κ is from (2.1) corresponding to $\alpha = 2/(2 - \beta)$ and C is a constant,

$$\Gamma_\mu := \sum_{m \in \mathcal{D}_\mu} \gamma_m \quad \textit{for} \quad \mu \in \mathbb{N}, \quad \textit{and} \quad \Gamma_{-1} := \Gamma_0 = \{\gamma_1\} \tag{3.1}$$

Corollary 3.2 *Under the hypothesis of Theorem 3.1, we have*

$$\sum(\gamma;f)_\beta \le \kappa C \sum_{m=1}^{\infty} m^{-\beta/2}\gamma_m \left(\frac{(\omega(f;\frac{\pi}{m}))}{\sum_{i=1}^{m}\frac{1}{\lambda_i}}\right)^{\beta/2},$$

In the case when $\gamma_m \equiv 1$, it follows from the above Corollary that $\sum(1;f)_\beta := \sum_{|m|\ge 1}|\hat{f}(m)|^\beta$

$$\le C\sum_{m=1}^{\infty} m^{-\beta/2}\left(\frac{(\omega(f;\frac{\pi}{m})}{\sum_{i=1}^{m}\frac{1}{\lambda_i}}\right)^{\beta/2}.$$

This gives the result [6, Theorem1,with$n_k = k$, forall k,] as a particular case.

Above corollary can easily follow from the Theorem 3.1.

Theorem 3.3 *If $f \in \Lambda BV^{(p)}(\overline{\mathbb{T}})$ and $\gamma = \{\gamma_m\} \in \mathcal{A}_{2/(2-\beta)}$ for some $\beta \in (0,2)$ then*

$$\sum(\gamma;f)_\beta \le \kappa C \sum_{\mu=0}^{\infty} 2^{-\mu\beta/2}\Gamma_{\mu-1}\left(\left(\frac{(\omega^{((2-p)s+p)}(f;\frac{\pi}{2^\mu}))^{2r-p}}{\sum_{i=1}^{2^\mu}\frac{1}{\lambda_i}}\right)^{1/r}\right)^{\beta/2},$$

where $\frac{1}{r}+\frac{1}{s}=1$, κ is from (2.1) corresponding to $\alpha = 2/(2-\beta)$ and C is a constant.

Corollary 3.4 *Under the hypothesis of Theorem 3.3, we have*

$$\sum(\gamma;f)_\beta \le \kappa C \sum_{m=1}^{\infty} m^{-\beta/2}\gamma_m \left(\left(\frac{(\omega^{((2-p)s+p)}(f;\frac{\pi}{m}))^{2r-p}}{\sum_{i=1}^{m}\frac{1}{\lambda_i}}\right)^{1/r}\right)^{\beta/2},$$

In the case when $\gamma_m \equiv 1$, it follows from the above Corollary that

$$\sum(1;f)_\beta := \sum_{|m|\ge 1}|\hat{f}(m)|^\beta$$
$$\le C\sum_{m=1}^{\infty} m^{-\beta/2}\left(\left(\frac{(\omega^{((2-p)s+p)}(f;\frac{\pi}{m}))^{2r-p}}{\sum_{i=1}^{m}\frac{1}{\lambda_i}}\right)^{1/r}\right)^{\beta/2}.$$

This gives the result [4, Theorem1, with $n_k = k$, forall k,] as a particular case.

Above Corollary 3.4 can be easily follows from the Theorem 3.3.

Proof of Theorem 3.1 $f \in \Lambda BV(\overline{\mathbb{T}})$ implies that f is bounded over $\overline{\mathbb{T}}$ and hence $f \in L^2(\overline{\mathbb{T}})$. For given $h > 0$, put $f_j = T_{jh}f - T_{(j-1)h}f$, then $\hat{f}_j(m) = 2i\hat{f}(m)e^{im(j-\frac{1}{2}h)}$ $\sin(\frac{mh}{2})$.

By Parseval's equality, we get

$$4\sum_{m\in\mathbb{Z}} |\hat{f}(m)|^2 \sin^2\left(\frac{mh}{2}\right) = O(||f_j||_2^2).$$

Putting $h = \frac{\pi}{2^\mu}$, $\mu \in \mathbb{N}$, and observing that

$$\frac{\pi}{4} < \frac{|m|\pi}{2^{\mu+1}} \leq \frac{\pi}{2} \quad \textit{for} \quad |m| \in \mathcal{D}_\mu, \quad \textit{implies} \quad \sin^2\left(\frac{mh}{2}\right) > \frac{1}{2}.$$

Thus, we have

$$\begin{aligned} B = \sum_{|m|\in\mathcal{D}_\mu} |\hat{f}(m)|^2 &= O\left(||f_j||_2^2\right) \\ &= O\left(\omega(f;h)\right)\left(\int_0^{2\pi} |f_j(x)|dx\right). \end{aligned} \tag{3.2}$$

Multiplying both the sides of the above inequality by $\frac{1}{\lambda_j}$ and then summing over $j = 1$ to $j = 2^\mu$, we have

$$B = O\left(\frac{\omega(f;h)}{\sum_{j=1}^{2^\mu}\frac{1}{\lambda_j}}\right)\left(\int_0^{2\pi}\sum_{j=1}^{2^\mu}\frac{(|f_j(x)|)}{\lambda_j}\,dx\right) = O\left(\frac{\omega(f;h)}{\sum_{j=1}^{2^\mu}\frac{1}{\lambda_j}}\right),$$

as $f \in \Lambda BV(\overline{\mathbb{T}})$ implies $\sum_{j=1}^{2^\mu}\frac{(|f_j(x)|)}{\lambda_j} = O(1)$.

Since $1 = \frac{\beta}{2} + \frac{2-\beta}{2}$, by Holder's inequality, for $\mu \geq 1$, we have

$$\begin{aligned} S_\mu := \sum_{|m|\in\mathcal{D}_\mu} \gamma_m|\hat{f}(m)|^\beta &\leq \left(\sum_{|m|\in\mathcal{D}_\mu}|\hat{f}(m)|^2\right)^{\beta/2}\left(\sum_{|m|\in\mathcal{D}_\mu}\gamma_m^{2/(2-\beta)}\right)^{(2-\beta)/2} \\ &\leq C\left(\frac{\omega(f;h)}{\sum_{j=1}^{2^\mu}\frac{1}{\lambda_j}}\right)^{\frac{\beta}{2}}\left(\sum_{|m|\in\mathcal{D}_\mu}\gamma_m^{2/(2-\beta)}\right)^{(2-\beta)/2}. \end{aligned} \tag{3.3}$$

Thus for $\mu \geq 1$,

$$S_\mu \leq C\kappa\left(2^{-\mu\beta/2}\,\Gamma_{\mu-1}\left(\frac{\omega(f;h)}{\sum_{j=1}^{2^\mu}\frac{1}{\lambda_j}}\right)^{\frac{\beta}{2}}\right).$$

If $\mu = 0$, then from (3.3) it follows that

$$S_0 := \gamma_1(|\hat{f}(-1)|^\beta + |\hat{f}(1)|^\beta) = O\left(\gamma_1\left(\frac{\omega(f;\pi)}{\frac{1}{\lambda_1}}\right)\right).$$

Hence, the result follows from

$$\sum_{|m|\geq 1} \gamma_m |\hat{f}(m)|^\beta = \sum_{\mu=0}^{\infty} S_\mu.$$

Proof of Theorem 3.3. $f \in \Lambda BV^{(p)}(\overline{\mathbb{T}})$ implies that f is bounded over $\overline{\mathbb{T}}$ [4, in view of Lemma 1, p.771] and hence $f \in L^2(\overline{\mathbb{T}})$. Proceeding as in the proof of Theorem 3.1, we get (3.2).

Since $2 = \frac{(2-p)s+p}{s} + \frac{p}{r}$, by using Holder's inequality, we have

$$||f_j||_2^2 \leq (||f_j||_p)^{p/r} \left(\int_0^{2\pi} |f_j|^{(2-p)s+p} dx\right)^{1/s} \leq (||f_j||_p)^{p/r}\, \Omega_h^{1/r},$$

where $\Omega_h^{1/r} = (\omega^{(2-p)s+p}(f;h))^{2r-p}$.

This together with (3.2) implies

$$B^r = \left(\sum_{|m|\in \mathcal{D}_\mu} |\hat{f}(m)|^2\right)^r = O\left(\Omega_h \int_0^{2\pi} |f_j(x)|^p dx\right).$$

Multiplying both the sides of the above inequality by $\frac{1}{\lambda_j}$ and then summing over $j = 1$ to $j = 2^\mu$, we have

$$B^r = O\left(\frac{\Omega_h}{\sum_{j=1}^{2^\mu} \frac{1}{\lambda_j}}\right)\left(\int_0^{2\pi} \sum_{j=1}^{2^\mu} \frac{(|f_j(x)|^p)}{\lambda_j}\, dx\right) = O\left(\frac{\Omega_h}{\sum_{j=1}^{2^\mu} \frac{1}{\lambda_j}}\right).$$

Thus

$$B = O\left(\frac{\Omega_h}{\sum_{j=1}^{2^\mu} \frac{1}{\lambda_j}}\right)^{1/r}.$$

Now, proceeding as in the proof of the Theorem 3.1 the result follows.

References

1. Gogoladze, L., Meskhia, R.: On the absolute convergence of trigonometric Fourier series. Proc. Razmadze. Math. Inst. **141**, 29–40 (2006)
2. Móricz, F., Veres, A.: Absolute convergence of multiple Fourier series revisited. Anal. Math. **34**(2), 145–162 (2008)
3. Ul'yanov, P.L.: Series with respect to a Haar system with monotone coefficients (in Russia). Izv. Akad. Nauk. SSSR Ser. Mat. **28**, 925–950 (1964)
4. Vyas, R.G.: On the absolute convergence of Fourier series of functions of $\Lambda BV^{(p)}$ and $\varphi\Lambda BV$. Georgian Math. J. **14**(4), 769–774 (2007)
5. Vyas, R.G., Darji, K.N.: On absolute convergence of multiple Fourier series. Math. Notes **94**(1), 71–81 (2013)
6. Vyas, R.G., Patadia, J.R.: On the absolute convergence of F ourier series of functions of generalized bounded variations. J. Indian Math. Soc. **62**(1–4), 129–136 (1996)

Some New Inequalities for the Ratio of Gamma Functions

Sourav Das and A. Swaminathan

Abstract In this work, certain new inequalities involving ratios of q-analogue of Euler gamma function are derived. Interesting generalizations and particular cases are also discussed.

Keywords Euler gamma function · Polygamma function · Inequalities · q-analogue

1 Introduction

In 1729, Euler introduced the gamma function for the generalization of $n!$ for non-integral values of n. He [4] discovered that

$$\int_0^1 (-\ln t)^x dt = \int_0^\infty t^x e^{-t} dt \quad (x > -1) \tag{1}$$

which gives $x!$ for $x \in \mathbb{N}$. Later, Legendre [12, Vol. 1, p. 221] denoted the integral (1) as $\Gamma(x+1)$ and is known as the gamma function. It can be noted that the integral on the right side of (1) converges for $x \in \mathbb{C}$ if $\Re(x) > -1$. For the brief history of the gamma function and its applications in various fields we refer to [2–9, 20–22] and references therein.

The problem of finding new inequalities for the ratio of gamma functions has attracted the attention of many researchers [1, 6, 10, 11, 13–19, 21]. In particular C. Alsina and M.S. Tomás [1] found the following inequality

S. Das (✉) · A. Swaminathan
Department of Mathematics, Indian Institute of Technology Roorkee, Uttarakhand 247667, Roorkee, India
e-mail: das90dma@iitr.ac.in

A. Swaminathan
e-mail: swamifma@iitr.ac.in

V.K. Singh et al. (eds.), *Modern Mathematical Methods and High Performance Computing in Science and Technology*, Springer Proceedings in Mathematics & Statistics 171, DOI 10.1007/978-981-10-1454-3_20

$$\frac{1}{n!} \leq \frac{\Gamma(1+x)^n}{\Gamma(1+nx)} \leq 1.$$

for all $x \in [0, 1]$ and all nonnegative integers n, which was generalized in [19].

There are several generalization of the above mentioned inequalities exist in the literature. For example [8, Theorem 2.1]

$$\frac{1}{\Gamma_q(1+a)} \leq \frac{\Gamma_q(1+x)^a}{\Gamma_q(1+ax)} \leq 1,$$

for all $a \geq 1$ and all $x \in [0, 1]$ with $q \in (0, 1)$.

For the case $q \to 1$, it was given by J. Sándor [19]. Similarly, the inequality [13, Theorem 2.3]

$$\frac{\Gamma_q(a)^c}{\Gamma_q(b)^d} \leq \frac{\Gamma_q(a+bx)^c}{\Gamma_q(b+ax)^d} \leq \frac{\Gamma_q(a+b)^c}{\Gamma_q(a+b)^d} \tag{2}$$

for $q \to 1$ was proved by A.Sh. Shabani [16, Theorem 2.4] using series representation of digamma function $\psi(x) = \frac{\Gamma'(x)}{\Gamma(x)}$.

A.Sh. Shabani [18] generalized [17] and unified many existing for q-analogue of gamma functions by

$$\frac{\Gamma_q(a+b)^c}{\Gamma_q(d+e)^f} \leq \frac{\Gamma_q(a+bx)^c}{\Gamma_q(d+ex)^f} \leq \frac{\Gamma_q(a)^c}{\Gamma_q(d)^f} \tag{3}$$

for $q \in (0, 1)$, $x \in [0, 1]$, $a + bx > 0$, $d + ex > 0$, $a + bx \leq d + ex$, $ef \geq bc > 0$ with $\psi_q(a + bx) > 0$ or $\psi_q(d + ex) > 0$.

In this work, we are interested in establishing some double inequalities for the polygamma function related to (2) and (3). This article is organized as follows. First we will establish the bounds for the ratio of $\psi_q^{(n)}(a + bx)^c$ in Sect. 2. In Sect. 3, we will generalize the results of Sect. 2 and find the bounds for the ratio of $\psi_q^{(n)}(a + bx)^c$. In Sect. 4 bounds for the ratio of $\Gamma(x + \alpha)$ and its q-analogue are established.

2 Inequalities Involving q-Analogue of Polygamma Functions

Note that digamma function is defined as $\psi(x) = \frac{\Gamma'(x)}{\Gamma(x)}$. Polygamma function is defined as the nth order derivative of digamma function. The q-analogue of digamma function and polygamma function are defined, respectively, as $\psi_q(x) = \frac{\Gamma_q'(x)}{\Gamma_q(x)}$ and $\psi_q^{(n)}(x) = \frac{d^n}{dx^n}\psi_q(x)$.

From [15, Eq. 1.17] we have, for $0 < q < 1$,

$$\psi_q(x) = -\ln(1-q) + \ln q \sum_{k=1}^{\infty} \frac{q^{kx}}{1-q^k}. \tag{4}$$

Differentiating (4) with respect to x, n times we have

$$\psi_q^{(n)}(x) = (\ln q)^{n+1} \sum_{k=1}^{\infty} \frac{k^n q^{kx}}{1-q^k}, \quad n \geq 1. \tag{5}$$

Since $\ln q < 0$ for $0 < q < 1$, it is easy to see that

$$\psi_q^{(n)}(x) = \begin{cases} > 0, & \text{if } n \text{ is odd;} \\ < 0, & \text{if } n \text{ is even.} \end{cases}$$

Lemma 1 *Let $x \in [0, 1]$, $q \in (0, 1)$ and a, b be any two real numbers such that $a \geq b > 0$. Then*

$$\psi_q^{(n)}(a+bx) \geq \psi_q^{(n)}(b+ax), \quad \textit{if } n \textit{ is even;}$$
$$\psi_q^{(n)}(a+bx) \leq \psi_q^{(n)}(b+ax), \quad \textit{if } n \textit{ is odd.}$$

Proof Let $x \geq y > 0$. Then, from (5);

$$\psi_q^{(n)}(x) - \psi_q^{(n)}(y) = (\ln q)^{n+1} \sum_{k=1}^{\infty} \frac{k^n}{(1-q^k)} (q^{kx} - q^{ky}).$$

Now,

$$x \geq y > 0 \Longrightarrow q^x \leq q^y \quad \text{for } 0 < q < 1.$$

Hence, $\psi_q^{(n)}(x) - \psi_q^{(n)}(y) \geq 0$ (≤ 0) if n is even (odd). Replacing x by $a + bx$ and y by $b + ax$ gives the result.

Lemma 2 *Let n be any odd natural number and $x \in [0, 1]$, $q \in (0, 1)$, a, b be any two real numbers such that $a \geq b > 0$. Let c, d be any two positive real numbers such that $ad \geq bc > 0$. Then*

$$bc\psi_q^{(n+1)}(a+bx)\psi_q^{(n)}(b+ax) - ad\psi_q^{(n)}(a+bx)\psi_q^{(n+1)}(b+ax) \geq 0.$$

Proof Let n be odd natural number. Then

$$\psi_q^{(n)}(a+bx) > 0,\ \psi_q^{(n)}(b+ax) > 0,\ \psi_q^{(n+1)}(a+bx) < 0 \text{ and } \psi_q^{(n+1)}(b+ax) < 0.$$

Now by Lemma 1, we have

$$\begin{aligned} bc\psi_q^{(n+1)}(a+bx)\psi_q^{(n)}(b+ax) &\geq ad\psi_q^{(n+1)}(a+bx)\psi_q^{(n)}(b+ax) \\ &\geq ad\psi_q^{(n+1)}(b+ax)\psi_q^{(n)}(a+bx) \end{aligned}$$

and the proof is complete.

Theorem 1 *Let $f(x)$ be a function defined as*

$$f(x) = \frac{\psi_q^{(n)}(a+bx)^c}{\psi_q^{(n)}(b+ax)^d}.$$

Let n be odd natural number, $x \in [0, 1]$, $q \in (0, 1)$, $a \geq b > 0$, $c, d > 0$ with $ad \geq bc > 0$. Then $f(x)$ is increasing on $[0, 1]$ and the following double inequality holds:

$$\frac{\psi_q^{(n)}(a)^c}{\psi_q^{(n)}(b)^d} \leq \frac{\psi_q^{(n)}(a+bx)^c}{\psi_q^{(n)}(b+ax)^d} \leq \frac{\psi_q^{(n)}(a+b)^c}{\psi_q^{(n)}(a+b)^d}.$$

Proof Let $g(x) = \log f(x)$. Then

$$g'(x) = \frac{bc\psi_q^{(n+1)}(a+bx)\psi_q^{(n)}(b+ax) - ad\psi_q^{(n+1)}(b+ax)\psi_q^{(n)}(a+bx)}{\psi_q^{(n)}(a+bx)\psi_q^{(n)}(b+ax)}.$$

By Lemma 2, $g'(x) > 0$, which implies $g(x)$ is increasing function on $[0, 1]$. Hence $f(x)$ is also increasing function on $[0, 1]$. Consequently, using $f(0) \leq f(x) \leq f(1)$, proves the theorem.

These results can be obtained for other cases like $x \geq 1$ and n being even natural numbers, by similar procedures. We state some of them without proof.

Lemma 3 *Let $x \geq 1$, $q \in (0, 1)$ and $a, b > 0$ with $b \geq a$. Then*

$$\begin{aligned} \psi_q^{(n)}(a+bx) &\geq \psi_q^{(n)}(b+ax), \quad \text{if } n \text{ is even;} \\ \psi_q^{(n)}(a+bx) &\leq \psi_q^{(n)}(b+ax), \quad \text{if } n \text{ is odd.} \end{aligned}$$

Lemma 4 *Let $x \geq 1$, $q \in (0, 1)$, $b \geq a > 0$ and c, d be any two real numbers such that $ad \geq bc > 0$. Then*

$$bc\psi_q^{(n+1)}(a+bx)\psi_q^{(n)}(b+ax) - ad\psi_q^{(n)}(a+bx)\psi_q^{(n+1)}(b+ax) \geq 0$$

for n as odd natural number.

Theorem 2 *Let $x \geq 1$, $q \in (0, 1)$ and $b \geq a > 0$ and c, d be two real numbers such that $ad \geq bc > 0$. Then $\frac{\psi_q^{(n)}(a+bx)^c}{\psi_q^{(n)}(b+ax)^d}$ is an increasing function on $[1, \infty)$ for any odd natural number n.*

For even numbers, the results follows in similar way. We omit the proof.

Lemma 5 *Let n be any even natural number, $x \in [0, 1]$, $q \in (0, 1)$, $a \geq b > 0$ and c, d be two real numbers such that $bc \geq ad > 0$. Then*

$$bc\psi_q^{(n+1)}(a+bx)\psi_q^{(n)}(b+ax) - ad\psi_q^{(n)}(a+bx)\psi_q^{(n+1)}(b+ax) \geq 0.$$

Lemma 6 *Let n any even natural number, $x \geq 1$, $q \in (0, 1)$, $b \geq a > 0$ and $c, d > 0$ be such that $bc \geq ad > 0$. Then*

$$bc\psi_q^{(n+1)}(a+bx)\psi_q^{(n)}(b+ax) - ad\psi_q^{(n)}(a+bx)\psi_q^{(n+1)}(b+ax) \leq 0.$$

Theorem 3 *Let $f(x)$ be a function defined as*

$$f(x) = \frac{\psi_q^{(n)}(a+bx)^c}{\psi_q^{(n)}(b+ax)^d}.$$

Let n be any even natural number, $x \in [0, 1]$, $q \in (0, 1)$, $a \geq b > 0$ and c, d be any two real numbers such that $bc \geq ad > 0$. Then $f(x)$ is decreasing on $[0, 1]$ and satisfy the following double inequality:

$$\frac{\psi_q^{(n)}(a+b)^c}{\psi_q^{(n)}(a+b)^d} \leq \frac{\psi_q^{(n)}(a+bx)^c}{\psi_q^{(n)}(b+ax)^d} \leq \frac{\psi_q^{(n)}(a)^c}{\psi_q^{(n)}(b)^d}.$$

Theorem 4 *Let n be any even natural number, $x \geq 1$, $q \in (0, 1)$, $b \geq a > 0$ and c, d be any real numbers such that $ad \geq bc > 0$. Then $\frac{\psi_q^{(n)}(a+bx)^c}{\psi_q^{(n)}(b+ax)^d}$ is an decreasing function on $[1, \infty)$.*

3 Some Generalizations

In this section we will generalize the results of Sect. 2. The following result is similar to Lemma 1.

Lemma 7 *Let $q \in (0, 1)$ and $y > x > 0$. Then*

$$\psi_q^{(n)}(x) < \psi_q^{(n)}(y), \quad \text{if } n \text{ is even};$$
$$\psi_q^{(n)}(x) > \psi_q^{(n)}(y), \quad \text{if } n \text{ is odd}.$$

With the help of Lemma 7 it is easy to prove the following lemma.

Lemma 8 *Let $q \in (0,1)$ and $a+bx > 0$, $d+ex > 0$ with $a+bx \le d+ex$. Then*

$$\psi_q^{(n)}(a+bx) - \psi_q^{(n)}(d+ex) \le 0 \quad \text{if } n \text{ is even;}$$
$$\psi_q^{(n)}(a+bx) - \psi_q^{(n)}(d+ex) \ge 0 \quad \text{if } n \text{ is odd.}$$

Lemma 9 *Let a, b, c, d, e, f be real numbers such that $a+bx > 0$, $d+ex > 0$, $a+bx \le d+ex$ and $bc \ge ef > 0 > 0$. Let $q \in (0,1)$ and n be odd. Then*

$$bc\psi_q^{(n+1)}(a+bx)\psi_q^{(n)}(d+ex) - ef\psi_q^{(n+1)}(d+ex)\psi_q^{(n)}(a+bx) \le 0.$$

Proof Let n be odd. Then $\psi_q^{(n+1)}(a+bx) < 0$ and $\psi_q^{(n)}(d+ex) > 0$. Now by Lemma 8 we have,

$$\begin{aligned} bc\psi_q^{(n+1)}(a+bx)\psi_q^{(n)}(d+ex) &\le ef\psi_q^{(n+1)}(a+bx)\psi_q^{(n)}(d+ex) \\ &\le ef\psi_q^{(n+1)}(d+ex)\psi_q^{(n)}(a+bx). \end{aligned}$$

which proves the result.

Theorem 5 *Let n be any odd natural number and $f(x)$ be a function defined as*

$$f(x) = \frac{\psi_q^{(n)}(a+bx)^c}{\psi_q^{(n)}(d+ex)^f}, \quad x \ge 0, \ q \in (0,1)$$

where a, b, c, d, e, f are real numbers such that $a+bx > 0$, $d+ex > 0$, $a+bx \le d+ex$ with $bc \ge ef > 0$. Then $f(x)$ is decreasing for $x \ge 0$ and for all $x \in [0,1]$, the following double inequality holds:

$$\frac{\psi_q^{(n)}(a+b)^c}{\psi_q^{(n)}(d+e)^f} \le \frac{\psi_q^{(n)}(a+bx)^c}{\psi_q^{(n)}(d+ex)^f} \le \frac{\psi_q^{(n)}(a)^c}{\psi_q^{(n)}(d)^f}. \tag{6}$$

Proof Let n be odd. Then $\psi_q^{(n)}(a+bx)$, $\psi_q^{(n)}(d+ex) > 0$. Let $g(x) = \log f(x)$. Then

$$g'(x) = \frac{bc\psi_q^{(n+1)}(a+bx)\psi_q^{(n)}(d+ex) - ef\psi_q^{(n+1)}(d+ex)\psi_q^{(n)}(a+bx)}{\psi_q^{(n)}(a+bx)\psi_q^{(n)}(d+ex)}.$$

Now by using Lemma 9, we have $g'(x) \le 0$. Hence $g(x)$ is decreasing for all $x \ge 0$, which implies that $f(x)$ is decreasing for all $x \ge 0$. Consequently, $f(x)$ is decreasing in $[0,1]$. Hence $f(1) \le f(x) \le f(0)$ for $x \in [0,1]$, which proves the theorem.

The results for even numbers follows in similar way. For clarity, the results are directly stated.

Lemma 10 *Let a, b, c, d, e, f be real numbers such that $a + bx > 0$, $d + ex > 0$, $a + bx \leq d + ex$ and $ef \geq bc > 0$. Let $q \in (0, 1)$ and n be even natural number. Then*

$$bc\psi_q^{(n+1)}(a + bx)\psi_q^{(n)}(d + ex) - ef\psi_q^{(n+1)}(d + ex)\psi_q^{(n)}(a + bx) \geq 0.$$

Theorem 6 *Let $f(x)$ be a function defined as*

$$f(x) = \frac{\psi_q^{(n)}(a + bx)^c}{\psi_q^{(n)}(d + ex)^f}, \quad x \geq 0, \ q \in (0, 1)$$

where a, b, c, d, e, f are real numbers such that $a + bx > 0$, $d + ex > 0$, $a + bx \leq d + ex$ and $ef \geq bc > 0$. If n is even natural number then $f(x)$ is increasing for all $x \geq 0$ and for all $x \in [0, 1]$, the following double inequality holds:

$$\frac{\psi_q^{(n)}(a)^c}{\psi_q^{(n)}(d)^f} \leq \frac{\psi_q^{(n)}(a + bx)^c}{\psi_q^{(n)}(d + ex)^f} \leq \frac{\psi_q^{(n)}(a + b)^c}{\psi_q^{(n)}(d + e)^f}. \tag{7}$$

Remark 1 For particular values of parameters given in this section, results of previous section and other results in the literature can be obtained.

4 Inequalities Involving the Gamma Function

The results given in this section are different from the inequalities of (2) and (3). However, the results are given there due to their importance in Asymptotic analysis.

In this section, we will find bounds for $\frac{\Gamma(x+\alpha)}{\Gamma(x+\beta)}$ and $\frac{\Gamma_q(x+\alpha)}{\Gamma_q(x+\beta)}$ for $\alpha \geq \beta > 0$ and $q \in (0, 1)$ using techniques given in [11].

Theorem 7 *Let $x > 0$ and $\alpha \geq \beta > 0$, then*

$$e^{(\alpha-\beta)\psi(x+\beta)} \leq \frac{\Gamma(x + \alpha)}{\Gamma(x + \beta)} \leq e^{(\alpha-\beta)\psi(x+\alpha)}$$

holds true and the equality holds if and only if $\alpha = \beta$.

Proof The case $\alpha = \beta$ is trivial.

Let $f(t) = \log \Gamma(t)$. Then for fixed $x > 0$, the classical mean value theorem on $[x + \beta, x + \alpha]$, gives

$$f'(d) = \frac{\log \frac{\Gamma(x+\alpha)}{\Gamma(x+\beta)}}{\alpha - \beta} \quad \text{for } d \in (x + \beta, x + \alpha).$$

Taking $d = c + x$ we have, $(\alpha - \beta)\psi(x + c) = \log \frac{\Gamma(x+\alpha)}{\Gamma(x+\beta)}, \quad c \in (\beta, \alpha)$.

Since $\psi(x)$ is increasing for $x > 0$, we get

$$(\alpha - \beta)\psi(x + \beta) < \log \frac{\Gamma(x + \alpha)}{\Gamma(x + \beta)} < (\alpha - \beta)\psi(x + \alpha)$$

$$\Longrightarrow e^{(\alpha-\beta)\psi(x+\beta)} < \frac{\Gamma(x + \alpha)}{\Gamma(x + \beta)} < e^{(\alpha-\beta)\psi(x+\alpha)}.$$

The following generalization is immediate.

Theorem 8 *Let $x > 0$, $q \in (0, 1)$ and $\alpha \geq \beta > 0$, then*

$$e^{(\alpha-\beta)\psi_q(x+\beta)} \leq \frac{\Gamma_q(x + \alpha)}{\Gamma_q(x + \beta)} \leq e^{(\alpha-\beta)\psi_q(x+\alpha)}$$

holds true and the equality holds if and only if $\alpha = \beta$.

Acknowledgments The authors wish to thank the reviewers for their useful suggestions.

References

1. Alsina, C., Tomás, M.S.: A geometrical proof of a new inequality for the gamma function. JIPAM. J. Inequal. Pure Appl. Math. **6**(2), Article 48, 3 p. (2005). (electronic)
2. Andrews, G.E., Askey, R., Roy, R.: Special functions. Encyclopedia of Mathematics and its Applications, vol. 71. Cambridge Univ. Press, Cambridge (1999)
3. Ernst, T.: A comprehensive treatment of q-calculus. Birkhäuser/Springer Basel AG, Basel (2012)
4. Euler, L.: De progressionibus transcendentibus seu quarum termini generales algebraice dari nequeunt. In: Böhm, C., Faber, G. (eds.) Opera Omnia (1), vol. 14, pp. 1–24. B. G. Tubner, Berlin (1925)
5. Gasper, G., Rahman, M.: Encyclopedia of Mathematics and its Applications. Basic Hypergeometric Series, vol. 96, 2nd edn. Cambridge Univ. Press, Cambridge (2004)
6. Gautschi, W.: Some elementary inequalities relating to the gamma and incomplete gamma function, J. Math. Phys. **38**, 77–81 (1959/60)
7. Koekoek, R., Lesky, P.A., Swarttouw, R.F.: Hypergeometric Orthogonal Polynomials and Their q-Analogues. Springer Monographs in Mathematics. Springer, Berlin (2010)
8. Kim, T., Adiga, C.: On the q-analogue of gamma functions and related inequalities, JIPAM. J. Inequal. Pure Appl. Math. **6**(4), Article 118, 4 p. (2005). (electronic)
9. Laforgia, A.: Further inequalities for the gamma function. Math. Comput. **42**(166), 597–600 (1984)
10. Laforgia, A., Natalini, P.: Some inequalities for the ratio of gamma functions. J. Inequal. Spec. Funct. **2**(1), 16–26 (2011)
11. Laforgia, A., Natalini, P.: On some inequalities for the gamma function. Adv. Dyn. Syst. Appl. **8**(2), 261–267 (2013)
12. Legendre, A.M.: Exercices de calcul intégral sur divers ordres de transcendantes et sur les quadratures, Mme Ve Courcier, Paris, I (1811), II (1817), III (1816)
13. Mansour, T.: Some inequalities for the q-gamma function, JIPAM. J. Inequal. Pure Appl. Math. **9**(1), Article 18, 4 p. (2008)
14. Nantomah, K.: Some Inequalities for the Ratios of Generalized Digamma Functions. Advances in Inequalities and Applications, vol. 2014, Article ID 28 (2014)

15. Qi, F.: Bounds for the ratio of two gamma functions, J. Inequal. Appl. **2010**, Article ID 493058, 84 p
16. Shabani, A.Sh.: Some inequalities for the gamma function. JIPAM. J. Inequal. Pure Appl. Math. **8**(2), Article 49, 4 p. (2007)
17. Shabani, A.Sh.: Generalization of some inequalities for the gamma function. Math. Commun. **13**(2), 271–275 (2008)
18. Shabani, A.Sh.: Generalization of some inequalities for the q-gamma function. Ann. Math. Inf. **35**, 129–134 (2008)
19. Sándor, J.: A note on certain inequalities for the Gamma function, JIPAM. J. Inequal. Pure Appl. Math. **6**(3), Article 61, 3 p. (2005). (electronic)
20. Srinivasan, G.K.: The gamma function: an eclectic tour. Am. Math. Mon. **114**(4), 297–315 (2007)
21. Watson, G.N.: A note on Gamma functions. Proc. Edinb. Math. Soc. **11**(2) (1958/1959), Edinb. Math. Notes No. 42, pp. 7–9 (misprinted 41) (1959)
22. Whittaker, E.T., Watson, G.N.: A course of modern analysis, reprint of the fourth (1927) edition, Cambridge Mathematical Library. Cambridge Univ. Press, Cambridge (1996)

Some New I-Lacunary Generalized Difference Sequence Spaces in n-Normed Space

Tanweer Jalal

Abstract In this paper, we introduce a new class of ideal convergent (briefly I-convergent) sequence spaces using, infinite matrix, lacunary sequences with respect to a sequence of modulus functions and difference operator defined on n-normed space. We study these spaces for some linear topological structures and algebraic properties. We also give some inclusion relations for these sequence spaces.

Keywords I-convergence · n-normed · Infinite matrix · Lacunary sequences · Modulus function

1 Introduction

The notion of ideal convergence (I-convergence) was first introduced by Kostyrko et al. [12] as a generalization of statistical convergence of sequences in a metric space and studied some properties of such convergence. Since then many researchers have studied these subjects and obtained various results (see [12, 13]). Note that I -convergence is an interesting generalization of statistical convergence.

Let X be a nonempty set, then a family of sets $I \subset 2^x$ (the class of all subsets of all X) is called an ideal if and only if for each $A, B \in I$, we have $A \cup B \in I$ and for each $A \in I$ and $B \subset A$, we have $B \in I$. A non-empty family of sets $F \subset 2^x$ is a filter on X if and only if $\phi \notin F$, for each $A, B \in F$, we have $A \cap B \in F$ and for each $A \in F$ and each $A \subset F$, we have $B \in F$. An ideal I is called nontrivial ideal if each $I \neq \phi$ and $X \notin I$. Evidently $I \subset 2^x$ is a nontrivial ideal if and only if $F = F(I) = \{X - A : A \in I\}$ is a filter on X. A nontrivial ideal $I \subset 2^x$ is called admissible if and only if $\{\{x\} : x \in X\} \subset I$. A non-trivial ideal I is maximal if there cannot exist any non-trivial $J \neq I$ containing I as a subset. Further details on ideals can be found in Kostyrko et al. [12].

T. Jalal (✉)
Dapartment of Mathematics, National Institute of Technology,
Hazratbal, Srinagar 190006, India
e-mail: tjalal@rediffmail.com

V.K. Singh et al. (eds.), *Modern Mathematical Methods and High Performance Computing in Science and Technology*, Springer Proceedings in Mathematics & Statistics 171, DOI 10.1007/978-981-10-1454-3_21

The space of lacunary strongly convergent sequences was defined by Freedman et al. [3]. By a lacunary sequence $\theta = (k_r); r = 0, 1, 2, \ldots$ where $k_0 = 0$, we shall mean an increasing sequence of non-negative integers with $k_r - k_{r-1} \to \infty$ as $r \to \infty$. The intervals determined by θ will be denoted by $I_r = (K_{r-1}, K_r]$ and $h_r = k_r - k_{r-1}$.

The notion of difference sequence spaces was introduced by Kizmaz [11]. It was further generalized by Et and Colak [1]. Later on Et and Esi [2] defined the sequence spaces

$$X(\Delta_m^s) = \{x = (x_k) \in w : (\Delta_m^s x_k) \in X\}.$$

for $X = l_\infty, c$ and c_0, where $s \in N$ and $m = (m_k)$ is any fixed sequence of non-zero complex numbers and

$$\Delta_m^0 = m_k x_k, \quad \Delta_m x = m_k x_k - m_{k+1} x_{k+1}$$

are non-negative,

$$\Delta_m^s x = (\Delta_m^s x_k) = (\Delta_m^{s-1} x_k - \Delta_m^{s-1} x_{k+1})$$

and so that

$$\Delta_m^s x_k = \sum_{\nu=0}^{s} (-1)^\nu (s\nu) m_{k+\nu} x_{k+\nu}.$$

Taking $m = s = 1$, we get the spaces of $l_\infty(\Delta), c(\Delta), c_0(\Delta)$, introduced and studied by Kizmaz [11].

The concept of 2-normed spaces was initially introduced by Gahler [4] in the mid 1960s. Since then this concept has been studied by many authors and generalized to the notion of n-normed spaces, see for instance [5, 10, 14]. Sahiner et al. [16] introduced the notion of I-convergence in 2-normed spaces. Later on it was extended to n-normed spaces by Gurdal and Sahiner [7], Savas [17] and Jalal [8, 9].

A modulus function is a function $f : [0, \infty) \to [0, \infty)$, such that

(i) $f(x) = 0$ if and only if $x = 0$.
(ii) $f(x + y) \leq f(x) + f(y)$,
(iii) f is increasing;
(iv) f is continuous from right at zero.

The following well-known inequality will be used throughout the article. Let $p = (p_k)$ be any sequence of positive real numbers with $0 \leq p_k \leq \sup_k p_k = G$, $D = \max\{1, 2^{G-1}\}$ then

$$|a_k + b_k|^{p_k} \leq D(|a_k|^{p_k} + |b_k|^{p_k})$$

for all $k \in N$ and $a_k, b_k \in C$. Also $|a|^{p_k} \leq \max\{1, |a|^G\}$ for all $a \in C$.

2 Definitions and Preliminaries

Let n be a non-negative integer and X be a real vector space of dimension $d \geq n$ (d may be infinite). A real-valued function $||., \ldots, .||$ on X^n satisfying the following conditions:

(1) $||(x_1, x_2, \ldots, x_n)|| = 0$ if and only if $x_1, x_2, \ldots, x_n$ are linearly dependent.
(2) $||(x_1, x_2, \ldots, x_n)||$ is invariant under permutation,
(3) $||\alpha(x_1, x_2, \ldots, x_n)|| = |\alpha|||(x_1, x_2, \ldots, x_n)||$, for any $\alpha \in R$.
(4) $||(x_1 + \overline{x}, x_2, \ldots, x_n)|| \leq ||(x_1, x_2, \ldots, x_n)|| + ||(\overline{x}, x_2, \ldots, x_n)||$.

is called an n-norm on X and the pair $(X, ||., \ldots, .||)$ is called an n-normed space (see [6]).

A trivial example of an n-normed space is $X = R^n$, equipped with the Euclidean n-norm $||(x_1, x_2, \ldots, x_n)||_E =$ volume of the n-dimensional parallelepiped spanned by the vectors $x_1, x_2, \ldots, x_n$ which may be given explicitly by the formula

$$||(x_1, x_2, \ldots, x_n)||_E = |det(x_{ij})| = abs(det(< x_i, x_j >))$$

where $x_i = (x_{i1}, x_{i2}, \ldots, x_{in}) \in R^n$ for each $i = 1, 2, \ldots, n$.

Let $(X, ||., \ldots, .||)$ be an n-normed space of dimension $d \geq n \geq 2$ and $\{a_1, a_2, \ldots, a_n\}$ be a linearly independent set in X. Then the function $||., \ldots, .||$ on X^{n-1} is defined by

$$||(x_1, x_2, \ldots, x_n)||_\infty = \max_{1 \leq i \leq n} \{||x_1, x_2, \ldots, x_{n-1}, a_i||\}$$

defines as (n − 1)-norm on X with respect to $\{a_1, a_2, \ldots, a_n\}$ and this is known as the derived (n − 1)-norm.

The standard n-norm on a real inner product space of dimension $d \geq n$ is as follows:

$$||(x_1, x_2, \ldots, x_n)||_s = [det(< x_i, x_j >)]^{\frac{1}{2}},$$

where $< ., . >$ denotes the inner product on X. If we take $X = R^n$ then this n-norm is exactly the same as the Euclidean n-norm $||(x_1, x_2, \ldots, x_n)||_E$ mentioned earlier. For n = 1, this n-norm is the usual norm $||x_1|| = \sqrt{< x_1, x_1 >}$ for further details (see [5]).

We first introduce the following definitions.

Definition 2.1 A sequence (x_k) in an n-normed space $(X, ||., \ldots, .||)$ is said to be convergent to some $L \in X$ with respect to the n-norm if for each $\epsilon > 0$ there exists a positive integer n_0 such that $||x_k - L, z_1, z_2, \ldots, z_{n-1}|| < \epsilon$, for all $k > n_0$ and for every $z_1, z_2, \ldots, z_{n-1} \in X$.

Definition 2.2 A sequence (x_k) in an n-normed space $(X, ||., \ldots, .||)$ is said to be I-convergent to some $L \in X$ with respect to the n-norm if for each $\epsilon > 0$

such that the set $\{k \in N : ||x_k - L, z_1, z_2, \ldots, z_{n-1}|| \geq \epsilon\}$ belongs to I for every $z_1, z_2, \ldots, z_{n-1} \in X$.

In this article, we study some new ideal convergent sequence spaces on n-normed spaces using an infinite matrix $A = (a_{nk})$, modulus functions and generalized difference operator.

3 Main Results

Before we state our main results, first we shall present some new ideal convergent sequence spaces by combining an infinite matrix $A = (a_{nk})$, lacunary sequences and modulus functions and study their linear topological structures. Also we give some relations related to these sequence spaces.

Let I be an admissible ideal of N, $p = (p_n)$ be a bounded sequence of positive real numbers for all $n \in N$ and $A = (a_{nk})$ be an infinite matrix. Let f be a modulus function and $(X, ||., \ldots, .||)$ be an n-normed space. $w(n-X)$ denotes the spaces of X-valued sequence spaces defined over $(X, ||., \ldots, .||)$. For every $z_1, z_2, \ldots, z_{n-1} \in X$ and for every $\epsilon > 0$ we define the following sequence spaces:

$$[N_\theta, A, \Delta^s_m, f, p, ||., \ldots, .||]^I$$
$$= \left\{x = (x_k) \in w(n-X) : \left\{r \in N : \frac{1}{h_r} \sum_{n \in I_r} f\left[\sum_{k=1}^{\infty} a_{nk}(||\Delta^s_m(x) - L, z_1, z_2, \ldots, z_{n-1}||)\right]^{p_n} \geq \epsilon\right\} \in I,\ for\ L \in X\right\}.$$

$$[N_\theta, A, \Delta^s_m, f, p, ||., \ldots, .||]^I_0$$
$$= \left\{x = (x_k) \in w(n-X) : \left\{r \in N : \frac{1}{h_r} \sum_{n \in I_r} f\left[\sum_{k=1}^{\infty} a_{nk}(||\Delta^s_m(x), z_1, z_2, \ldots, z_{n-1}||)\right]^{p_n} \geq \epsilon\right\} \in I\right\}.$$

Theorem 1 *If $p = (p_n)$ is bounded then the spaces $\left[N_\theta, A, \Delta^s_m, p, \|., \ldots, .\|\right]^I$ and $\left[N_\theta, A, \Delta^s_m, p, \|., \ldots, .\|\right]^I_0$ are linear.*

Proof We shall prove the theorem for the space $\left[N_\theta, A, \Delta^s_m, p, \|., \ldots, .\|\right]^I_0$ only and the other can be proved in a similar manner. Let $x = (x_k)$ and $y = (y_k)$ be two elements in $\left[N_\theta, A, \Delta^s_m, p, \|., \ldots, .\|\right]^I_0$ and let α, β be scalars in $\Re$. Therefore

$$\left\{r \in N : \frac{1}{h_r} \sum_{n \in I_r} f\left[\sum_{k=1}^{\infty} a_{nk}\left(\|\Delta^s_m x_k, z_1, z_2, \ldots, z_{n-1}\|\right)\right]^{p_n} \geq \epsilon\right\} \in I$$

and

$$\left\{r \in N : \frac{1}{h_r} \sum_{n \in I_r} f\left[\sum_{k=1}^{\infty} a_{nk}\left(\|\Delta^s_m y_k, z_1, z_2, \ldots, z_{n-1}\|\right)\right]^{p_n} \geq \epsilon\right\} \in I.$$

Since f is a modulus function and Δ_m^s is linear the following inequality holds:

$$\frac{1}{h_r}\sum_{n\in I_r} f\left[\sum_{k=1}^{\infty} a_{nk}\left(\|\Delta_m^s(\alpha x_k+\beta y_k), z_1, z_2, \dots, z_{n-1}\|\right)\right]^{p_n}$$
$$\le D\frac{1}{h_r}T_\alpha^{\sup p_n}\sum_{n\in I_r} f\left[\sum_{k=1}^{\infty} a_{nk}\left(\|\Delta_m^s x_k, z_1, z_2, \dots, z_{n-1}\|\right)\right]^{p_n}$$
$$+ D\frac{1}{h_r}T_\beta^{\sup p_n}\sum_{n\in I_r} f\left[\sum_{k=1}^{\infty} a_{nk}\left(\|\Delta_m^s y_k, z_1, z_2, \dots, z_{n-1}\|\right)\right]^{p_n}$$

where T_α and T_β are positive integers such that $|\alpha| \le T_\alpha$ and $|\beta| \le T_\beta$. On the other hand from the above inequality, we get

$$\left\{r\in N : \frac{1}{h_r}\sum_{n\in I_r} f\left[\sum_{k=1}^{\infty} a_{nk}\left(\|\Delta_m^s(\alpha x_k+\beta y_k), z_1, z_2, \dots, z_{n-1}\|\right)\right]^{p_n} \ge \epsilon\right\}$$
$$\subseteq \left\{r\in N : D\frac{1}{h_r}T_\alpha^{\sup p_n}\sum_{n\in I_r} f\left[\sum_{k=1}^{\infty} a_{nk}\left(\|\Delta_m^s x_k, z_1, z_2, \dots, z_{n-1}\|\right)\right]^{p_n} \ge \epsilon\right\}$$
$$\bigcup \left\{r\in N : D\frac{1}{h_r}T_\beta^{\sup p_n}\sum_{n\in I_r} f\left[\sum_{k=1}^{\infty} a_{nk}\left(\|\Delta_m^s y_k, z_1, z_2, \dots, z_{n-1}\|\right)\right]^{p_n} \ge \epsilon\right\}.$$

The last two sets on the right hand side belong to I and this completes the proof.

Lemma 1 *Let f be the modulus function and let $0 < \delta < 1$. Then for each $x > \delta$ we have $f(x) \le 2f(1)\delta^{-1}x$ (see [15]).*

Theorem 2 *Let f be a modulus function. Then $\left[N_\theta, A, \Delta_m^s, p, \|., \dots, .\|\right]^I \subseteq \left[N_\theta, A, \Delta_m^s, f, p, \|., \dots, .\|\right]^I$.*

Proof If $x \in \left[N_\theta, A, \Delta_m^s, p, \|., \dots, .\|\right]^I$ then for some $L > 0$ and each $z_1, z_2, \dots, z_{n-1} \in X$

$$\left\{r\in N : \frac{1}{h_r}\sum_{n\in I_r} f\left[\sum_{k=1}^{\infty} a_{nk}\left(\|\Delta_m^s x_k - L, z_1, z_2, \dots, z_{n-1}\|\right)\right]^{p_n} \ge \epsilon\right\} \in I.$$

Now let $\epsilon > 0$ be given. We can choose $0 < \delta < 1$ such that for every t with $0 \le t \le \delta$ we have $f(t) < \epsilon$. Now assuming Lemma 1 we get

$$\frac{1}{h_r}\sum_{n\in I_r} f\left[\sum_{k=1}^{\infty} a_{nk}\left(\|\Delta_m^s x_k, z_1, z_2, \ldots, z_{n-1}\|\right)\right]^{p_n}$$
$$= \frac{1}{h_r}\sum_{n\in I_r, \|\Delta_m^s x_k - L, z_1, z_2, \ldots, z_{n-1}\|\le\delta} f\left[\sum_{k=1}^{\infty} a_{nk}\left(\|\Delta_m^s, z_1, z_2, \ldots, z_{n-1}\|\right)\right]^{p_n}$$
$$+ \frac{1}{h_r}\sum_{n\in I_r, \|\Delta_m^s x_k - L, z_1, z_2, \ldots, z_{n-1}\|>\delta} f\left[\sum_{k=1}^{\infty} a_{nk}\left(\|\Delta_m^s, z_1, z_2, \ldots, z_{n-1}\|\right)\right]^{p_n}$$
$$\le \frac{1}{h_r}\left(h_r \max\left\{\epsilon^{\inf p_n}, \epsilon^{\sup p_n}\right\}\right) + \frac{1}{h_r}\max\{a_1, a_2\}\sum_{n\in I_r} f\left[\sum_{k=1}^{\infty} a_{nk}\left(\|\Delta_m^s x_k, z_1, z_2, \ldots, z_{n-1}\|\right)\right]^{p_n}$$

where $a_1 = (2f(1)\delta^{-1})^{\inf p_n}$ and $a_2 = (2f(1)\delta^{-1})^{\sup p_n}$.

So

$$\left\{r \in N : \frac{1}{h_r}\sum_{n\in I_r} f\left[\sum_{k=1}^{\infty} a_{nk}\left(\|\Delta_m^s x_k, z_1, z_2, \ldots, z_{n-1}\|\right)\right]^{p_n} \ge \epsilon\right\}$$
$$= \left\{r \in N : \frac{1}{h_r}(h_r \max\{\epsilon^{\inf p_n}, \epsilon^{\sup p_n}\}) \ge \epsilon\right\}$$
$$\bigcup\left\{r \in N : \frac{1}{h_r}\max\{a_1, a_2\}\sum_{n\in I_r}\left[\sum_{k=1}^{\infty} a_{nk}\left(\|\Delta_m^s x_k, z_1, z_2, \ldots, z_{n-1}\|\right)\right]^{p_n} \ge \epsilon\right\}$$

and this completes the proof.

Theorem 3 *Let f be a modulus function. If $\lim_{t\to\infty}\sup \frac{f(t)}{t} = A > 0$, then $\left[N_\theta, A, \Delta_m^s, f, p, \|., \ldots, .\|\right]^I = \left[N_\theta, A, \Delta_m^s, p, \|., \ldots, .\|\right]^I$.*

Proof It is sufficient only to show that $\left[N_\theta, A, \Delta_m^s, p, \|., \ldots, .\|\right]^I \subset \left[N_\theta, A, \Delta_m^s, f, p, \|., \ldots, .\|\right]^I$. If we have $\lim_{t\to\infty}\sup \frac{f(t)}{t} = A > 0$ then there exists a constant $B > 1$ such that $f(t) \ge Bt$ for all $t \ge 0$. Hence

$$\frac{1}{h_r}\sum_{n\in I_r} f\left[\sum_{k=1}^{\infty} a_{nk}\left(\|\Delta_m^s x_k - L, z_1, z_2, \ldots, z_{n-1}\|\right)\right]^{p_n}$$
$$\ge \frac{1}{h_r}B^{\sup p_n}\sum_{n\in I_r} f\left[\sum_{k=1}^{\infty} a_{nk}\left(\|\Delta_m^s x_k - L, z_1, z_2, \ldots, z_{n-1}\|\right)\right]^{p_n}$$

and this inequality gives the result.

More generally we have the following:

Theorem 4 *Let f_1 and f_2 be a modulus functions. If $\lim_{t\to\infty}\sup \frac{f_1(t)}{f_2(t)} = A > 0$, Then $[N_\theta, A, f_1(t), p, \|., \ldots, .\|]^I = [N_\theta, A, f_2(t), p, \|., \ldots, .\|]^I$.*

Theorem 5 *Let* $(X, \|., \ldots, .\|_S)$ *and* $(X, \|., \ldots, .\|_E)$ *be standard and Euclid n-normed spaces respectively, then* $\left[N_\theta, A, \Delta_m^s, f, p, \|., \ldots, .\|_E\right]^I \bigcap \left[N_\theta, A, \Delta_m^s, f, p, \|., \ldots, .\|_S\right]^I \subseteq \left[N_\theta, A, \Delta_m^s, f, p, (\|., \ldots, .\|_E + \|., \ldots, .\|_S)\right]^I$

Proof The following inequality that gives us the desired inclusion

$$\begin{aligned}
&\frac{1}{h_r}\sum_{n\in I_r} f\left[\sum_{k=1}^{\infty} a_{nk}(\|., \ldots, .\|_E + \|., \ldots, .\|_S)\left(\Delta_m^s x_k - L, z_1, z_2, \ldots, z_{n-1}\right)\right]^{p_n}\\
&\quad \le D\frac{1}{h_r}\sum_{n\in I_r} f\left[\sum_{k=1}^{\infty} a_{nk}\left(\|\Delta_m^s x_k - L, z_1, z_2, \ldots, z_{n-1}\|_E\right)\right]^{p_n}\\
&\quad + D\frac{1}{h_r}\sum_{n\in I_r} f\left[\sum_{k=1}^{\infty} a_{nk}\left(\|\Delta_m^s x_k - L, z_1, z_2, \ldots, z_{n-1}\|_S\right)\right]^{p_n}
\end{aligned}$$

Theorem 6 *Let* $(X, \|., \ldots, .\|)$ *be a n-normed space and* f_1, f_2 *be modulus functions. Then*

(i) $\left[N_\theta, A, \Delta_m^s, f_1, p, \|., \ldots, .\|\right]_0^I \bigcap \left[N_\theta, A, \Delta_m^s, f_2, p, \|., \ldots, .\|\right]_0^I \subseteq \left[N_\theta, A, \Delta_m^s, f_1 + f_2, p, \|., \ldots, .\|\right]_0^I$

(ii) $\left[N_\theta, A, \Delta_m^s, f_1, p, \|., \ldots, .\|\right]^I \bigcap \left[N_\theta, A, \Delta_m^s, f_2, p, \|., \ldots, .\|\right]^I \subseteq \left[N_\theta, A, \Delta_m^s, f_1 + f_2, p, \|., \ldots, .\|\right]^I$

We will prove (i) only.

Proof Let $x \in \left[N_\theta, A, \Delta_m^s, f_1, p, \|., \ldots, .\|\right]_0^I \bigcap \left[N_\theta, A, \Delta_m^s, f_2, p, \|., \ldots, .\|\right]_0^I$. The fact

$$\begin{aligned}
&\frac{1}{h_r}\sum_{n\in I_r}(f_1 + f_2)\left[\sum_{k=1}^{\infty} a_{nk}\left(\|\Delta_m^s x_k, z_1, z_2, \ldots, z_{n-1}\|\right)\right]^{p_n}\\
&\quad \le D\frac{1}{h_r}\sum_{n\in I_r} f_1\left[\sum_{k=1}^{\infty} a_{nk}\left(\|\Delta_m^s x_k, z_1, z_2, v, z_{n-1}\|\right)\right]^{p_n}\\
&\quad + D\frac{1}{h_r}\sum_{n\in I_r} f_2\left[\sum_{k=1}^{\infty} a_{nk}\left(\|\Delta_m^s x_k, z_1, z_2, \ldots, z_{n-1}\|\right)\right]^{p_n}
\end{aligned}$$

gives the result.

Theorem 7 *Let* $s \ge 1$, *Then the following inclusions hold.*

(i) $\left[N_\theta, A, \Delta_m^{s-1}, f, p, \|., \ldots, .\|\right]_0^I \subseteq \left[N_\theta, A, \Delta_m^s, f, p, \|., \ldots, .\|\right]_0^I$

(ii) $\left[N_\theta, A, \Delta_m^{s-1}, f, p, \|., \ldots, .\|\right]^I \subseteq \left[N_\theta, A, \Delta_m^s, f, p, \|., \ldots, .\|\right]^I$

Proof (i) If $x \in \left[N_\theta, A, \Delta_m^{s-1}, f, p, \|., \ldots, .\|\right]_0^I$ then we have

$$\left\{n \in N : \frac{1}{h_r} \sum_{n \in I_r} f\left[\sum_{k=1}^{\infty} a_{nk}\left(\|\Delta_m^s x_k, z_1, z_2, \ldots, z_{n-1}\|\right)\right]^{p_n} \geq \epsilon\right\} \in I.$$

Hence the following inequalities gives the result

$$\begin{aligned}
&\frac{1}{h_r} \sum_{n \in I_r} f\left[\sum_{k=1}^{\infty} a_{nk}\left(\|\Delta_m^s x_k, z_1, z_2, \ldots, z_{n-1}\|\right)\right]^{p_n} \\
&\leq D \frac{1}{h_r} \sum_{n \in I_r} f\left[\sum_{k=1}^{\infty} a_{nk}\left(\|\Delta_m^{s-1} x_k, z_1, z_2, \ldots, z_{n-1}\|\right)\right]^{p_n} \\
&+ D \frac{1}{h_r} \sum_{n \in I_r} f\left[\sum_{k=1}^{\infty} a_{nk}\left(\|\Delta_m^{s-1} x_{k+1}, z_1, z_2, \ldots, z_{n-1}\|\right)\right]^{p_n}.
\end{aligned}$$

(ii) can be proved similarly.

4 Sequence Spaces Defined by Sequences of Modulus Functions

Let G be the space of sequences of modulus functions $F = (f_n)$ such that $\limsup_{t \to 0} \sup_n f_n(t) = 0$ and $(X, ||., \ldots, .||)$ is an n-Banach space. We define the following spaces:

$$\begin{aligned}
&N_\theta, A, \Delta_m^s, F, p, ||., \ldots, .||]^I \\
&= \left\{x = (x_k) \in w(n-X) : \left\{r \in N : \frac{1}{h_r} \sum_{n \in I_r} f_n\left[\sum_{k=1}^{\infty} a_{nk}(||\Delta_m^s x_k - L, z_1, z_2, \ldots, z_{n-1}||)\right]^{p_n} \geq \epsilon\right\} \in I,\ for\ L \in X\right\} \\
&[N_\theta, A, \Delta_m^s, F, p, ||., \ldots, .||]_0^I \\
&= \left\{x = (x_k) \in w(n-X) : \left\{r \in N : \frac{1}{h_r} \sum_{n \in I_r} f_n\left[\sum_{k=1}^{\infty} a_{nk}(||\Delta_m^s x_k, z_1, z_2, \ldots, z_{n-1}||)\right]^{p_n} \geq \epsilon\right\} \in I\right\}.
\end{aligned}$$

Theorem 8 $\left[N_\theta, A, \Delta_m^{s-1}, F, p, \|., \ldots, .\|\right]^I$ *and* $\left[N_\theta, A, \Delta_m^s, F, p, \|., \ldots, .\|\right]_0^I$ *are linear spaces.*

Proof The proof is similar to Theorem 1.

Theorem 9 *Let $F = (f_n)$ be a sequence of modulus functions, $(X, \|., \ldots, .\|)$ in an n-Banach space and (x_k) is lacunary strongly convergent to L in $\left[N_\theta, A, \Delta_m^s, p, \|., \ldots, .\|\right]^I$ then (x_k) is lacunary strongly convergent to L in $\left[N_\theta, A, \Delta_m^s, F, p, \|., \ldots, .\|\right]^I$.*

Proof Let $\epsilon > 0$ be given. We can choose $0 < \delta < 1$, such that for every t with $0 \le t \le \delta$ we have $f(t) < \epsilon$. Now using the Lemma 1 we get

$$\begin{aligned}
&\frac{1}{h_r}\sum_{n\in I_r} f_n\left[\sum_{k=1}^{\infty} a_{nk}\left(\|\Delta_m^s x_k - L, z_1, z_2, \ldots, z_{n-1}\|\right)\right]^{p_n} \\
&= \frac{1}{h_r}\sum_{n\in I_r, \|\Delta_m^s x_k - L, z_1, z_2, \ldots, z_{n-1}\| \le \delta} f_n\left[\sum_{k=1}^{\infty} a_{nk}\left(\|\Delta_m^s x_k - L, z_1, z_2, \ldots, z_{n-1}\|\right)\right]^{p_n} \\
&\quad + \frac{1}{h_r}\sum_{n\in I_r, \|\Delta_m^s x_k - L, z_1, z_2, \ldots, z_{n-1}\| > \delta} f_n\left[\sum_{k=1}^{\infty} a_{nk}\left(\|\Delta_m^s x_k - L, z_1, z_2, \ldots, z_{n-1}\|\right)\right]^{p_n} \\
&\le \frac{1}{h_r}\left(h_r \max\left\{\epsilon^{\inf p_n}, \epsilon^{\sup p_n}\right\}\right) + \frac{1}{h_r}\max\{a_1, a_2\}\sum_{n\in I_r} f_n\left[\sum_{k=1}^{\infty} a_{nk}\left(\|\Delta_m^s x_k - L, z_1, z_2, \ldots, z_{n-1}\|\right)\right]^{p_n}
\end{aligned}$$

where $a_1 = \left(2\sup_n p_n f_n(1)\right)^{\inf p_n}$, $a_2 = \left(2\sup_n p_n f_n(1)\right)^{\sup p_n}$

Thus we have

$$\begin{aligned}
&\left\{r \in N : \frac{1}{h_r}\sum_{n\in I_r} f_n\left[\sum_{k=1}^{\infty} a_{nk}\left(\|\Delta_m^s x_k - L, z_1, z_2, \ldots, z_{n-1}\|\right)\right]^{p_n}\right\} \\
&= \left\{r \in N : \frac{1}{h_r}\left(h_r \max\left\{\epsilon^{\inf p_n}, \epsilon^{\sup p_n}\right\}\right) \ge \epsilon\right\} \\
&\bigcup \left\{r \in N : \frac{1}{h_r}\max\{a_1, a_2\}\sum_{n\in I_r} f_n\left[\sum_{k=1}^{\infty} a_{nk}\left(\|\Delta_m^s - L, z_1, z_2, \ldots, z_{n-1}\|\right)\right]^{p_n} \ge \epsilon\right\}
\end{aligned}$$

This completes the proof.

Theorem 10 *Let* $(X, \|., \ldots, .\|)$ *in an n-Banach space and* $F = (F_n)$ *be a sequence of modulus functions with* $\lim_{t\to\infty}\inf_n \frac{f_n(t)}{t} > 0$, *then* $\left[N_\theta, A, \Delta_m^s, F, p, \|., \ldots, .\|\right]^I = \left[N_\theta, A, \Delta_m^s, p, \|., \ldots, .\|\right]^I$

Proof The following inequality gives us the required result

$$\begin{aligned}
&\left\{r \in N : \frac{1}{h_r}\sum_{n\in I_r} f_n\left[\sum_{k=1}^{\infty} a_{nk}\left(\|\Delta_m^s x_k - L, z_1, z_2, \ldots, z_{n-1}\|\right)\right]^{p_n}\right\} \\
&\supseteq \left\{r \in N : \frac{B}{h_r}\sum_{n\in I_r} f_n\left[\sum_{k=1}^{\infty} a_{nk}\left(\|\Delta_m^s x_k - L, z_1, z_2, \ldots, z_{n-1}\|\right)\right]^{p_n}\right\}
\end{aligned}$$

where B is a positive number such that $\frac{f_n(t)}{t} > Bu$ for $u > 0$ and each $n \in N$.

Acknowledgments The author would like to record his gratitude to his referees for their careful reading and making some useful corrections which improve the presentation of the paper.

References

1. Et, M., Colak, R.: On some generalized difference sequence spaces. Soochow J. Math. **21**(4), 377–386 (1995)
2. Et, M., Esi, A.: On Köthe-Toeplitz duals of generalized difference sequence spaces. Bull. Malays. Math. Sci. Soc. **23**, 1–8 (2000)
3. Freedman, A.R., Sember, J.J., Raphael, M.: Some Cesro- type summability spaces. Proc. Lond. Math. Soc. **37**(3), 508–520 (1978)
4. Gahler, S.: 2-metrische Raume und ihre topologische Struktur. Math. Nachr. **26**, 115–148 (1963)
5. Gunawa, H., Mashadi, M.: On finite dimensional 2-normed spaces. Soochow J. Math. **27**(3), 321–329 (2001)
6. Gunawan, H., Mashadi, M.: On n-normed spaces. Int. J. Math. Sci. **27**(10), 631–639 (2001)
7. Gurdal, M., Sahiner, A.: Ideal convergence in n-normed spaces and some new sequence spaces via n-norm. J. Fundam. Sci. **4**, 233–244 (2008)
8. Jalal, T.: Some new I- convergent sequence spaces defined by using a sequence of modulus functions in n-normed spaces. Int. J. Math. Arch. **5**(9), 202–209 (2014)
9. Jalal, T.: New sequence spaces in multiple normed space through lacunary sequences. Int. Bull. Math. Res. **2**(1), 173–179 (2015)
10. Kim, S.S., Cho, Y.J.: Strict convexity in linear n-normed spaces. Demonstr. Math. **29**, 739–744 (1996)
11. Kizmaz, H.: On certain sequence spaces. Canad. Math. Bull. **24**, 169–176 (1981)
12. Kostyrko, P., Macaj, M., Salat, T.: I-convergence. Real Anal. Exch. **26**(2), 669–686 (2000)
13. Kostyrko, P., Macaj, M., Salat, T., Sleziak, M.: I-convergence and extremal I-limit points. Math. Slovaca **55**, 443–464 (2005)
14. Malceski, A.: Strong n-convex n-normed spaces. Mat. Bilt. **21**, 81–102 (1997)
15. Pehlivan, S., Fisher, B.: Lacunary strong convergence with respect to a sequence of modulus functions. Comment. Math. Univ. Carol. **36**(1), 71–78 (1995)
16. Sahiner, A., Gurdal, M., Saltan, S., Gunawan, H.: Ideal convergence in 2-normed spaces. Taiwan. J. Math. **11**(5), 1477–1484 (2007)
17. Savas, E: Some new double defined by Orlicz function in n-normed spaces, J. Inequal. Appl. 1–9 (2011)

GPU-Accelerated Simulation of Maxwell's Equations

Tony W.H. Sheu

Abstract In this study an explicit finite difference scheme is developed to solve the Maxwell's equations in time domain for a lossless medium. From the physical point of view, the hyperbolic system of Maxwell equations shall be discretized explicitly. From the computational point of view, this developed three-dimensional explicit scheme can be more effectively implemented in parallel in CPU/GPU with the Nvidia K-20 card. From the mathematical point of view, symplectic scheme is adopted for the approximation of temporal derivative terms so that all Hamiltonians in Maxwell's equations can be conserved at all times. Moreover, to predict the long-time accurate solution a phase velocity preserving scheme is developed for the spatial derivative terms so that the chosen time increment and grid spacing can be excellently paired following the employed theoretical guideline. Computational performance will be assessed based on the results obtained from the computed results in one GPU card and in one I7-4820K CPU card.

Keywords Hamiltonians · Symplecticity · CPU/GPU · Three dimensional maxwell's equations · Dispersion relation equation · In parallel

1 Introduction

While approximating the derivative terms in Maxwell's equations, dissipation error can more or less smear the solution profiles. Dispersion error can moreover worse destabilize the discrete system and yield, therefore, an erroneously predicted phase

T.W.H. Sheu (✉)
Department of Engineering Science and Ocean Engineering,
National Taiwan University, No. 1, Sec. 4, Roosevelt Road, Taipei, Taiwan, ROC
e-mail: twhsheu@ntu.edu.tw

T.W.H. Sheu
Center for Advanced Study in Theoretical Sciences, National Taiwan University,
Taipei, Taiwan, ROC

V.K. Singh et al. (eds.), *Modern Mathematical Methods and High Performance Computing in Science and Technology*, Springer Proceedings in Mathematics & Statistics 171, DOI 10.1007/978-981-10-1454-3_22

speed or group velocity. To reduce dispersion and dissipation errors at the same time is therefore essential while approximating the first-order spatial derivative terms in EM wave equations.

After a long calculation of the electromagnetic wave equations, the solution quality may be deteriorated substantially due to the accumulation of numerical error resulting from the applied non-symplectic time-stepping scheme. How to numerically preserve the symplectic property existing in Maxwell's equations motivates us to accurately approximate the time derivative terms in the Faraday's and Ampère's equations. When simulating Maxwell's equations, the quality of the solution predicted by the finite difference time domain (FDTD) method can be deteriorated as well by the introduced anisotropy error. Dispersion and anisotropy errors are both accumulative with time and can seriously contaminate the true propagation characteristics. All the above three types of numerical error will be reduced as much as possible in this study through different underlying theories in a domain of three dimensions.

Due to the advance in computer architectures, an important aspect one shall not ignore is related to the implementation of computer code on vector/parallel CPUs and GPUs (Graphics Processing Units). In comparison with CPU programming, hardware-oriented GPU programming executed in machines with much larger number of processing cores is now known to be able to reach a highly parallelized level. More importantly, modern GPUs can now offer us a relative low-cost computing power in different parallel applications. Thanks to the advent of GPU hardwares, Maxwell's equations have been effectively solved in parallel more recently in graphics processors. Such a new implementation in GPUs has been evidenced to reduce a large amount of computing time [1–6]. Owing to the potential of GPU parallelization, in this study the explicit FDTD scheme capable of yielding dispersive error-minimization will be implemented on GPU.

This paper is organized as follows. In Sect. 2, some of the distinguished physical and fruitful mathematical features in the ideal (or lossless) Maxwell's equations that are related to the scheme development and code verification are presented together with the two indispensable divergence-free constraint equations. In Sect. 3, the first-order spatial derivative terms in Faraday's and Ampère's equations are discretized in non-staggered grids rather than in conventional Yee's staggered grids. In this paper, the difference between the exact and numerical phase velocities is minimized to achieve a higher dispersive accuracy. Maxwell's equations belong to the class of integrable equations [7]. A symplectic structure-preserving time integrator shall be applied to conserve symplecticity numerically. For this reason, the explicit symplectic partitioned Runge–Kutta (SPRK) scheme is applied. Derivation of the corresponding stability condition for the proposed explicit scheme is also given in Sect. 3. In Sect. 4, the Nvidia K-20 GPU and the CPU/GPU hybrid architecture are briefly reviewed. To accelerate the speed of computation in time domain, the method described in Sect. 3 is further parallelized through a proper arrangement of the global and shared memories built in GPU. The proposed second-order accurate temporal scheme and the fourth-order accurate spatial scheme will be verified and validated through the respective problems. In addition, the performance of GPU implementation is also detailed. Finally, we will draw some conclusions in Sect. 5 based on the solutions computed in parallel in non-staggered grids on a single Nvidia K-20 GPU card.

2 Maxwell's Equations

Maxwell's equations in lossless media are represented below in terms of the dependent variables $\underline{E} = (E_x, E_y, E_z)^T$ and $\underline{H} = (H_x, H_y, H_z)^T$

$$\frac{\partial \underline{H}}{\partial t} = -\frac{1}{\mu} \nabla \times \underline{E}, \tag{1}$$

$$\frac{\partial \underline{E}}{\partial t} = \frac{1}{\varepsilon} \nabla \times \underline{H}. \tag{2}$$

The above set of equations is coupled with the Gauss's law which consists of the divergence-free equations $\nabla \cdot \underline{B} = 0$ and $\nabla \cdot \underline{D} = 0$. These two divergence-free constraint equations can be derived directly from the Faraday's law and Ampère's law, respectively, for a linear, isotropic, lossless material provided that the electric current density and electric charge density are neglected. Within the differential context, the Gauss's law is unconditionally satisfied in case two vectors $\underline{B}$ and $\underline{D}$ are initially divergence-free [8]. The differential set of the Maxwell's equations becomes however over-determined. Two divergence-free equations need to be neglected so that the numbers of unknowns and field equations are equal. Equations (1) and (2) are derived under the conditions of $\underline{D} = \varepsilon \underline{E}$ and $\underline{B} = \mu \underline{H}$, where $\underline{D}$ denotes the electric flux density and $\underline{E}$ is the electric field density. In the proportional constants, ε is known as the electric permittivity and μ is known as the magnetic permeability. The values of ε and μ determine the light speed c $(\equiv (\varepsilon\ \mu)^{-1/2})$.

The first Hamiltonian in the bi-Hamiltonian differential system of Eqs. (1) and (2) has association with the helicity Hamiltonian H_1 given below [9]

$$H_1 = \frac{1}{2} \int \frac{1}{\varepsilon} \underline{H} \cdot \nabla \times \underline{H} + \frac{1}{\mu} \underline{E} \cdot \nabla \times \underline{E}\, d\Omega. \tag{3}$$

The second quadratic Hamiltonian (or energy density) is expressed as follows [10].

$$H_2 = \frac{1}{2} \int \mu \underline{H} \cdot \underline{H} + \varepsilon \underline{E} \cdot \underline{E}\, d\Omega. \tag{4}$$

Two Hamiltonians given above will be used in this study to indirectly justify the proposed numerical scheme.

Numerical errors computed solely from the Faraday's and Ampère's equations may make the solutions computed from the magnetic and electric equations no longer divergence free. To overcome the difficulty owing to the omission of the Gauss's law, two gradient terms for the scalar variables Φ_1 and Φ_2 are introduced into the Eqs. (1) and (2), respectively. The resulting modified equations to be applied are expressed as $\frac{\partial \underline{E}}{\partial t} - \frac{1}{\varepsilon}\nabla \times \underline{H} + \nabla\Phi_1 = 0$ and $\frac{\partial \underline{H}}{\partial t} + \frac{1}{\mu}\nabla \times \underline{E} + \nabla\Phi_2 = 0$. These equations responsible for the two introduced correction potentials can be seen in [11].

3 Numerical Method

Unlike most of other EM wave solvers, in this study the Maxwell's equations are solved in non-staggered grids so that it is comparatively easy for us to execute the computer code in parallel.

3.1 Explicit Symplectic Partitioned Runge–Kutta Temporal Scheme

Maxwell's equations are mathematically separable. The explicit symplectic partitioned Runge–Kutta time-stepping scheme is therefore applied to integrate Faraday's and Ampère's equations [12]. Calculation of the $\underline{E}^{n+1}$ and $\underline{H}^{n+1}$ solutions from the solutions computed at time nΔt is split into the following steps by using the second-order accurate explicit partitioned Runge–Kutta scheme presented in [13]

$$\underline{H}^{n+\frac{1}{2}} = \underline{H}^n - \frac{dt}{2\mu}\nabla \times \underline{E}^n, \tag{5}$$

$$\underline{E}^{n+1} = \underline{E}^n + \frac{dt}{\varepsilon}\nabla \times \underline{H}^{n+\frac{1}{2}}, \tag{6}$$

$$\underline{H}^{n+1} = \underline{H}^{n+\frac{1}{2}} - \frac{dt}{2\mu}\nabla \times \underline{E}^{n+1}. \tag{7}$$

3.2 Numerical Scheme on Spatial Derivative Terms

In addition to the development of above symplecticity-preserving scheme, we are also aimed to reduce the dispersion error in space. To this end, the difference between the numerical and exact phase velocities shall be minimized in the space of wavenumbers [14]. The spatial derivative terms shown in (5)–(7) are therefore approximated by the methods of modified equation analysis, dispersion analysis, and the grid-anisotropy analysis.

At $t = n\Delta t$, we can get $\underline{H}^{n} = \underline{H}^{n-\frac{1}{2}} - \frac{dt}{2\mu}\nabla \times \underline{E}^{n}$ and, then, $\underline{H}^{n+\frac{1}{2}} = \underline{H}^{n-\frac{1}{2}} - \frac{dt}{2\mu}\nabla \times \underline{E}^{n}$ according to the following equations

$$E_z^{n+\frac{1}{2}} = E_z^{n-\frac{1}{2}} + \frac{\Delta t}{\varepsilon}\left(\frac{\partial H_y^n}{\partial x} - \frac{\partial H_x^n}{\partial y}\right), \tag{8}$$

$$E_x^{n+\frac{1}{2}} = E_x^{n-\frac{1}{2}} + \frac{\Delta t}{\varepsilon}\left(\frac{\partial H_z^n}{\partial y} - \frac{\partial H_y^n}{\partial z}\right), \tag{9}$$

$$E_y^{n+\frac{1}{2}} = E_y^{n-\frac{1}{2}} + \frac{\Delta t}{\varepsilon}\left(\frac{\partial H_x^n}{\partial z} - \frac{\partial H_z^n}{\partial x}\right). \tag{10}$$

From Eq. (6), we can get $\underline{E}^{n+\frac{1}{2}} = \underline{E}^{n-\frac{1}{2}} + \frac{dt}{\varepsilon}\nabla \times \underline{H}^{n}$.

To get a higher accuracy at a reasonable computational cost, we can apply either a compact or a combined compact difference scheme to effectively reduce numerical errors at small wavelengths [15]. We are aimed particularly at reducing not only the phase error but also the amplitude error [16]. In this study our goal of reducing dispersive error is to minimize the error of numerical dispersion relation equation [17, 18]. It is therefore necessary to derive the explicit form of the numerical dispersion relation equation.

The first-order derivative terms $\frac{\partial H_y^n}{\partial x}$ and $\frac{\partial H_x^n}{\partial y}$ shown in Eq. (8) are approximated in non-staggered grids so that programming takes becomes simplified without suffering checkerboarding oscillations. These derivative terms at an interior node (i, j, k) are approximated by the scheme given below

$$\begin{aligned}\frac{\partial H_y}{\partial x}\Big|_{i,j,k}^{n} =& \frac{1}{h}\Big[a_1\left(H_y|_{i+3,j,k}^{n} - H_y|_{i-3,j,k}^{n}\right) + a_2\left(H_y|_{i+2,j,k}^{n} - H_y|_{i-2,j,k}^{n}\right)\\ &+ a_3\left(H_y|_{i+1,j,k}^{n} - H_y|_{i-1,j,k}^{n}\right)\Big],\end{aligned} \tag{11}$$

$$\begin{aligned}\frac{\partial H_x}{\partial y}\Big|_{i,j,k}^{n} =& \frac{1}{h}\Big[a_1\left(H_x|_{i,j+3,k}^{n} - H_x|_{i,j-3,k}^{n}\right) + a_2\left(H_x|_{i,j+2,k}^{n} - H_y|_{i,j-2,k}^{n}\right)\\ &+ a_3\left(H_x|_{i,j+1,k}^{n} - H_x|_{i,j-1,k}^{n}\right)\Big].\end{aligned} \tag{12}$$

After substituting (11) and (12) into (8) and then expanding the resulting terms in Taylor series with respect to E_z, the following equation at an interior point (i, j, k) is derived as

$$\frac{\partial E_z}{\partial t}|^n_{i,j,k} + \frac{dt^2}{24}\frac{\partial^3 E_z}{\partial t^3}|^n_{i,j,k} + \frac{dt^4}{1920}\frac{\partial^5 E_z}{\partial t^5}|^n_{i,j,k} + \frac{dt^6}{322560}\frac{\partial^7 E_z}{\partial t^7}|^n_{i,j,k}$$
$$+\cdots = \frac{1}{\varepsilon}\Bigg\{\Bigg[\Big(6a_1+4a_2+2a_3\Big)\frac{\partial H_y}{\partial x}|^n_{i,j,k} + \Big(9a_1+\frac{8}{3}a_2+\frac{1}{3}a_3\Big)dx^2\frac{\partial^3 H_y}{\partial x^3}|^n_{i,j,k}$$
$$+\Big(\frac{81}{20}a_1+\frac{8}{15}a_2+\frac{1}{60}a_3\Big)dx^4\frac{\partial^5 H_y}{\partial x^5}|^n_{i,j,k} + \Big(\frac{243}{280}a_1+\frac{16}{315}a_2+\frac{1}{2520}a_3\Big)dx^6\frac{\partial^7 H_y}{\partial x^7}|^n_{i,j,k}$$
$$+\cdots\Bigg] - \Bigg[\Big(6a_1+4a_2+2a_3\Big)\frac{\partial H_y}{\partial x}|^n_{i,j,k} + \Big(9a_1+\frac{8}{3}a_2+\frac{1}{3}a_3\Big)dx^2\frac{\partial^3 H_y}{\partial x^3}|^n_{i,j,k}$$
$$+\Big(\frac{81}{20}a_1+\frac{8}{15}a_2+\frac{1}{60}a_3\Big)dx^4\frac{\partial^5 H_y}{\partial x^5}|^n_{i,j,k} + \Big(\frac{243}{280}a_1+\frac{16}{315}a_2+\frac{1}{2520}a_3\Big)dx^6\frac{\partial^7 H_y}{\partial x^7}|^n_{i,j,k}$$
$$+\cdots\Bigg]\Bigg\}. \tag{13}$$

The three introduced weighting coefficients a_1, a_2 and a_3 are determined by performing the rigorous modified equation analysis and the dispersion analysis described below.

All the time derivative terms $\frac{\partial^3 E_z}{\partial t^3}, \frac{\partial^5 E_z}{\partial t^5}, \frac{\partial^7 E_z}{\partial t^7}$...shown in (13) are replaced first by their equivalent spatial derivative terms through the Ampère's equations to get the corresponding equations for $\frac{\partial^i E_j}{\partial t^3}$ ($i = 3$ and 5, $j = x, y, z$). After replacing the high-order temporal derivative terms $\frac{\partial^3 E_z}{\partial t^3}$ and $\frac{\partial^5 E_z}{\partial t^5}$ with the corresponding spatial derivative terms, the equation equivalent to (13) is derived. By comparing the resulting equation with the equation $\frac{\partial E_z}{\partial t} = \frac{1}{\varepsilon}\left(\frac{\partial H_y}{\partial x} - \frac{\partial H_x}{\partial y}\right)$, the equations for a_1, a_2 and a_3 are derived as $3a_1 + 2a_2 + a_3 = \frac{1}{2}$ and $9a_1 + \frac{8}{3}a_2 + \frac{1}{3}a_3 - \frac{Cr^2}{12}(3a_1 + 2a_2 + a_3) = 0$. In the above, $Cr = \frac{c\Delta t}{h}$ denotes the Courant number and h denotes the grid spacing.

Determination of the undetermined coefficients shown above needs to derive the third algebraic equation. Substitution of the plane wave solution $\underline{E} = \underline{E}_0 exp\,(I\,(k_x i\,\Delta x + k_y j\Delta y + k_z k\Delta z - \omega n\Delta t))$, where $I = -1^{1/2}$, into the equation given by $\frac{\partial \underline{E}}{\partial t}|^n_{i,j,k} = \frac{\underline{E}^{n+\frac{1}{2}}|_{i,j,k} - \underline{E}^{n-\frac{1}{2}}|_{i,j,k}}{\Delta t}$ and the equations given by
$\frac{\partial \underline{E}}{\partial x}|^n_{i,j,k} = \frac{1}{h}\left[a_1\left(\underline{E}|^n_{i+3,j,k} - \underline{E}|^n_{i-3,j,k}\right) + a_2\left(\underline{E}|^n_{i+2,j,k} - \underline{E}|^n_{i-2,j,k}\right) + a_3\left(\underline{E}|^n_{i+1,j,k} - \underline{E}|^n_{i-1,j,k}\right)\right]$,
$\frac{\partial \underline{E}}{\partial y}|^n_{i,j,k} = \frac{1}{h}\left[a_1\left(\underline{E}|^n_{i,j+3,k} - \underline{E}|^n_{i,j-3,k}\right) + a_2\left(\underline{E}|^n_{i,j+2,k} - \underline{E}|^n_{i,j-2,k}\right) + a_3\left(\underline{E}|^n_{i,j+1,k} - \underline{E}|^n_{i,j-1,k}\right)\right]$ and
$\frac{\partial \underline{E}}{\partial z}|^n_{i,j,k} = \frac{1}{h}\left[a_1\left(\underline{E}|^n_{i,j,k+3} - \underline{E}|^n_{i,j,k-3}\right) + a_2\left(\underline{E}|^n_{i,j,k+2} - \underline{E}|^n_{i,j,k-2}\right) + a_3\left(\underline{E}|^n_{i,j,k+1} - \underline{E}|^n_{i,j,k-1}\right)\right]$, we can get $\frac{\partial \underline{E}}{\partial t}, \frac{\partial \underline{E}}{\partial x}, \frac{\partial \underline{E}}{\partial y}$, and $\frac{\partial \underline{E}}{\partial z}$ and then the equations for $\frac{\partial^2 \underline{E}}{\partial t^2}\left(= c^2\left(\frac{\nabla^2 \underline{E}}{\partial x^2} + \frac{\nabla^2 \underline{E}}{\partial y^2} + \frac{\nabla^2 \underline{E}}{\partial z^2}\right)\right)$, $\frac{\partial^2 \underline{E}}{\partial x^2}, \frac{\partial^2 \underline{E}}{\partial y^2}$ and $\frac{\partial^2 \underline{E}}{\partial z^2}$. Numerical dispersion relation equation can be derived as follows by substituting Eqs. (12) and (13) into the second-order wave equation for $\underline{E}$

$$\frac{1}{c^2}\frac{\omega^2}{4}\left(\frac{\sin(\omega\Delta t/2)}{\omega\Delta t}\right)^2 = k_x^2\left(3a_1\frac{\sin(3k_x\Delta x)}{3k_x\Delta x} + 2a_2\frac{\sin(2k_x\Delta x)}{2k_x\Delta x} + a_3\frac{\sin(k_x\Delta x)}{k_x\Delta x}\right)^2$$
$$+ k_y^2\left(3a_1\frac{\sin(3k_y\Delta y)}{3k_y\Delta y} + 2a_2\frac{\sin(2k_y\Delta y)}{2k_y\Delta y} + a_3\frac{\sin(k_y\Delta y)}{k_y\Delta y}\right)^2$$
$$+ k_z^2\left(3a_1\frac{\sin(3k_z\Delta z)}{3k_z\Delta z} + 2a_2\frac{\sin(2k_z\Delta z)}{2k_z\Delta z} + a_3\frac{\sin(k_z\Delta z)}{k_z\Delta z}\right)^2. \tag{14}$$

The wavenumber vector is defined as $\underline{k} = (k_x, k_y, k_z)$. The exact dispersion relation equation can be similarly derived as $\left(\frac{\omega}{c}\right)^2 = k_x^2 + k_y^2 + k_z^2$ by substituting the plane wave solution into the second-order wave equation $\frac{\partial^2 \underline{E}}{\partial t^2} = c^2 \nabla^2 \underline{E}$.

To get a correct propagation characteristics while solving the Maxwell's equations in time domain, we need to develop a scheme whose numerical phase velocity $\underline{v}_p$ $(\equiv \frac{\omega_{num}}{\underline{k}})$ matches perfectly with its exact counterpart. To this end, the error function defined as $\left[|\frac{\omega_{num}}{\underline{k}}| - |\frac{\omega_{exact}}{\underline{k}}| \right]^2$ shall be minimized in a weak sense. The function to be minimized within the integral range of $-m_p \pi \le h\underline{k} \le m_p \pi$ is as follows

$$E_p = \int_{-m_p\pi}^{m_p\pi} \left[|\frac{\omega_{num}}{\underline{k}}| - |\frac{\omega_{exact}}{\underline{k}} (\equiv c)| \right]^2 W_p \, d(k_x \Delta x) \, d(k_y \Delta y) \, d(k_z \Delta z). \quad (15)$$

In the above, $k_x \Delta x$, $k_y \Delta y$ and $k_z \Delta z$ denote the scaled (or modified) wavenumbers along the x, y and z directions, respectively. Application of the above weighting function W_p enables us to integrate E_p analytically for the value of m_p in between 0 and $\frac{1}{2}$. By enforcing the limiting condition given by $\frac{\partial E_p}{\partial a_3} = 0$, the third algebraic equation for a_1, a_2 and a_3 is derived as

$$\begin{aligned} &- 0.00946472 \, a_1 - 0.00787899 \, a_2 + 0.224744 \, a_1^3 + 0.0948775 \, a_2^3 + 0.367829 \, a_2^2 a_1 \\ &+ 0.0166091 \, a_3^3 + 0.107206 \, a_3^2 a_1 + 0.261056 \, a_1^2 a_3 + 0.156637 \, a_2^2 a_3 - 0.00453852 \, a_3 \\ &+ 0.492672 \, a_1^2 a_2 + 0.395351 \, a_3 a_2 a_1 + 0.0875208 \, a_3^2 a_2 = 0 \end{aligned} \quad (16)$$

The stability condition of the proposed explicit scheme, which conserves not only the symplecticity but also preserves the dispersion relation equation, is derived by considering the equivalent eigenvalue equations. The proposed conditionally stable explicit scheme is subject to $\Delta t \le \frac{1}{c} \left(\frac{\max(F_x^2)}{\Delta x^2} + \frac{\max(F_y^2)}{\Delta y^2} + \frac{\max(F_z^2)}{\Delta z^2} \right)^{-\frac{1}{2}}$. By substituting the previously derived coefficients a_1, a_2 and a_3 into the above inequality equation, the stability condition for the current scheme developed to solve the three dimensional Maxwell's equations is $\Delta t \le 0.673844 \frac{h}{c}$.

4 GPU Calculation of Maxwell's Equations

The explicit dispersion-relation-equation preserving scheme developed in Sect. 3 for solving the Maxwell's equations is suitable for parallel implementation. The reason is that the update of electric field components requires only the available magnetic field values, and vice versa. A parallel algorithm is therefore executed on the Nvidia K-20 GPU card aiming at reducing computing time while solving the three-dimensional Maxwell's equations.

Table 1 The predicted L_2 errors and the corresponding spatial rates of convergence (sroc) for the analytical test problem investigated in Sect. 4

Meshes	L_2- error norm of E_z	sroc
$10 \times 10 \times 10$	1.8366E-05	–
$20 \times 20 \times 20$	1.2339E-06	3.8957
$30 \times 30 \times 30$	8.5169E-08	3.8567
$40 \times 40 \times 40$	5.3609E-09	3.9897

Fig. 1 The computed and exact energy densities, shown in (3) and (4), are plotted with respect to time for the analytical problem in Sect. 4 using the proposed phase velocity optimized compact difference scheme. **a** Hamiltonian function; **b** Energy density

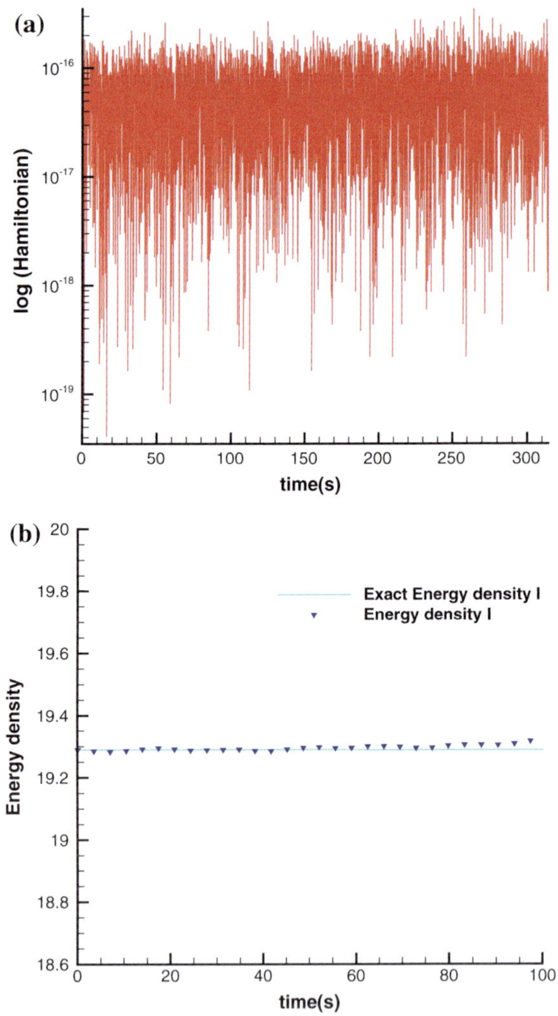

A parallelized Fortran code is executed mostly in the blocks of streaming multiprocessors (SMXs). The threads in a block are grouped into several wraps during the calculation. In Nvidia Tesla K20c, one block has 1024 threads distributed in 32 wraps. One SMX can execute several thread block calculations. Several SMXs constitute a TPC (Texture Processing Cluster). Only one wrap in a block is permitted to use.

The proposed explicit symplectic PRK scheme developed in non-staggered grids is verified first by solving the three dimensional Maxwell's equations amenable to the exact solution in a cube of $-\pi \leq x \leq \pi$, $-\pi \leq y \leq \pi$ and $-\pi \leq z \leq \pi$. The solution sought at $\mu = 1$ and $\varepsilon = 1$ is subject to the initial solenoidal solutions $E_x(x, y, z, 0) = E_y(x, y, z, 0) = E_z(x, y, z, 0) = 0$, $H_x(x, y, z, 0) = \cos(x + y + z)$, $H_y(x, y, z, 0) = \frac{1}{2}(-1 + \sqrt{3})\cos(x + y + z)$ and $H_z(x, y, z, 0) = -\frac{1}{2}(1 + \sqrt{3})$ $\cos(x + y + z)$. The exact electric and magnetic field solutions to Eqs. (1) and (2) are given by $E_x(x, y, z, t) = \sin(\sqrt{3}t)\sin(x + y + z)$, $E_x(x, y, z, t) = \sin(\sqrt{3}t)$ $\sin(x + y + z)$, $E_y(x, y, z, t) = -\frac{1}{2}(1 + \sqrt{3})\sin(\sqrt{3}t)\sin(x + y + z)$, $E_z(x, y, z, t) = \frac{1}{2}(-1 + \sqrt{3})\sin(\sqrt{3}t)\sin(x + y + z)$, $H_x(x, y, z, t) = \cos(\sqrt{3}t)\cos(x + y + z)$, $H_y(x, y, z, t) = \frac{1}{2}(-1 + \sqrt{3})\cos(\sqrt{3}t)\cos(x + y + z)$, $H_z(x, y, z, t) = -\frac{1}{2}(1 + \sqrt{3})\cos(\sqrt{3}t)\cos(x + y + z)$.

The spatial rate of convergence is computed first at $\Delta t = 10^{-5}$, which is much smaller than the grid sizes chosen as $\Delta x = \Delta y = \Delta z = \pi/5, \pi/10, \pi/15$ and $\pi/20$ in this study. From the predicted L_2-error norms tabulated in Table 1 one can see that there is only a very small difference between the predicted spatial rate of convergence and the theoretically derived fourth-order accuracy.

The Hamiltonian defined in (3) and the energy density given in (4) are computed from the predicted solutions of $\underline{E}$ and $\underline{H}$ for making an additional theoretical justification of the proposed scheme. One can clearly find from Fig. 1 that the computed

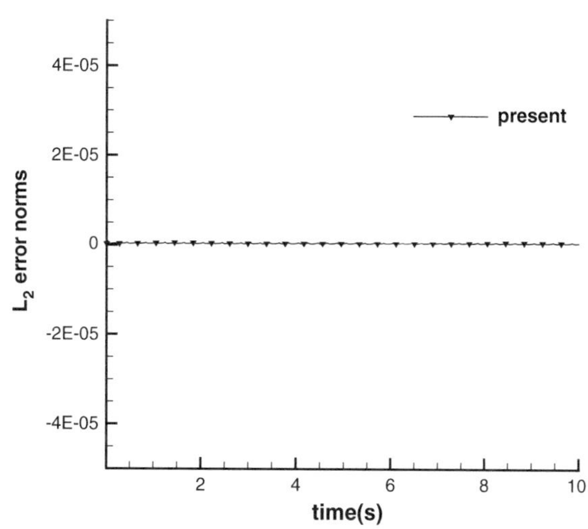

Fig. 2 The computed L_2-norm of $\nabla \cdot \underline{H}$ is plotted with respect to time using the present explicit partitioned Runge–Kutta symplectic scheme

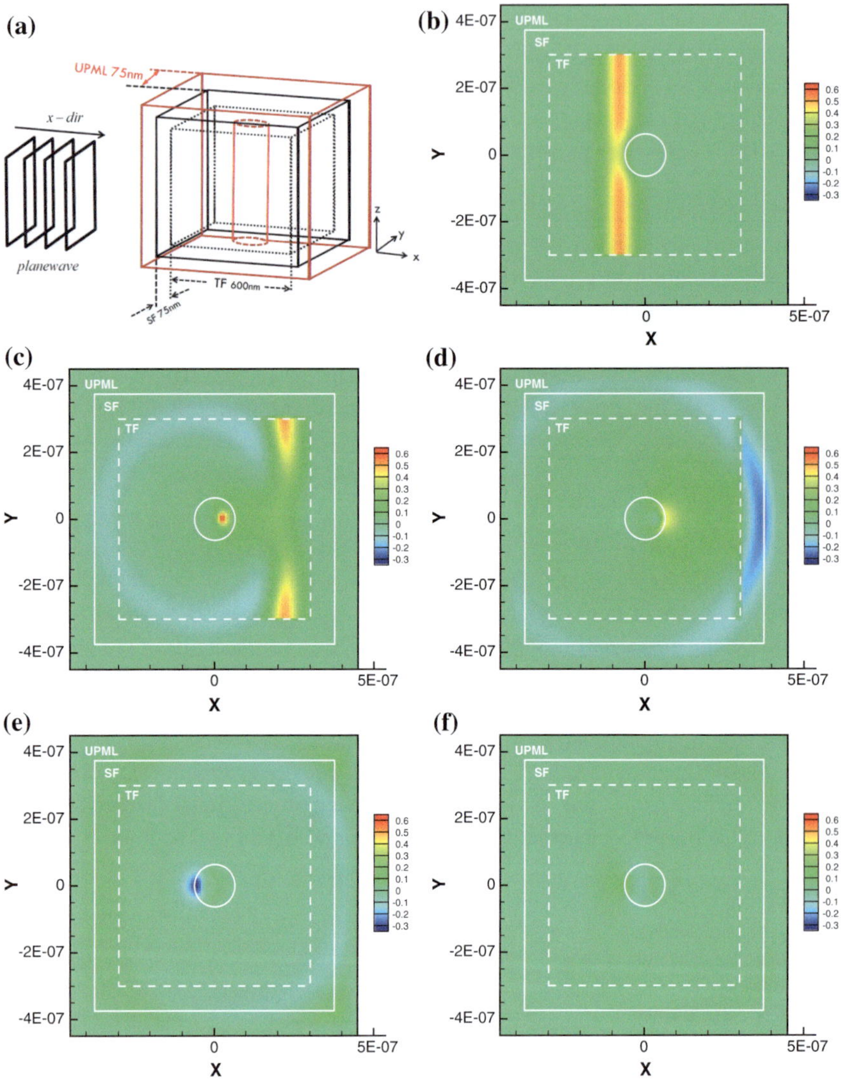

Fig. 3 **a** Schematic of the 3D Mie scattering problem; The predicted time-varying contours E_z ($z = 0$) at the cutting plane containing a cylindrical scatter. **b** time step = 560 (2.8 fs); **c** time step = 760 (3.8 fs); **d** time step = 1350 (4.25 fs); **e** time step = 1600 (5.8 fs); **f** time step = 1900 (9 fs)

Hamiltonian and energy density are not changed too much with time. The predicted norms of $\nabla \cdot \underline{H}$ and $\nabla \cdot \underline{E}$ are also plotted with respect to time to assure that the Gauss's law is indeed satisfied discretely. In Fig. 2, the predicted magnetic field predicted by the proposed scheme is essentially divergence-free.

The diameter of the dielectric cylinder under current study is 126.56 nm. In Fig. 3a, this isotropic cylinder located at the center of a cube with the volume of 720^3 nm^3 has $\varepsilon_r = 12.1104$. The cross-section area is $760 \times 760\,\text{nm}^2$. The incident x-polarized plane wave with the amplitude of 0.5 $\frac{\text{v}}{\text{m}}$ and the angular frequency of 13.263 $\frac{\text{rad}}{\text{s}}$ propagates rightward according to the one-dimensional Maxwell's equations $\frac{\partial E_z}{\partial t} = \frac{1}{\varepsilon}\nabla \times H$, $\frac{\partial H}{\partial t} = -\frac{1}{\mu}\nabla \times \underline{E}$.

In the presence of a single dielectric cylinder, the incident wave is scattered so that the total field/scattered field formulation is adopted. The physical domain is divided into the regions known as the total field, scattered field, and uniaxial perfectly matched layer to absorb waves.

The results are calculated at the same Courant number $Cr = 0.2$, which corresponds to the specified time increment $\Delta t = 0.0026685\ fs$. The three-dimensional results of E_z are plotted in Fig. 3b–f at the cutting plane $z = 0$ nm.

5 Conclusions

A high-order FDTD scheme has been developed in three-point grid stencil to solve the three-dimensional Maxwell's solutions in non-staggered grids. Our first aim is to numerically preserve symplecticity and conserve Hamiltonian. To retain these theoretical properties at all times, the explicit partitioned Runge–Kutta symplectic time integrator is applied together with the space-centered scheme. To increase the dispersive accuracy that is essential to predict wave propagation correctly, the discrepancy between the numerical and exact phase velocities is minimized. The numerically verified temporally second-order and spatially fourth-order accurate compact finite difference scheme is also shown to satisfy the discrete Gauss' law. The solutions computed on a single Nvidia K-20 card for the analytical and benchmark problems for the verification and validation purposes have been shown to agree very well with the exact and the benchmark numerical solutions.

Acknowledgments This work was supported by the Ministry of Science and Technology (MOST) of the Republic of China under the Grants NSC96-2221-E-002-293-MY2, NSC96-2221-E-002-004, and CQSE97R0066-69.

References

1. Chi, J., Liu, F., Weber, E., Li, Y., Crozier, S.: GPU-accelerated FDTD modeling of radio-frequency field-tissue interactions in high-field MRI. IEEE Trans. Biomed. Eng. **58**(6), 1789–96 (2011)
2. Zunoubi, M.R., Payne, J., Roach, W.P.: CUDA implementation of TE-FDTD solution of Maxwell's equations in dispersive media. IEEE Antennas and Propagation Society **9**, 756–759 (2010)

3. Lee, K.H., Ahmed, I., Goh, R.S.M., Khoo, E.H., Li, E.P., Hung, T.G.G.: Implementation of the FDTD method based on Lorentz-Drude dispersive model on GPU for plasmonics applications. Progr. Electromagn. Res. **116**, 441–456 (2011)
4. Zygiridis, T.T.: High-order error-optimized FDTD algorithm with GPU implementation. IEEE Trans. Magnetics **49**(5), 1809–1813 (2013)
5. Micikevicius, P.: 3D Finite Difference Computation on GPUs Using CUDA. ACM New York **79–84**, (2009)
6. Zhang, B., Xue, Z.H., Ren, W., Li, W.M.: X, pp. 410–413. Q. Sheng, Accelerating FDTD algorithm using GPU computing, IEEE (ICMTCE) (2011)
7. Bridges, T.J., Reich, S.: Multi-symplectic integration numerical scheme for Hamiltonian PDEs that conserves symplecticity. Phys. Lett. A **284**, 184–193 (2001)
8. Cockburn, B., Li, F., Shu, C.-W.: Locally divergence-free discontinuous Galerkin methods for the Maxwell equations. J. Comput. Phys. **194**, 588–610 (2004)
9. Anderson, N., Arthurs, A.M.: Helicity and variational principles for Maxwell's equations. Int. J. Electron. **54**, 861–864 (1983)
10. Marsden, J.E., Weinstein, A.: The Hamiltonian structure of Maxwell-Vlasov equations. PhysicalD **4**, 394–406 (1982)
11. Sheu, T.W.H., Hung, Y.W., Tsai, M.H., Li, J.H.: On the development of a triple-preserving Maxwell's equations solver in non-staggered grids. Int. J. Numer. Methods Fluids. **63**, 1328–1346 (2010)
12. Sanz-Serna, J.M.: Symplectic Runge-Kutta and related methods: recent results. Physica D **293–302**, (1992)
13. Jiang, L.L., Mao, J.F., Wu, X.L.: Symplectic finite-difference time-domain method for Maxwell equations. IEEE Trans. Magn. **42**(8), 1991–1995 (2006)
14. Sha, W., Huang, Z.X., Chen, M.S., Wu, X.L.: Survey on symplectic finite-difference time-domain schemes for Maxwell's equations. IEEE T. Antenn. Propag. **56**, 493–510 (2008)
15. Lele, S.K.: Compact finite difference schemes with spectral-like resolution. J. Comput. Phys. **17**, 328–346 (1996)
16. Zingy, D.W., Lomax, H., Jurgens, H.: High-accuracy finite-difference schemes for linear wave propagation. SIAM J. Sci. Comput. **17**, 328–346 (1996)
17. Spachmann, H., Schuhmann, R., Weiland, T.: High order spatial operators for the finite integration theory. ACES Journal **17**(1), 11–22 (2002)
18. Kashiwa, T., Sendo, Y., Taguchi, K., Ohtani, T., Kanai, Y.: Phase velocity errors of the nonstandard FDTD method and comparison with other high-accuracy FDTD methods. IEEE Transactions on Magnetics **39**(4), 2125–2128 (2003)

RETRACTED CHAPTER: A Collocation Method for Integral Equations in Terms of Generalized Bernstein Polynomials

Vinai K. Singh and A.K. Singh

Abstract In this study, a collocation method based on generalized Bernstein polynomials is presented for approximate solution of Fredholm–Volterra integral equations. While this method is applicable directly to linear integral equations of the first, second and third kinds, it is applicable iteratively to nonlinear integral equations using method of quasilinearzation. Error bounds are demonstrated for the Bernstein collocation method, and the convergence of this method is shown. Moreover, some numerical examples are given to illustrate the accuracy, efficiency and applicability of the method.

Keywords Bernstein polynomials linear and nonlinear integral equations · Quasilinearzation technique · Collocation method

2000 Mathematics Subject Classification 42A60 · 42A10

1 Introduction

Integral equations are closely related to a number of different areas of mathematics. For instance, many problems included to ordinary and partial differential equations can be converted to the integral equations. In addition, these equations are often used in the engineering, mathematical physics, potential theory, electrostatic and

---

The original version of this chapter was revised: The chapter was retracted as it contains significant parts plagiarizing another publication. The erratum to this chapter is available at 10.1007/978-981-10-1454-3_26

V.K. Singh (✉)
Department of Applied Mathematics, Raj Kumar Goel Institute of Technology, NH-58, Delhi-Meerut Road, Ghaziabad 201003, India
e-mail: drvinaiksingh@rkgit.edu.in

A.K. Singh
Department of Science and Technology,
Government of India, Technology Bhavan, New Mehrauli Road, New Delhi 110016, India
e-mail: ashokk.singh@nic.in

V.K. Singh et al. (eds.), *Modern Mathematical Methods and High Performance Computing in Science and Technology*, Springer Proceedings in Mathematics & Statistics 171, DOI 10.1007/978-981-10-1454-3_23

radioactive heat transfer. Therefore, many researchers are interested in numerical methods to get the solution of integral equations.

Quasilinearzation [1, 2] is an effective method that solves the nonlinear equations recursively by a sequence of linear equations. The main advantage of this method is that it converges quadratically to the solution of the original equation. Besides, since many problems in system identification and optimization can be reduced to this format, quasilinearzation is a useful computational technique in modern control applications. This method has also been applied the variety of nonlinear equations such as ordinary differential equations [3–7], functional differential equations [8–10], integral equations [11, 12], integro-differential equations [13].

Bernstein polynomials have many useful properties, such as the positivity, continuity, recursion's relation, symmetry, unity partition of the basis set over the interval [0, 1], and the polynomials are differentiable and integrable. For this reason, these polynomials have been used to numerical solution of Volterra [14–17, 25] and Fredholm [18, 19] integral equations.

The definitions of the Bernstein polynomials and their basis from that can be easily generalized on the interval [a, b], are given as follows.

Definition 1.1 Generalized Bernstein basis polynomials can be defined on the interval [a, b] by

$$p_{i,n}(x) = \frac{1}{(b-a)^n}\binom{n}{i}(x-a)^i(b-x)^{n-i},\ i = 0, 1, 2, \ldots, n.$$

Definition 1.2 Let $y : [a, b] \to R$ be continuous function on the interval $[a, b]$. Bernstein polynomials of nth-degree are defined by

$$B_n(y; x) = \sum_{i=0}^{n} y\left(a + \frac{(b-a)i}{n}\right) p_{i,n}(x).$$

Theorem 1.1 *If $y \in C^m[a, b]$ and $m \geq 0$, for some integer then*

$$\lim_{n\to\infty} B_n^{(k)}(y; x) = y^{(k)}(x);\ k = 0, 1, 2, \ldots, m$$

converges uniformly to y for as $n \to \infty$. For more information about Bernstein polynomials, see [4, 20, 21].

Definition 1.3 A linear Fredholm–Volterra integral equation of the third kind is given by:

$$a(x)y(x) = g(x) + \lambda_1 \int_a^b f(x,t)y(t)dt + \lambda_2 \int_a^x k(x,t)y(t)dt \quad (1)$$

such that $a(x) \neq 0$ and $a(x) \neq 1$. Here $a(x)$, $g(x)$, $f(x, t)$ and $k(x, t)$ are given functions. λ_1 and λ_2 are constants, $y(x)$ is unknown function. Privately Eq. (1) is called as linear Fredholm–Volterra integral equations of the first or second kind for $a(x) \equiv 0$ and $a(x) \equiv 1$, respectively.

Definition 1.4 A nonlinear Fredholm–Volterra integral equation is defined as follows:

$$a(x)y(x) = g(x) + \lambda_1 \int_a^b f(x,t,y(t))dt + \lambda_2 \int_a^x k(x,t,y(t))dt \tag{2}$$

where $g(x)$, $f(x,t,y(t))$ and $k(x,t,y(t))$ are continuous functions, λ_1 and λ_2 are constants, $y(x)$ is unknown function.

The reminder of this paper is organized as follows: In Sect. 2, a collocation method is developed directly and iteratively to get the solution of the integral equations by means of the generalized Bernstein polynomials. In Sect. 3, error bounds and convergence analysis are given for the proposed method. Section 4 is devoted to the applicability of the presented method. In this part, some linear and nonlinear examples are solved and compared with different methods. Finally, the paper is ended with conclusions.

2 Method of Solution

In this paper, the purpose is to approximate the solution of the linear Fradholm–Volterra integral equation (1) directly and nonlinear Fredholm–Volterra integral equation (2) via the quasilinearization method iteratively with the generalized Bernstein polynomials:

$$y(x) \cong B_n(y;x) = \sum_{i=0}^{n} y\left(a + \frac{(b-a)i}{n}\right) p_{i,n}(x). \tag{3}$$

Theorem 2.1 *Let $x_s \in [a,b]$ be collocation points. Linear Fredholm–Volterra equation (1) has following matrix form:*

$$[\mathbf{AP} - \lambda_1 \mathbf{F} - \lambda_2 \mathbf{K}]\mathbf{Y} = \mathbf{G} \tag{4}$$

Here $\mathbf{A} = diag[a(x_s)]$, $\mathbf{F} = [F_{s,i}]$, $\mathbf{K} = [K_{s,i}]$, $\mathbf{P} = [p_{i,n}(x_s)]$ are $(n+1) \times (n+1)$ matrices, and $\mathbf{Y} = [y(a + \frac{(b-a)i}{n})]$, $\mathbf{G} = [g(x_s)]$ are $(n+1) \times 1$ matrices for $i, s = 0, 1, \ldots, n$.

Proof The expression (3) can be written as

$$y(x) \cong \mathbf{P}(x)\mathbf{Y} \tag{5}$$

such that

$$\mathbf{P}(x) = [p_{0,n}(x)\ p_{1,n}(x)....P_{n,n}(x)],\ \mathbf{Y} = [y(a),\ y\left(a + \frac{(b-a)}{n}\right), ..., y(b)]^T.$$

Substituting the collocation points and relation (5) into Eq. (1), we obtain the linear algebraic equation system

$$a(x_s)\mathbf{P}(x_s)\mathbf{Y} = g(x_s) + \lambda_1 \int_a^b f(x_s,t)\mathbf{P}(t)dt\mathbf{Y} + \lambda_2 \int_a^{x_s} k(x_s,t)\mathbf{P}(t)dt\mathbf{Y} \quad (6)$$

such that $y(x_s) = B_n(y; x_s)$; $s = 0, 1, \ldots, n$. If the integrals at the sides of λ_1 and λ_2 are called respectively $\mathbf{F}(x_s)$ and $\mathbf{K}(x_s)$, then for $i = 0, 1 \ldots, n$, the elements of these matrices can be written as

$\mathbf{F}(x_s) = [F_{s,0}\ F_{s,1} \ldots \ldots F_{s,n}]$; $F_{s,i} = \int_a^b f(x_s,t)p_{i,n}(t)dt$
$\mathbf{K}(x_s) = [K_{s,0}\ K_{s,1} \ldots \ldots K_{s,n}]$; $K_{s,i} = \int_a^{x_s} k(x_s,t)p_{i,n}(t)dt$,

Therefore the Eq. (6) becomes

$[a(x_s)\mathbf{P}(x_s) - \lambda_1\mathbf{F}(x_s) - \lambda_2\mathbf{K}(x_s)]\mathbf{Y} = g(x_s)$.

For $s = 0, 1, \ldots, n$. This system is equivalent to matrix equation (4), completed the proof.

The Eq. (4) can be written in the compact form

$$\mathbf{WY} = \mathbf{F} \text{ or } [\mathbf{W}; \mathbf{F}],$$

So that $\mathbf{W} = \mathbf{AP} - \lambda_1\mathbf{F} - \lambda_2\mathbf{K}$. If $rank(\mathbf{W}) = rank[\mathbf{W}; \mathbf{F}] = n + 1$, then solution of this system is uniquely determined

Theorem 2.2 *Let $x_s \in [a, b]$ be collocation points. Nonlinear Fredholm–Volterra integral equation (2) has the iteration matrix in the form:*

$$[\mathbf{AP} - \lambda_1\mathbf{F_r}\lambda_2\mathbf{K_r}]\mathbf{Y_{r+1}} = \mathbf{H_r};\ r = 0, 1 \ldots.. \quad (7)$$

*Here matrices **A** and **P** are as given in the Theorem 2.1, $\mathbf{F_r} = [F_{r,s,i}]$ and $\mathbf{K_r} = [K_{r,s,i}]$ are $(n+1) \times (n+1)$ matrices, $\mathbf{Y_{r+1}} = [y_{r+1}(a + \frac{(b-a)^i}{n})]$ and $\mathbf{H_r} = [h_r(x_s)]$ are $(n+1) \times 1$ matrices for $i, s = 0, 1, \ldots, n$.*

Proof Let $y_0(x)$ be arbitrary chosen function for starting iteration. By considering the quasilinearization method, Eq. (2) is expressed as a sequence of linear equations for $r = 0, 1 \ldots..$

$$a(x)y_{r+1}(x) = g(x) + \lambda_1 \int_a^b [f(x,t,y_r(t)) + f_y(x,t,y_r(t))(y_{r+1}(t) - y_r(t))]dt$$
$$+ \lambda_2 \int_a^x [k(x,t,y_r(t)) + k_y(x,t,y_r(t))(y_{r+1}(t) - y_r(t))]dt, \quad (8)$$

and then Bernstein collocation method for solving a sequence of linear equations (8) is applied. From expression (5), we have

$$y_{r+1}(x) \cong \mathbf{P}(x)\mathbf{Y_{r+1}};\ r = 0, 1 \ldots.. \tag{9}$$

Substituting the collocation points and relation (9) into Eq. (8), we obtain linear algebraic system

$$a(x_s)\mathbf{P}(x_s)\mathbf{Y_{r+1}} - \lambda_1 \mathbf{F_r}(x_s)\mathbf{Y_{r+1}} - \lambda_2 \mathbf{K_r}(x_s)\mathbf{Y_{r+1}} = h_r(x_s) \tag{10}$$

Here $\mathbf{F_r}(x_s)$, $\mathbf{K_r}(x_s)$ and $h_r(x_s)$ are given by

$$\mathbf{F_r}(x_s) = [F_{r,s,0}\ \ F_{r,s,1} \ldots F_{r,s,n}],\ \mathbf{K_r}(x_s) = [K_{r,s,0}\ \ K_{r,s,1} \ldots K_{r,s,n}].$$

$$h_r(x_s) = g(x_s) + \lambda_1 \int_a^b [f(x_s, t, y_r(t)) - f_y(x_s, t, y_r(t))y_r(t)]dt$$
$$+ \lambda_2 \int_a^{x_s} [k(x_s, t, y_r(t)) + k_y(x_s, t, y_r(t))y_r(t)]dt$$

such that

$$F_{r,s,i} = \int_a^b f(x_s, t, y_r(t))p_{i,n}(t)dt,\ K_{r,s,i} = \int_a^{x_s} k(x_s, t, y_r(t))p_{i,n}(t)dt$$

Considering the matrices

$$\mathbf{F_r} = \begin{bmatrix} \mathbf{F_r}(x_0) \\ \mathbf{F_r}(x_1) \\ \ldots \\ \ldots \\ \mathbf{F_r}(x_n) \end{bmatrix}$$

$$\mathbf{K_r} = \begin{bmatrix} \mathbf{K_r}(x_0) \\ \mathbf{K_r}(x_1) \\ \ldots \\ \ldots \\ \mathbf{K_r}(x_n) \end{bmatrix}$$

and

$$\mathbf{H_r} = \begin{bmatrix} \mathbf{h_r}(x_0) \\ \mathbf{h_r}(x_1) \\ \ldots \\ \ldots \\ \mathbf{h_r}(x_n) \end{bmatrix}$$

the Eq. (10) can be written as matrix form (7). This completed the proof.

The Eq. (7) can be written in the compact form

$$\mathbf{W_r}\,\mathbf{Y_{r+1}} = \mathbf{H_r} \quad or \quad [\mathbf{W_r}; \mathbf{H_r}];\ r = 0, 1, \ldots$$

so that $\mathbf{W_r} = \mathbf{AP} - \lambda_1 \mathbf{F_r} - \lambda_2 \mathbf{K_r}$.

3 Convergence and Error Analysis

Definition 3.1 Error is denoted by $e_n(x) = y(x) - B_n(y; x)$ such that $y(x)$ is an exact solution and $B_n(y; x)$ is a generalized Bernstein approximate solution. Then the maximum error can be defined as

$$E_n(y) = \| e_n \|_\infty = \max_{a \le x \le b} |e_n(x)|,$$

and on the collocation points; maximum, mean and root of the mean square errors are defined by

$$E_{\max} = \max_{x_s \in [a,b]} |e_n(x_s)|,\ E_{\text{mean}} = \frac{1}{n+1}\sum_{s=0}^{n} |e_n(x_s)|,\ E_{\text{root}} = \sqrt{\frac{1}{n+1}\sum_{s=0}^{n}(e_n(x_s))^2}$$

Let $y(x_s) \neq 0$ and $B_n(y; x_s)$ be scalars, then the absolute relative error in $B_n(y; x_s)$ as an approximation to $y(x_s)$ is the number

$$E_{rel} = \frac{|e_n(x_s)|}{|y(x_s)|}$$

Let f be a continuous function on the square $[a, b] \times [a, b]$. Then the maximum norm of f can be denoted by

$$\| f \|_\infty = \max_{x,t \in [a,b]} |f(x, t)|$$

Residual error can be defined for the presented method on the following:

$$R_n(x) = a(x)B_n(y; x) - \lambda_1 \int_a^b f(x, t)B_n(y; t)dt - \lambda_2 \int_a^x k(x, t)B_n(y; t)dt - g(x). \tag{11}$$

Definition 3.2 ([1]) Let y_0 be initial approximation to root m, with $y_0 < m$ and $e_r(x) = y_{r+1}(x) - y_r(x)$ be rth iteration error. Then the following relation obtained for error in quasilinearization is called quadratic convergence:

$$|e_r(x)| \le M|e_{r-1}(x)|^2$$

where

$$M = \max_{y_0 \le \theta \le m} \left[\frac{|y''(\theta)|}{|y'(\theta)|} + \frac{|\phi''(\theta)|}{2} \right]; \phi(\theta) = \theta - \frac{y(\theta)}{y'(\theta)}, y_{r-1} \le \theta \le y_r.$$

Definition 3.3 ([22]) Let $y(x)$ be defined on $[a, b]$, the modulus of continuity of $y(x)$ on $[a, b]$, $\omega(\delta)$,is defined for $\delta > 0$ by

$$\omega(\delta) = \sup_{x_1, x_2 \in [a,b]; |x_1 - x_2| \le \delta} |y(x_1) - y(x_2)|$$

Lemma 3.1 ([22]) *If* $\lambda > 0$, *then* $\omega(\lambda\delta) = (1 + \lambda)(\omega(\delta)$.

Lemma 3.2 ([22]) $y(x)$ *is uniformly continuous on* $[a, b]$ *iff*

$$\lim_{\delta \to 0} \omega(\delta) = 0.$$

Lemma 3.3 *The generalized Bernstein basis polynomials have the following relation:*

$$\sum_{i=0}^{n} \left(x - \left(a + \frac{b-a}{n} i \right) \right)^2 p_{i,n}(x) = \frac{(x-a)(b-x)}{n}.$$

Proof Consider the following relation given in Ref. [22] for Bernstein basis polynomials defined on the interval [0,1]:

$$\sum_{i=0}^{n} \left(t - \frac{i}{n} \right)^2 p_{i,n}(x) = \frac{t(1-t)}{n}.$$

Applying the transformation $t = \frac{x-a}{b-a}$ to this expression and multiplying both sides with $(b-a)^2$,, then we obtain the desired relation.

Theorem 3.1 *Let* $B_n y$ *be generalized Bernstein approximate solution on* $[a, b]$. *If exact solution* $y(x)$ *is continuous on* $[a, b]$, *then the error is*

$$|e_n(x)| \le \omega(n^{-1/2})(1 + \sqrt{(x-a)(b-x)})$$

and

$$\lim_{n \to \infty} \| e_n \|_\infty = 0$$

Proof This theorem is easily proved with similar way in Ref. [22]. However, Lemma 3.3 is used for proof different from [22].

Theorem 3.2 *Consider the linear Fredholm–Volterra integral equation (1). Let $a(x)$, $y(x)$ be continuous functions on the interval $[a, b]$ and $f(x, t)$, $k(x, t)$ be continuous functions on the square $[a, b] \times [a, b]$. Then residual error of the generalized Bernstein polynomials approach holds the following:*

$$\| R_n \|_\infty \leq c \| e_n \|_\infty, \lim_{n \to \infty} \| R_n \|_\infty = 0$$

such that c is a positive constant.

Proof Considering absolute value of residual error (11) and substituting $g(x)$ given in Eq. (1), the residual error can be written by

$$|R_n(x)| \leq |a(x)||B_n(y; x) - y(x)| + |\lambda_1| \int_a^b |f(x, t)||B_n(y; t) - y(t)|dt$$
$$+ |\lambda_2| \int_a^x |k(x, t)| \; |B_n(y; t) - y(t)|dt. \qquad (12)$$

Moreover, from the definition of the maximum error and properties of the norm, we further get

$$\begin{aligned} \| R_n \|_\infty &\leq \| a \|_\infty \| e_n \|_\infty + (b - a)|\lambda_1| \; \| f \|_\infty \| e_n \|_\infty + (x - a)|\lambda_2| \; \| k \|_\infty \| e_n \|_\infty \\ &\leq (\| a \|_\infty + (b - a)|\lambda_1| \; \| f \|_\infty + (x - a)|\lambda_2| \; \| k \|_\infty) \| e_n \|_\infty \\ &\leq (\| a \|_\infty + (b - a)|\lambda_1| \; \| f \|_\infty + \max_{x \in [a,b]} (x - a)|\lambda_2| \; \| k \|_\infty) \| e_n \|_\infty \\ &\leq (\| a \|_\infty + (b - a)|\lambda_1| \; \| f \|_\infty + (b - a)|\lambda_2| \; \| k \|_\infty) \| e_n \|_\infty \\ &\leq c \| e_n \|_\infty \end{aligned}$$

such that $c = \| a \|_\infty + (b - a)|\lambda_1| \; \| f \|_\infty + (b - a)|\lambda_2| \; \| k \|_\infty$. Since $y(x)$ is continuous on the interval $[a, b]$ as follows from Theorem 3.1 $\| R_n \|_\infty \to 0$, as $n \to \infty$.

Theorem 3.3 *Let y be a continuous function on the interval $[a, b]$ and $x_s; s = 0, 1, ., n$ be collocation points. Then, the residual error bound at the collocation points for linear Fredholm–Volterra integral equation (1), is*

$$|R_n(x_s)| < k(b - a) \left(1 + \frac{b - a}{2}\right) \omega(sn^{-3/2}),$$

and

$$\lim_{n \to \infty} |R_n(x_s)| = 0.$$

Here k is positive constant.

Proof Substituting the collocation points into the absolute residual error (12), we have

$$|R_n(x_s)| \le |a(x_s)||B_n(y;x_s) - y(x_s)| + |\lambda_1| \int_a^b |f(x_s,t)||B_n(y;t) - y(t)|dt$$
$$+ |\lambda_2| \int_a^{x_s} k(x_s,t)||B_n(y;t) - y(t)|dt$$

From Theorem 3.1 and considering $B_n(y;x_s) = y(x_s)$ for the generalized Bernstein polynomials approach on the collocation points, the residual error becomes

$$R_n(x_s)| \le \omega(n^{-1/2})[|\lambda_1| \int_a^b |f(x_s,t)|[1 + \sqrt{(t-a)(b-t)}]dt$$
$$+ |\lambda_2| \int_a^{x_s} |k(x_s,t)|[1 + \sqrt{(t-a)(b-t)}]]dt$$

Denoting

$$\epsilon = |\lambda_1||f(x_s,t)|,\ \delta = |\lambda_2| \max_{t\in[a,b]} |k(x_s,t)|,\ \max_{t\in[a,b]} [1 + \sqrt{(t-a)(b-t)}] = 1 + \frac{b-a}{2},$$

and considering the Lemma 3.1, desirable inequality is obtained as:

$$\begin{aligned}
|R_n(x_s)| &\le ((b-a)\epsilon + (x_s - a)\delta)\left(1 + \frac{b-a}{2}\right)\omega(n^{-1/2}) \\
&\le \left(\epsilon + \delta\frac{s}{n}\right)(b-a)\left(1 + \frac{b-a}{2}\right)\omega(n^{-1/2}) \\
&\le k(b-a)\left(1 + \frac{b-a}{2}\right)\left(1 + \frac{s}{n}\right)\omega(n^{-1/2}) \\
&\le k(b-a)\left(1 + \frac{b-a}{2}\right)\omega(sn^{-3/2})
\end{aligned}$$

Such that k is bigger than ϵ and δ. Since $y(x)$ is continuous on the interval $[a,b]$, in view of Lemma 3.2, $|R_n(x_s)| \to 0$ is $n \to \infty$. This is completes the proof.

4 Numerical Results

Four linear and two nonlinear numerical examples are given using the presented method on the collocation points $x_s = a + \frac{(b-a)s}{n}$; $x_s = -\cos(\frac{\pi s}{n}$)and $x_s = -\frac{1-\cos(\frac{\pi s}{n})}{2}$; $s = 0, 1, \ldots\ldots, n$. Numerical results computed in MATLAB 7.1 with 32 digits are compared with the other methods.

Table 1 Mean errors of Example 1

n	$x_s = a + \frac{(b-a))s}{n}$	$x_s = -\cos(\frac{\pi s}{n})$	n	$x_s = a + \frac{(b-a)s}{n}$	$x_s = -\cos(\frac{\pi s}{n})$
2	0	3.7e−017	11	2.7e−015	3.3e−016
3	4.9e−017	1.1e−016	12	1.4e−015	3.6e−016
4	2.5e−017	1.8e−016	13	4.4.e−015	2.5e−016
5	3.6e−016	5.4e−017	14	2.6e−015	1.7e−016
6	5.4e−016	9.8e−017	16	1.5e−015	3.5e−016
7	4.2e−016	9.8e−017	17	2.1e−014	2.7e−016
8	2.8e−016	2.4e−016	18	2.5e−014	2.2e−016
9	1.2e−016	2.0e−016	19	8.8e−014	3.2e−016
10	2.4e−016	2.0e−016	20	1.5e−013	4.0e−016

Example 1 Consider the

$$y(x) = 1 + \int_{-1}^{1} (xt + x^2t^2)y(t)dt; \quad -1 \leq x \leq 1$$

linear Fredholm integral equation of the second kind that the exact solution is $y(x) = 1 + \frac{10}{9}x^2$.

The mean errors of proposed method with increasing n are given in Table 1. Mean error obtained for $n = 4$ by using the numerical method based on the Bernstein basis polynomials is nearby 10^{-13}. Whereas mean error of the presented method is nearby 10^{-17} on the collocation points $x_s = a + \frac{(b-a)s}{4}$; $s = 0, 1, \ldots, 4$. Therefore, we can say that our method is more effective than the other method given by Ahmad et al. [8].

Example 2 Consider the

$$y(x) = \cos(x) - e^x \sin(x) + \int_0^x e^x y(t)dt; \quad 0 \leq x \leq 1$$

Linear Volterra integral equation of the second kind that analytic solution is $y(x) = \cos x$.

The root of the mean square errors are compared with the numerical method based on the Bersetein polynomials [15] in Table 2. If collocation points $x_s = \frac{(1-\cos(\frac{\pi s}{n}))}{2}$; $s = 0, 1, \ldots, n$ are considered, the presented method converges more rapidly than the other method for $n \geq 4$.

Example 3 Consider the linear Fredholm integral equation of the second kind:

$$y(x) = e^x + 2\int_0^1 e^{x+t} y(t)dt; \quad 0 \leq x \leq 1$$

Table 2 Comparison of the root of the mean square errors for Example 2

n	Presented method	K. Maleknejad
2	4.7e−003	2.0e−003
3	3.1e−004	2.3e−004
4	4.7e−006	6.2e−006
5	1.8e−007	5.4e−007
6	2.0e−009	1.1e−008
7	1.0e−010	1.1e−009
8	1.3e−012	3.4e−010
9	5.2e−014	3.3e−010
10	6.6e−016	3.4e−010

Table 3 Mean errors of Example 3

n	$x_s = a + \frac{(b-a)s}{n}$	$x_s = \frac{1-\cos(\frac{\pi s}{n})}{2}$	n	$x_s = a + \frac{(b-a)s}{n}$	$x_s = \frac{1-\cos(\frac{\pi s}{n})}{2}$
2	6.0e−004	6.0e−004	12	1.1e−016	1.6e−016
3	5.8e−005	2.5e−005	17	3.4e−016	1.9e−016
4	1.2e−006	4.7e−007	18	8.[illegible]e−01[illegible]	1.9e−016
5	1.1e−007	1.2e−008	19	3.0e−015	3.3e−016
6	2.0e−009	1.3e−010	20	1.6e−014	2.5e−016
7	1.5e−010	3.9e−012	21	1.2e−013	1.8e−016
8	2.2e−012	3.8e−014	22	1.9e−013	3.0e−016
9	1.4e−013	1.4e−015	23	9.0e 015	2.2e−016
10	1.6e−015	1.6e−016	24	6.4e−013	2.0e−016
11	1.7e−016	2.3e−016	25	5.9e−014	2.4e−016

Exact solution of the above equation is $y(x) = \frac{e^x}{2-e^2}$.

Mean errors obtained by using the presented method are given in Table 3. Besides, when the collocation points are not equally spaced, Table 3 shows that results obtained on these points are much better than the other results. The absolute relative errors can be compared with the Galerkin method based on the Bernstein basis polynomials [19] in Table 4. It shows that the results obtained on the collocation points $x_s = a + \frac{(b-a)s}{n}$; $s = 0, 1, \ldots, n$ are better than the results given by other method.

Example 4 Consider the following linear Fredholm–Volterra integral equation of the third kind.

$$3y(x) = 3x^2 - \sin x(x^2 \sin x + 2x \cos x - 2\sin x - \sin 1 + 2\cos 1) + \int_0^1 \sin x \cos t\, y(t)dt + \int_0^x \sin x \cos t\, y(t)dt$$

Analytic solution of the above equation is $y(x) = x^2$.

Table 4 Comparison of the absolute relative error for Example 3

n = 5			n = 7	
x	Presented method	A. Shirin	Presented method	A. Shirin
0.1	1.5e−006	1.9e−005	1.4e−009	1.3e−005
0.2	3.4e−007	7.2e−006	1.2e−009	3.8e−006
0.3	8.6e−007	1.0e−005	3.4e−010	6.1e−006
0.4	3.4e−007	1.1e−005	3.1e−010	3.5e−066
0.5	2.1e−008	4.8e−007	6.4e−010	7.6e−006
0.6	3.4e−007	9.2e−006	3.4e−010	1.4e−006
0.7	7.1e−007	7.8e−006	3.7e−010	5.9e−006
0.8	3.4e−007	3.6e−006	8.8e−010	3.3e−006
0.9	6.0e−007	9.3e−006	4.5e−010	5.5e−006
1	3.4e−007	2.1e−005	4.5e−010	[illegible]

Table 5 Mean errors of Example 4

n	$x_s = a + \frac{(b-a)s}{n}$	$x_s = \frac{1-\cos(\frac{\pi s}{n})}{2}$	n	$x_s = a + \frac{(b-a)s}{n}$	$x_s = \frac{1-\cos(\frac{\pi s}{n})}{2}$
2	9.8e−018	9.3e−018	10	3.5e−017	8.1e−017
3	2.3e−017	4.3e−017	14	7.6e−017	7.0e−017
4	0	7.4e−018	15	7.7e−017	6.7e−017
5	3.7e−017	1.1e−016	16	4.7e−016	1.1e−016
6	8.2e−017	1.9e−017	18	5.5e−016	5.4e−017
7	1.3e−016	1.8e−016	20	5.0e−015	9.3e−017
8	2.6e−017	6.2e−017	25	4.8 e−014	7.6e−017
9	5.9e−017	7.0e−017	30	2.0e−012	7.3e−017

Table 6 Comparison of the maximum errors for Example 4

n	Presented method	Taylor expansion method	Collocation method	Fixed point method
14	4.4e−016	2.0e−015	–	–
16	9.7e−016	–	7.8e−005	3.8e−004
32	2.8e−012	–	4.7e−005	9.5e−005

The mean errors of the presented method with increasing n are given in Table 5. The maximum errors are compared with the Taylor expansion method [23], collocation method and fixed point method [11] in Table 6. It shows that the presented method on the collocation points $x_s = a + \frac{(b-a)s}{n}$; $s = 0, 1, \ldots, n$ is more effective than the other methods. Moreover, the results of proposed method obtained without iteration are better than the results of collocation and fixed point methods [11] obtained with eigth iteration.

Example 5 Consider the following nonlinear Volterra integral equation:

$$y(x) = 2 - e^x + \int_0^x e^{x-t} y^2(t)dt; \quad 0 \le t \le 1$$

Exact solution of the above equation is $y(x) = 1$. Let $y_0(x) = 2 - e^x$ be the first iteration function.

Mean errors of the presented method with increasing n and r are given in Table 7. The absolute errors are compared with the results given by Malekeknajad and Najafi [24] in Table 8. They have used the method combining collocation method and iterations of the quasilinear technique [15]. Table 8 shows that presented method gives the exact solution $y(x) = 1$ for r = fifth iteration, and the proposed method has better numerical solutions than the other method.

Example 6 Consider the following nonlinear Fredholm–Volterra integral equation:

$$y(x) = \frac{2x + 7 - 7x^4}{3} + \int_{-1}^{x} (x + t)y^2(t)dt + \int_{-1}^{1} (x - t)y(t)dt; \quad 0 \le x \le 1$$

Table 7 Mean errors of Example 5

n	r = 1	r = 2	r = 3	r = 4	r = 5
2	2.3e−002	6.8e−00[illegible]	1.6e−007	3.6e−014	0
4	2.5e−002	1.9e−004	9.6e−009	0	0
8	2.0e−002	1.3e−004	4.1e−009	2.0e−016	5.3e−016
16	1.7e−002	9.2e−005	2.3e−009	1.1e−014	1.3e−015

Table 8 Comparison of the absolute error of Example 5

Presented method			Presented method		Collocation method [15]	
$y_0(x) = 2\text{-}e^x$, n = 3			$y_0(x) = 2\text{-}e^x$, n = 4		$y_o(x) = 2\text{-}e^x$, n = 4	
x	r = 2	r = 5	r = 2	r = 5	r = 2	r = 5
0.1	2.3e−004	0	5.8e−005	0	2.5e−003	1.8e−015
0.2	2.8e−004	0	3.9e−005	0	7.9e−004	6.6e−016
0.3	2.2e−004	0	8.7e−006	0	1.4e−003	1.5e−015
0.4	1.3e−004	0	3.9e−006	0	1.8e−003	2.1e−015
0.5	1.1e−004	0	3.5e−005	0	9.0e−004	1.2e−015
0.6	2.2e−004	0	8.4e−005	0	3.6e−003	1.1e−015
0.7	5.5e−004	0	1.1e−004	0	1.8e−002	2.4e−015
0.8	1.2e−003	0	3.5e−005	0	5.5e−002	1.3e−015
0.9	2.2e−003	0	2.3e−004	0	1.3e−001	1.5e−014

Table 9 Mean errors of Example 6

n	$r = 2$	$r = 3$	$r = 4$	$r = 5$	$r = 6$
2	1.6e−001	3.7e−002	3.1e−003	2.6e−005	7.1e−008
4	8.2e−002	6.4e−003	4.2e−005	1.7e−007	6.5e−008
8	5.0e−002	1.6e−003	1.5e−006	2.6e−007	1.1e−007
16	4.1e−002	1.1e−003	4.2e−007	3.5e−006	9.5e−007

Exact solution of the above equation is $y(x) = 2x$. Let $y_0(x) = 0$ be the first iteration function.

Mean errors of the presented method with Increasing n are given on the collocation points $x_s = a + \frac{(b-a)s}{n}$; $s = 0, 1, \ldots, n$ in Table 9. We can say that the numerical results of proposed method converge more rapidly for increasing iterations r.

5 Conclusions

In this work, a collocation method based on the generalized Bernstein polynomials has been developed for the numerical solution of linear and nonlinear Fredholm–Volterra integral equations directly and iteratively using the quasilinear technique. Since, the generalized Bernstein polynomials approximation is valid for continuous functions on the interval [a, b],the presented method can be applied to solve the integral equations. The error bounds and convergence of the presented method have been presented by considering the generalized Bernstein polynomials approach. Some numerical examples have been given to show the applicability and accuracy of the proposed method. This method has much better numerical results obtained directly on collocation points with equally and not equally spaced for increasing values n, and it is more effective than the other methods given in examples (1), (2), (3) and (4). Moreover, the proposed method derived iteratively converges more rapidly for increasing iterations r as is seen from example (5) and (6), because of the quadratic convergence of this method. Consequently; all these positive reasons are encouraging for an application of the more used method to the the linear and nonlinear equations.

References

1. Belmann, R.E., Kalaba, R.E.: Quasilinearization and Nonlinear Boundary Value Problems. American Elsevier Publishing Company Inc., New York (1965)
2. Stanley, E.L.: Quasilinearization and Invariant Imbedding. Academic Press Inc., New York (1968)

3. Agarwal, R.P.: Iterative methods for a fourth order boundary value problems. J. Comput. Appl. Math. **10**, 203–217 (1984)
4. Dascioglu, A.A., Isler, N.: Bersntein collocation method for solving nonlinear differential equations. Math. Comput. Appl. **18**(3), 293–300 (2013)
5. Baird, Jr. A.C.: Modified quasilinearization technique for the solution of boundary value problems for ordinary differential equations. J. Optim. Theory Appl. **3**, 227–242 (1969)
6. Mandelzweig, V.B., Tabakin, F.: Quasilinearzation approach to nonlinear problems in physics with application to nonlinear ODEs. Comput. Phys. Commun. **141**, 268–281 (2001)
7. Ramos, J.I.: Piecewise quasilinearization techniques for singular boundary problems. Comput. Phys. Commun. **158**, 12–25 (2004)
8. Ahmad, B., Khan, R.A., Sivasundaram, S.: Generalized quasilinearization method for nonlinear functional differential equations. J. Appl. Math. Stoch. Anal. **16**, 33–43 (2003)
9. Dricia, Z., McRaeb, F.A., Devic, J.V: Quasilinearization for functional differential equations with Retardation and anticipation. Nonlinear Anal. **70**, 1763–1775 (2009)
10. Pandey, R.K., Bhardwaj, A., Syam, M.: An efficient method for solving fractional differential equations using Bernstein polynomials. J. Fract. Calc. Appl. **5**(1), 129–145 (2014)
11. Calio, F., Munoz, F., Marchetti, E.: Direct and iterative methods for the numerical solution of mixed integral equations. Appl. Math. Comput. **216**, 3739–3746 (2010)
12. Pandit, S.G.: Quadratically converging iterative schemes for nonlinear Volterra integral equations and an application. J. Appl. Math. Stoch. Anal. **10**, 169–178 (1997)
13. Wang, P., Wu, Y., Wiwatanapaphee, B.: An extension of method of quasilinearization for integro-differential equations. Int. J. Pure Appl. Math. **54**, 27–37 (2009)
14. Bhattacharya, S., Mandal, B.N.: Use of Bernstein polynomials in numerical solutions of Volterra integral equations. Appl. Math. Sci. **2**, 1773–1787 (2008)
15. Maleknejad, K., Hashemizadeh, E., Ezzati, R.: A new approach to the numerical solution of Volterra integral equations by using Bernstein approximation. Commun. Nonlinear Sci. Numer. Simul. **16**, 647–655 (2011)
16. Singh, V.K., Pandey, R.K., Singh, O.P.: New stable numerical solutions of singular integral equations of Abel type by using normalized Bernstein polynomials. Appl. Math. Sci. **3**, 241–255 (2009)
17. Syam, M.I., Anwar, M.N.: A computational method for solving a class of non-linear singularly perturbed Volterra Integro-differential boundary-value problems. J. Math. Comput. Sci. **3**(1), 73–86 (2013)
18. Bhattacharya, S., Mandal, B.N.: Numerical solution of some classes of integral equations using Bernstein polynomials. Appl. Math. Comput. **190**, 1707–1716 (2007)
19. Shirin, A., Islam, M.S.: Numerical solutions of Fredholm integral equations using Bernstein polynomials. J. Sci. Res. **2**, 264–272 (2010)
20. Farouki, R.T., Rajan, V.T.: Algorithms for polynomials in Bernstein form. Comput. Aided Geom. Des. **5**, 1–26 (1988)
21. Lorentz, G.G.: Bernstein Polynomials. Chelsea Publishing, New York (1986)
22. Rivlin, T.J.: An Introduction to the Approximation of Functions. Dover Publications, New York (1969)
23. Chen, Z., Jiang, W.: An approximate solution for a mixed linear Volterra–Fredholm integral equation. Appl. Math. Lett. **25**, 1131–1134 (2012)
24. Maleknejad, K., Najafi, E.: Numerical solution of nonlinear volterra integral equations using the idea of quasilinearization. Commun. Nonlinear Sci. Numer. Simul. **16**, 93–100 (2011)
25. Al-Mdallal, Q.M., Syam, M.I.: The Chebyshev collocation-path following method for solving sixth-order Sturm - Liouville problems. Appl. Math. Comput. **232**(1), 391–398 (2014)

Convergence Estimates in Simultaneous Approximation for Certain Generalized Baskakov Operators

Vijay Gupta and Vinai K. Singh

Abstract In the present article, we consider the Durrmeyer type integral modification of the generalized Baskakov operators. The special cases of our operators provides the well-known Baskakov–Durrmeyer and Szász–Durrmeyer operators. We estimate convergence estimates in simultaneous approximation.

Keywords Linear positive operators · Baskakov operators · Pointwise estimation · Asymptotic formula

1 Introduction

For $a \geq 0$, we consider the modified form of linear positive operators due to [16], depending on certain parameter $c \geq 0$ as

$$V_n^{a,c}(x) = \sum_{k=0}^{\infty} b_{n,k}^{a,c}(x) f\left(\frac{k}{n}\right),$$

where the kernel is given by

$$b_{n,k}^{a,c}(x) = \frac{e^{-acx/(1+cx)}}{k!} \sum_{i=0}^{k} \binom{k}{i} (n/c)_i \cdot a^{k-i} \frac{(cx)^k}{(1+cx)^{n/c+k}}.$$

V. Gupta (✉)
Department of Mathematics, Netaji Subhas Institute of Technology, Sector 3 Dwarka, New Delhi 110078, India
e-mail: vijaygupta2001@hotmail.com

V.K. Singh
Raj Kumar Goel Institute of Technology, NH-58, Delhi- Meerut Road, Ghaziabad 201003, India

V.K. Singh et al. (eds.), *Modern Mathematical Methods and High Performance Computing in Science and Technology*, Springer Proceedings in Mathematics & Statistics 171, DOI 10.1007/978-981-10-1454-3_24

It can easily be seen that $\sum_{k=0}^{\infty} b_{n,k}^{a,c}(x) = 1$. For the special value $a = 0, c = 1$ the above operators become Baskakov operators and $a = 0, c \to 0$ the above operators reduce to the classical Szász–Mirakyan operators. These operators preserve only the constant functions, but for special value $a = 0$ these preserve linear functions also.

In order to consider generalization of the operators discussed in [4], Erencin [2] considered the certain other form of such operators and established some direct results, we consider the Durrmeyer variant in the following way:

$$D_n^{a,c}(f,x) = (n-c)\sum_{k=0}^{\infty} b_{n,k}^{a,c}(x) \int_0^{\infty} b_{n,k}^{0,c}(t) f(t)dt, x \in [0,\infty) \qquad (1)$$

where the kernel is given by

$$b_{n,k}^{a,c}(x) = \frac{e^{-acx/(1+cx)}}{k!} \sum_{i=0}^{k} \binom{k}{i} (n/c)_i \cdot a^{k-i} \frac{(cx)^k}{(1+cx)^{n/c+k}}, b_{n,k}^{0,c}(x) = \frac{(n/c)_k}{k!} \frac{(cx)^k}{(1+cx)^{n/c+k}}.$$

We may point out here that for the general $a > 0$ the basis $b_{n,k}^{a,c}(t)$ under integral sign are not possible to handle due to technical difficulties in finding the moments. In the past two decades, simultaneous approximation properties have been discussed by many researchers on different operators, we mention few of them as [1, 3, 4, 6–15, 17, 18]. Recently Agarwal and Gupta presented some of them in the recent book [5]. The operators defined by (1) produce rational functions so for asymptotic formula in simultaneous approximation, one cannot find the exact form as of those considered in above mentioned papers.

In the present article, we discuss some direct estimates in simultaneous approximation for the operators (1), which include point-wise estimation and the asymptotic formula. We also present the exact expressions of the asymptotic formulae in ordinary approximation and for first derivatives.

2 Basic Results

Lemma 1 *For $m \in N^0$, $a \geq 0$, if we define*

$$T_{n,m}^{a,c}(x) = \sum_{k=0}^{\infty} b_{n,k}^{a,c}(x) \left(\frac{k}{n}\right)^m,$$

then the following recurrence relation holds:

$$T_{n,m+1}^{a,c}(x) = \frac{x(1+cx)}{n}[T_{n,m}^{a,c}(x)]' + \left[\frac{acx}{n(1+cx)} + x\right] T_{n,m}^{a,c}(x).$$

Proof Using the identity

$$x(1+cx)^2[b_{n,k}^{a,c}(x)]' = [(k-nx)(1+cx)-acx]b_{n,k}^{a,c}(x)$$

we have

$$\begin{aligned} x(1+cx)^2[T_{n,m}^{a,c}(x)]' &= \sum_{k=0}^{\infty}[(k-nx)(1+cx)-acx]b_{n,k}^{a,c}(x)\left(\frac{k}{n}\right)^m \\ &= (1+cx)nT_{n,m+1}^{a,c}(x) - [acx+nx(1+cx)]T_{n,m}^{a,c}(x). \end{aligned}$$

□

Lemma 2 *If the mth order ($m \in N^0$) of the operators (1) is defined as*

$$U_{n,m}^{a,c}(x) := D_n^{a,c}(t^m, x) = (n-c)\sum_{k=0}^{\infty} b_{n,k}^{a,c}(x)\int_0^{\infty} b_{n,k}^{0,c}(t)t^m dt,$$

then there holds the following recurrence relation:

$$\begin{aligned}[n-c(m+2)](1+cx)U_{n,m+1}^{a,c}(x) &= x(1+cx)^2(U_{n,m}^{a,c}(x))' \\ &\quad + [acx+(m+1+nx)(1+cx)]U_{n,m}^{a,c}(x). \end{aligned}$$

Proof Using the identity $x(1+cx)^2[b_{n,k}^{a,c}(x)]' = [(k-nx)(1+cx)-acx]b_{n,k}^{a,c}(x)$, we may write

$$\begin{aligned} x(1+cx)^2(U_{n,m}^{a,c}(x))' &= (n-c)\sum_{k=0}^{\infty}[(k-nx)(1+cx)-acx]b_{n,k}^{a,c}(x)\int_0^{\infty} b_{n,k}^{0,c}(t)t^m dt \\ &= (n-c)(1+cx)\sum_{k=0}^{\infty}(k-nx)b_{n,k}^{a,c}(x)\int_0^{\infty} b_{n,k}^{0,c}(t)t^m dt - acxU_{n,m}^{a,c}(x) \\ &= (n-c)(1+cx)\sum_{k=0}^{\infty} b_{n,k}^{a,c}(x)\int_0^{\infty}(k-nt)b_{n,k}^{0,c}t^m dt - acxU_{n,m}^{a,c}(x). \\ &\quad - nx(1+cx)U_{n,m}^{a,c}(x) + n(1+cx)U_{n,m+1}^{a,c}(x). \end{aligned}$$

Using the identity $t(1+ct)[b_{n,k}^{0,c}(t)]' = (k-nt)b_{n,k}^{0,c}(t)$, we have

$$\begin{aligned} & x(1+cx)^2(U_{n,m}^{a,c}(x))' + [acx+nx(1+cx)]U_{n,m}^{a,c}(x) - n(1+cx)U_{n,m+1}^{a,c}(x) \\ &= (n-c)(1+cx)\sum_{k=0}^{\infty} b_{n,k}^{a,c}(x)\int_0^{\infty} t(1+ct)[b_{n,k}^{0,c}(t)]'t^m dt \\ &\quad - (m+1)(1+cx)U_{n,m}^{a,c}(x) - c(m+2)(1+cx)U_{n,m+1}^{a,c}(x), \end{aligned}$$

which is the required recurrence relation. □

Remark 1 By Lemma 2, we have

(i) $U^{a,c}_{n,0}(x) = 1$,

(ii) $U^{a,c}_{n,1}(x) = x + \dfrac{acx}{(n-2c)(1+cx)} + \dfrac{1+2cx}{n-2c}$,

(iii) $U^{a,c}_{n,2}(x) = x^2 + \dfrac{6c(n-c)x^2+4nx+2}{(n-2c)(n-3c)} + \dfrac{2acx(2+nx)}{(n-2c)(n-3c)(1+cx)} + \dfrac{a^2c^2x^2}{(n-2c)(n-3c)(1+cx)^2}$,

(iv) $U^{a,c}_{n,3}(x) = x^3 + \dfrac{12c(n^2-2n+2c^2)x^3+9n(n+c)x^2+18nx+6}{(n-2c)(n-3c)(n-4c)}$
$+\dfrac{3acn(n+c)x^3+18acnx(1+x)}{(n-2c)(n-3c)(n-4c)(1+cx)} + \dfrac{3a^2c^2x^2(n+nx)}{(n-2c)(n-3c)(n-4c)(1+cx)^2}$
$+\dfrac{a^3c^3x^3}{(n-2c)(n-3c)(n-4c)(1+cx)^3}$,

(v) for each $x \in (0,\infty)$ $U^{a,c}_{n,m}(x) = x^m + n^{-1}(q_m(x,a,c)+o(1))$, where $q_m(x, a, c)$ is a rational function of x depending on a, c and m.

Lemma 3 *If the mth order* ($m \in N^0$) *central moment for the operators (1) is defined as*

$$\mu^{a,c}_{n,m}(x) := D^{a,c}_n((t-x)^m, x) = (n-c)\sum_{k=0}^{\infty} b^{a,c}_{n,k}(x)\int_0^{\infty} b^{0,c}_{n,k}(t)(t-x)^m dt,$$

then there holds the following recurrence relation:

$$[n-c(m+2)](1+cx)\mu^{a,c}_{n,m+1}(x) = x(1+cx)^2[(\mu^{a,c}_{n,m}(x))' + 2m\mu^{a,c}_{n,m-1}(x)] + [acx+(m+1)(1+cx)(1+2cx)]\mu^{a,c}_{n,m}(x)$$

Proof By using the identity $x(1+cx)^2[b^{a,c}_{n,k}(x)]' = [(k-nx)(1+cx)-acx]\, b^{a,c}_{n,k}(x)$, we may write

$$x(1+cx)^2(\mu^{a,c}_{n,m}(x))' = (n-c)\sum_{k=0}^{\infty}[(k-nx)(1+cx)-acx]b^{a,c}_{n,k}(x)\int_0^{\infty} b^{0,c}_{n,k}(t)(t-x)^m dt - mx(1+cx)^2\mu^{a,c}_{n,m-1}(x).$$

Thus,

$$\begin{aligned} & x(1+cx)^2\left[(\mu^{a,c}_{n,m}(x))' + m\mu^{a,c}_{n,m-1}(x)\right] + acx\mu^{a,c}_{n,m}(x) \\ &= (n-c)(1+cx)\sum_{k=0}^{\infty}(k-nx)b^{a,c}_{n,k}(x)\int_0^{\infty} b^{0,c}_{n,k}(t)(t-x)^m dt \\ &= (n-c)(1+cx)\sum_{k=0}^{\infty} b^{a,c}_{n,k}(x)\int_0^{\infty}[(k-nt)+n(t-x)]b^{0,c}_{n,k}(t-x)^m dt \end{aligned}$$

$$= (n-c)(1+cx)\sum_{k=0}^{\infty} b_{n,k}^{a,c}(x)\int_0^{\infty}(k-nt)b_{n,k}^{0,c}(t-x)^m dt + n(1+cx)\mu_{n,m+1}^{a,c}(x)$$

Using the identity $t(1+ct)[b_{n,k}^{0,c}(t)]' = (k-nt)b_{n,k}^{0,c}(t)$, we have

$$x(1+cx)^2\left[(\mu_{n,m}^{a,c}(x))' + m\mu_{n,m-1}^{a,c}(x)\right] + acx\mu_{n,m}^{a,c}(x) - n(1+cx)\mu_{n,m+1}^{a,c}(x)$$
$$= (n-c)(1+cx)\sum_{k=0}^{\infty} b_{n,k}^{a}(x)\int_0^{\infty} t(1+ct)[b_{n,k}^{0,c}]'(t-x)^m dt$$

Finally using $t(1+ct) = c(t-x)^2 + (1+2cx)(t-x) + x(1+cx)$ and integrating by parts, we have

$$x(1+cx)^2\left[(\mu_{n,m}^{a,c}(x))' + m\mu_{n,m-1}^{a,c}(x)\right] + acx\mu_{n,m}^{a,c}(x) - n(1+cx)\mu_{n,m+1}^{a,c}(x)$$
$$= (n-c)(1+cx)\sum_{k=0}^{\infty} b_{n,k}^{a,c}(x)\int_0^{\infty}[c(t-x)^2 + (1+2cx)(t-x) + x(1+cx)][b_{n,k}^{0,c}(t)]'(t-x)^m dt$$
$$- (m+2)c(1+cx)\mu_{n,m+1}^{a,c}(x) - (m+1)(1+cx)(1+2cx)\mu_{n,m}^{a,c}(x) - mx(1+cx)^2\mu_{n,m-1}^{a,c}(x),$$

which is the required recurrence relation. □

Corollary 1 *For the function $\mu_{n,m}^{a,c}(x)$, from Lemma 3, we have*

(i) $\mu_{n,1}^{a,c}(x) = \dfrac{acx}{(n-2c)(1+cx)} + \dfrac{1+2cx}{n-2c},$

(ii) $\mu_{n,2}^{a,c}(x) = \dfrac{2x(1+cx)}{n-3c} + \dfrac{2[5c^2x^2+5cx+1]}{(n-2c)(n-3c)} + \dfrac{a^2c^2x^2}{(n-2c)(n-3c)(1+cx)^2} + \dfrac{2acx(2+3cx)}{(n-2c)(n-3c)(1+cx)};$

(iii) $\mu_{n,m}^{a,c}(x)$ *is a rational function of* x;

(iv) *For every* $x \in (0,\infty)$, $\mu_{n,m}^{a,c}(x) = O\left(n^{-[(m+1)/2]}\right),$

where $[\alpha]$ *denotes the integer part of* α.

Corollary 2 *Let γ and δ be any two positive real numbers and $[c,d] \subset (0,\infty)$ be any bounded interval. Then, for any $m > 0$ there exists a constant M' depending on m only such that*

$$\left\|(n-c)\sum_{k=0}^{\infty} b_{n,k}^{a,c}(x)\int_{|t-x|\geq\delta} b_{n,k}^{0,c}(t)t^{\gamma}dt\right\| \leq M'n^{-m},$$

where $\|.\|$ is the sup-norm over $[c,d]$.

Lemma 4 *For each $x \in (0, \infty)$ and $r \in N^0$, there exist polynomials $q_{i,j,r}(x)$ in x independent of n and k such that*

$$\frac{d^r}{dx^r} b^{a,c}_{n,k}(x) = b^{a,c}_{n,k}(x) \sum_{\substack{2i+j \le r \\ i,j \ge 0}} n^i (k - nx)^j \frac{q_{i,j,r}(x)}{(p(x))^r},$$

where $p(x) = x(1 + cx)^2$.

The proof of this lemma follows using the lines of [18].

3 Simultaneous Approximation

We consider the following space of functions defined as

$$C_\gamma[0, \infty) = \{f \in C[0, \infty) : |f(t)| \le Ct^\gamma, \text{for some } \gamma > 0,\ t \in [0, \infty)\},$$

it is observed that the operators $D^{a,c}_n(f, x)$ are well defined for $a, c \ge 0$.

Theorem 1 *Let $f \in C_\gamma[0, \infty)$. If $f^{(r)}$ exists at a point $x \in (0, \infty)$, then we have*

$$\lim_{n \to \infty} \left(\frac{d^r}{dw^r} D^{a,c}_n(f, w) \right)_{w=x} = f^{(r)}(x). \tag{2}$$

Further, if $f^{(r)}$ is continuous on $(c - \eta, d + \eta)$, $\eta > 0$, then the limit in (2) holds uniformly in $[c, d]$.

Proof Using Taylor's formula, we can write

$$f(t) = \sum_{i=0}^{r} \frac{f^{(i)}(x)}{i!} (t - x)^i + \psi(t, x)(t - x)^r, \quad t \in [0, \infty), \tag{3}$$

where $\psi(t, x) \to 0$ as $t \to x$. Applying (3) to the operator, we have

$$\begin{aligned} \left(\frac{d^r}{dw^r} D^{a,c}_n(f(t), w) \right)_{w=x} &= \sum_{i=0}^{r} \frac{f^{(i)}(x)}{i!} \left(\frac{d^r}{dw^r} D^{a,c}_n((t - x)^i, w) \right)_{w=x} + \left(\frac{d^r}{dw^r} D^{a,c}_n(\psi(t, x)(t - x)^r, w) \right)_{w=x} \\ &:= I_1 + I_2. \end{aligned}$$

First

$$\begin{aligned} I_1 &= \sum_{i=0}^{r} \frac{f^{(i)}(x)}{i!} \left\{ \frac{d^r}{dw^r} \left(\sum_{v=0}^{i} \binom{i}{v} (-x)^{i-v} D^{a,c}_n(t^v, w) \right) \right\}_{w=x} \\ &= \sum_{i=0}^{r} \frac{f^{(i)}(x)}{i!} \sum_{v=0}^{i} \binom{i}{v} (-x)^{i-v} \left(\frac{d^r}{dw^r} D^{a,c}_n(t^v, w) \right)_{w=x} \end{aligned}$$

$$= \sum_{i=0}^{r-1} \frac{f^{(i)}(x)}{i!} \sum_{v=0}^{i} \binom{i}{v} (-x)^{i-v} \left(\frac{d^r}{dw^r} D_n^{a,c}(t^v, w) \right)_{w=x} + \frac{f^{(r)}(x)}{r!} \sum_{v=0}^{r} \binom{r}{v} (-x)^{r-v} \left(\frac{d^r}{dw^r} D_n^{a,c}(t^v, w) \right)_{w=x}$$
$$:= I_3 + I_4.$$

Now, we may write

$$\begin{aligned} I_4 &= \frac{f^{(r)}(x)}{r!} \sum_{v=0}^{r-1} \binom{r}{v} (-x)^{r-v} \left(\frac{d^r}{dw^r} D_n^{a,c}(t^v, w) \right)_{w=x} + \frac{f^{(r)}(x)}{r!} \left(\frac{d^r}{dw^r} D_n^{a,c}(t^r, w) \right)_{w=x} \\ &:= I_5 + I_6. \end{aligned}$$

Making use of Remark 1 (iii), we obtain

$$I_6 = f^{(r)}(x) + O\left(\frac{1}{n}\right), I_3 = O\left(\frac{1}{n}\right) \text{ and } I_5 = O\left(\frac{1}{n}\right), \text{ as } n \to \infty.$$

From the above estimates, for each $x \in (0, \infty)$ we have $I_1 \to f^{(r)}(x)$ as $n \to \infty$. In view of Lemma 4, we have

$$|I_2| \leq (n-c) \sum_{k=0}^{\infty} \sum_{\substack{2i+j \leq r \\ i,j \geq 0}} n^i |k - nx|^j \frac{|q_{i,j,r}(x)|}{(p(x))^r} b_{n,k}^{a,c}(x) \int_0^{\infty} b_{n,k}^{0,c}(t) \psi(t,x) |t-x|^r dt \qquad (4)$$

Since $\psi(t,x) \to 0$ as $t \to x$, for a given $\varepsilon > 0$ there exists a $\delta > 0$ such that $|\psi(t,x)| < \varepsilon$ whenever $|t-x| < \delta$. For $|t-x| \geq \delta$, we have $|(t-x)^r \psi(t,x)| \leq Mt^{\gamma}$, for some $M > 0$. Thus, from Eq. (4) we may write

$$\begin{aligned} |I_2| &\leq (n-c) \sum_{k=0}^{\infty} \sum_{\substack{2i+j \leq r \\ i,j \geq 0}} n^i |k - nx|^j \frac{|q_{i,j,r}(x)|}{(p(x))^r} b_{n,k}^{a,c}(x) \left(\varepsilon \int_{|t-x|<\delta} b_{n,k}^{0,c}(t) |t-x|^r dt \right. \\ &\quad \left. + M \int_{|t-x| \geq \delta} b_{n,k}^{0,c}(t) t^{\gamma} dt \right) \\ &:= J_1 + J_2. \end{aligned}$$

Let $K = \sup\limits_{\substack{2i+j \leq r \\ i,j \geq 0}} \dfrac{|q_{i,j,r}(x)|}{(p(x))^r}$.

Using Schwarz inequality and Corollary 1, we have

$$\begin{aligned} J_1 &= (n-c)\varepsilon\ K \sum_{k=0}^{\infty} \sum_{\substack{2i+j \leq r \\ i,j \geq 0}} n^i |k - nx|^j b_{n,k}^{a,c}(x) \left(\int_0^{\infty} b_{n,k}^{0,c}(t) dt \right)^{1/2} \\ &\quad \left(\int_0^{\infty} b_{n,k}^{0,c}(t) |t-x|^{2r} dt \right)^{1/2} \end{aligned}$$

$$\leq \varepsilon\ K \sum_{\substack{2i+j\leq r\\ i,j\geq 0}} n^{i+j}\left(\sum_{k=0}^{\infty} b_{n,k}^{a,c}(x)\left(\frac{k}{n}-x\right)^{2j}\right)^{1/2}$$

$$\left(D_n^{a,c}(t-x)^{2r},x\right)^{1/2}$$

$$=\varepsilon \sum_{\substack{2i+j\leq r\\ i,j\geq 0}} n^{i+j}\left\{O\left(\frac{1}{n^j}\right)+O\left(\frac{1}{n^s}\right)\right\}^{1/2}$$

$$\times\left\{O\left(\frac{1}{n^r}\right)+O\left(\frac{1}{n^p}\right)\right\}^{1/2} \text{ for any } s,p>0.$$

Choosing s and p such that $s > j$, and $p > r$

$$J_1 \leq \varepsilon \sum_{\substack{2i+j\leq r\\ i,j\geq 0}} n^{i+j} O\left(\frac{1}{n^{j/2}}\right) O\left(\frac{1}{n^{r/2}}\right) = \varepsilon.O(1).$$

Since $\varepsilon > 0$ is arbitrary, $J_1 \to 0$ as $n \to \infty$.

Again, using Schwarz inequality, Lemma 1 and Corollary 2, we obtain

$$J_2 \leq (n-c)M_1 \sum_{\substack{2i+j\leq r\\ i,j\geq 0}} n^{i+j}\left(\sum_{k=0}^{\infty}\left(\frac{k}{n}-x\right)^{2j} b_{n,k}^{a,c}(x)\right)^{1/2}$$

$$\left((n-c)\sum_{k=0}^{\infty} b_{n,k}^{a,c}(x)\int_{|t-x|\geq\delta} b_{n,k}^{0,c}(t)t^{2\gamma}\,dt\right)^{1/2}$$

$$\leq M_1 \sum_{\substack{2i+j\leq r\\ i,j\geq 0}} n^{i+j}\left\{O\left(\frac{1}{n^j}\right)+O\left(\frac{1}{n^p}\right)\right\}^{1/2}\left\{O\left(\frac{1}{n^m}\right)\right\}^{1/2} \text{ for any } p>0.$$

Choosing p such that $p > j$

$$J_2 \leq M_1 \sum_{\substack{2i+j\leq r\\ i,j\geq 0}} n^{i+j} O\left(\frac{1}{n^{j/2}}\right) O\left(\frac{1}{n^{m/2}}\right)$$

$$= M_1 O\left(\frac{1}{n^{(m-r)/2}}\right)$$

which implies that $J_2 \to 0$, as $n \to \infty$ choosing $m > r$.

Thus, from the estimates of I_1 and I_2, the required result follows.
To prove the uniformity assertion, it is sufficient to remark that $\delta(\varepsilon)$ in the above

proof can be chosen to be independent of $x \in [c, d]$ and also that the other estimates hold uniformity in $x \in [c, d]$. This completes the proof.

Next, we establish a Voronovskaja type asymptotic formula in simultaneous approximation.

Theorem 2 (Voronovskaja type result) *Let* $f \in C_\gamma[0, \infty)$. *If* $f^{(r)}$ *exists at a point* $x \in (0, \infty)$, *then we have*

$$\lim_{n\to\infty} n\left(\left(\frac{d^r}{dw^r} D_n^{a,c}(f, w)\right)_{w=x} - f^{(r)}(x)\right) = \sum_{v=1}^{r+2} Q(v, r, a, c, x) f^{(v)}(x), \tag{5}$$

where $Q(v, r, a, c, x)$ *are certain rational functions of* x *depending on the parameter* a.

Further, if $f^{(r+2)}$ *is continuous on* $(c - \eta, d + \eta)$, $\eta > 0$, *then the limit in (5) holds uniformly in* $[c, d]$.

Proof From the Taylor's theorem, we may write

$$f(t) = \sum_{v=0}^{r+2} \frac{f^{(v)}(x)}{v!} (t - x)^v + \psi(t, x)(t - x)^{r+2}, \quad t \in [0, \infty), \tag{6}$$

where the function $\psi(t, x) \to 0$ as $t \to x$. From equation (6), we obtain

$$\begin{aligned}
\left(\frac{d^r}{dw^r} D_n^{a,c}(f(t), w)\right)_{w=x} &= \sum_{v=0}^{r+2} \frac{f^{(v)}(x)}{v!} \left(\frac{d^r}{dw^r} D_n^{a,c}((t - x)^v, w)\right)_{w=x} \\
&\quad + \left(\frac{d^r}{dw^r} D_n^{a,c}(\psi(t, x)(t - x)^{r+2}, w)\right)_{w=x} \\
&= \sum_{v=0}^{r+2} \frac{f^{(v)}(x)}{v!} \sum_{j=0}^{v} \binom{v}{j} (-x)^{v-j} \left(\frac{d^r}{dw^r} D_n^{a,c}(t^j, w)\right)_{w=x} \\
&\quad + \left(\frac{d^r}{dw^r} D_n^{a,c}(\psi(t, x)(t - x)^{r+2}, w)\right)_{w=x} \\
&:= I_1 + I_2.
\end{aligned} \tag{7}$$

Proceeding along the lines of the estimate of I_2 of Theorem 1, it follows that for each $x \in (0, \infty)$

$$\lim_{n\to\infty} n\left(\frac{d}{dw}(D_n^{a,c}(\psi(t, x)(t - x)^{r+2}, w)\right)_{w=x} = 0.$$

Now, we estimate I_1.

$$\begin{aligned} I_1 =& \sum_{v=0}^{r-1}\frac{f^{(v)}(x)}{v!}\sum_{j=0}^{v}\binom{v}{j}(-x)^{v-j}\left(\frac{d^r}{dw^r}D_n^{a,c}(t^j,w)\right)_{w=x} + \frac{f^{(r)}(x)}{r!}\sum_{j=0}^{r}\binom{r}{j}(-x)^{r-j}\left(\frac{d^r}{dw^r}D_n^{a,c}(t^j,w)\right)_{w=x} \\ &+ \frac{f^{(r+1)}(x)}{(r+1)!}\sum_{j=0}^{r+1}\binom{r+1}{j}(-x)^{r+1-j}\left(\frac{d^r}{dw^r}D_n^{a,c}(t^j,w)\right)_{w=x} \\ &+ \frac{f^{(r+2)}(x)}{(r+2)!}\sum_{j=0}^{r+2}\binom{r+2}{j}(-x)^{r+2-j}\left(\frac{d^r}{dw^r}D_n^{a,c}(t^j,w)\right)_{w=x}. \end{aligned} \tag{8}$$

In view of Remark 1 (3), we have

$$\begin{aligned} I_1 =& \sum_{v=1}^{r-1} f^{(v)}(x)O\left(\frac{1}{n}\right) + f^{(r)}(x)\left(1+O\left(\frac{1}{n}\right)\right) + f^{(r+1)}(x)O\left(\frac{1}{n}\right) \\ &+ f^{(r+2)}(x)O\left(\frac{1}{n}\right) \\ =& f^{(r)}(x) + n^{-1}\left(\sum_{v=1}^{r+2} Q(v,r,a,c,x)f^{(v)}(x) + o(1)\right). \end{aligned}$$

Combining the estimates of I_1 and I_2, we get the required result.
The uniformity assertion follows as in proof of Theorem 1. Hence the proof is completed. □

Corollary 3 *Let* $f \in C_\gamma[0,\infty)$ *for some* $\gamma > 0$. *If* f'' *exists at a point* $x \in [0,\infty)$ *then, we have*

$$\lim_{n\to\infty} n(D_n^{a,c}(f,x) - f(x)) = \left[\frac{acx}{1+cx} + (1+2cx)\right]f'(x) + x(1+cx)f''(x).$$

Proof From the Taylor's theorem, we may write

$$f(t) = f(x) + (t-x)f'(x) + \frac{1}{2}f''(x)(t-x)^2 + \psi(t,x)(t-x)^2, \quad t \in [0,\infty) \tag{9}$$

where the function $\psi(t,x) \to 0$ as $t \to x$.

Applying $D_n^{a,c}(.,x)$ and taking the limit as $n \to \infty$ on both sides of (9), we have

$$\begin{aligned} \lim_{n\to\infty} n(D_n^{a,c}(f,x) - f(x)) =& \lim_{n\to\infty} nD_n^{a,c}((t-x),x)f'(x) + \frac{f''(x)}{2}\lim_{n\to\infty} nD_n^{a,c}((t-x)^2,x) \\ &+ \lim_{n\to\infty} nD_n^{a,c}(\psi(t,x)(t-x)^2,x). \end{aligned}$$

In view of Corollary 1, we get

$$\lim_{n\to\infty} nD_n^{a,c}((t-x),x) = \frac{acx}{1+cx} + (1+2cx) \tag{10}$$

and

$$\lim_{n\to\infty} nD_n^{a,c}((t-x)^2,x) = 2x(1+cx). \tag{11}$$

Now, we prove that $nD_n^{a,c}(\psi(t,x)(t-x)^2,x) \to 0$, as $n\to\infty$. From the Cauchy–Schwarz inequality, we have

$$D_n^{a,c}(\psi(t,x)(t-x)^2,x) \le \sqrt{D_n^{a,c}(\psi^2(t,x),x)}\sqrt{D_n^{a,c}((t-x)^4,x)}. \tag{12}$$

Since $\psi(t,x)\to 0$ as $t\to x$, for a given $\varepsilon>0$ there exists $\delta>0$ such that $|\psi(t,x)|<\varepsilon$ whenever $|t-x|<\delta$. For $|t-x|\ge\delta$, there exists M_1 such that $|\psi(t,x)|\le M_1 t^{\gamma}$.

Let $\chi_\delta(t)$ denote the characteristic function of $(x-\delta,x+\delta)$. Then

$$\begin{aligned} D_n^{a,c}(\psi^2(t,x),x) &\le D_n^{a,c}(\psi^2(t,x)\chi_\delta(t),x,c) + D_n^{a,c}(\psi^2(t,x)(1-\chi_\delta(t)),x,c)\\ &\le \varepsilon^2 D_n^{a,c}(1,x) + M_1^2 D_n^{a,c}(t^{2\gamma}(1-\chi_\delta(t)),x)\\ &\le \varepsilon^2 + M_2 n^{-m}, \end{aligned}$$

in view of Corollary 2.

Hence, we have

$$\lim_{n\to\infty} D_n^{a,c}(\psi^2(t,x),x) = 0. \tag{13}$$

Further from Corollary 1,

$$D_n^{a,c}((t-x)^4,x) = O(n^{-2}), \tag{14}$$

which is a finite quantity for each fixed $x\in[0,\infty)$ thus from (12) to (14), we get

$$\lim_{n\to\infty} nD_n^{a,c}\left(\psi(t,x)(t-x)^2,x\right) = 0. \tag{15}$$

Combining (10), (11) and (15), we obtain the desired result. □

Next, we prove Voronovskaja type asymptotic formula for $\left(\frac{d}{d\omega}D_n^{a,c}(f,\omega,c)\right)_{\omega=x}$.

Corollary 4 *Let* $f\in C_\gamma[0,\infty)$ *admitting the derivative of third order at a fixed point* $x\in(0,\infty)$, *we have*

$$\lim_{n\to\infty} n\left(\left(\frac{d}{dw}D_n^{a,c}(f,w)\right)_{w=x} - f'(x)\right) = \left(2c + \frac{ac}{(1+cx)^2}\right)f'(x)$$
$$+\left(4cx + 2 + \frac{acx + ac^2x^2}{(1+cx)^2}\right)f''(x)$$
$$+\frac{1}{3!}\left(6x(1+cx) + \frac{15acx^2 + 6ac^2x^3}{(1+cx)^2}\right)f'''(x).$$

Proof From the Taylor's theorem, we may write

$$f(t) = \sum_{k=0}^{3} \frac{(t-x)^k}{k!} f^{(k)}(x) + \psi(t,x)(t-x)^3, t \in [0,\infty), \tag{16}$$

where $\lim_{t\to x} \psi(t,x) = 0$.
From Eq. (16), we obtain

$$\left(\frac{d}{d\omega}D_n^{a,c}(f(t),\omega)\right)_{\omega=x} = f'(x)\left(\frac{d}{d\omega}(D_n^{a,c}(t,\omega) - x)\right)_{\omega=x}$$
$$+\frac{f''(x)}{2}\left(\frac{d}{d\omega}(D_n^{a,c}(t^2,\omega) - 2xD_n^{a,c}(t,\omega) + x^2)\right)_{\omega=x}$$
$$+\frac{f'''(x)}{3!}\left(\frac{d}{d\omega}(D_n^{a,c}(t^3,\omega) - 3xD_n^{a,c}(t^2,\omega) + 3x^2D_n^{a,c}(t,\omega) - x^3)\right)_{\omega=x}$$
$$+\left(\frac{d}{d\omega}(D_n^{a,c}(\psi(t,x)(t-x)^3,\omega)\right)_{\omega=x}.$$

Using Lemma 2, we get

$$\left(\frac{d}{d\omega}D_n^{a,c}(f(t),\omega,c)\right)_{\omega=x} = f'(x)\left\{1 + \frac{ac}{(n-2c)(1+cx)^2} + \frac{2c}{n-2c}\right\}$$
$$+\frac{f''(x)}{2}\left\{\frac{12c(n-c)x + 4n}{(n-2c)(n-3c)} + \frac{4ac + 2acnx + 2ac^2nx^2 + 6ac^2x}{(n-2c)(n-3c)(1+cx)^2}\right.$$
$$\left.+\frac{2(x+cx^2)a^2c^2}{(n-2c)(n-3c)(1+cx)^4} - \frac{4cx}{n-2c}\right\}$$
$$+\frac{f'''(x)}{3!}\left\{\frac{36c(n^2-2n+2c^2)x^2 + 18n(n+c)x + 18n}{(n-2c)(n-3c)(n-4c)}\right.$$
$$+\frac{3acn(n+c)(3x^2+2cx^3) + 18acn + 18acn(2x+cx^2)}{(n-2c)(n-3c)(n-4c)(1+cx)^2}$$
$$+\frac{6a^2c^2n(x+x^2) + 3a^2c^2n(3x^2+c^2x^4+4cx^3) + 3a^3c^3x^2}{(n-2c)(n-3c)(n-4c)(1+cx)^4}$$
$$-\frac{36c(n-c)x^2 + 12nx}{(n-2c)(n-3c)} - \frac{12acx + 6acn(2x^2+cx^3)}{(n-2c)(n-3c)(1+cx)^2}$$
$$\left.-\frac{6(x^2+cx^3)a^2c^2}{(n-2c)(n-3c)(1+cx)^4} + \frac{3acx^2}{(n-2c)(1+cx)^2} + \frac{6cx^2}{n-2c}\right\}$$
$$+\left(\frac{d}{d\omega}(D_n^{a,c}(\psi(t,x)(t-x)^3,\omega,c)\right)_{\omega=x}.$$

Taking limit as $n \to \infty$ on both sides of the above equation, we have

$$\lim_{n\to\infty} n\left(\left(\frac{d}{d\omega}D_n^{a,c}(f,\omega,c)\right)_{\omega=x} - f'(x)\right) = f'(x)\left(2c + \frac{ac}{(1+cx)^2}\right)$$
$$+ f''(x)\left(4cx + 2 + \frac{acx + ac^2x^2}{(1+cx)^2}\right)$$
$$+ \frac{f'''(x)}{3!}\left(6x + 6cx^2 + \frac{15acx^2 + 6ac^2x^3}{(1+cx)^2}\right)$$
$$+ \lim_{n\to\infty} n\left(\frac{d}{d\omega}(D_n^{a,c}(\psi(t,x)(t-x)^3,\omega,c)\right)_{\omega=x}.$$

Proceeding in the same manner as in Corollary 3, we can easily show that

$$\lim_{n\to\infty} n\left(\frac{d}{d\omega}(D_n^{a,c}(\psi(t,x)(t-x)^3,\omega)\right)_{\omega=x} = 0,$$

since $\lim_{n\to\infty} n^3(D_n^{a,c}(t-x)^6, x)$ is finite for each $x \in [0,\infty)$ in view of Lemma 3. Thus, the proof is completed.

References

1. Agrawal, P.N., Gupta, V.: Simultaneous approximation by linear combination of modified Bernstein polynomials. Bull. Greek Math. Soc. **39**, 29–43 (1989)
2. Erencin, A.: Durrmeyer type modification of generalized Baskakov operators. Appl. Math. Comput. **218**(3), 4384–4390 (2011)
3. Gupta, V.: Simultaneous approximation by Szász-Durrmeyer operators. Math. Stud.-India **64**(1), 27–36 (1995)
4. Gupta, V.: A note on modified Baskakov type operators. Approx. Theory Appl. **10**(3), 74–78 (1994)
5. Gupta, V., Agarwal, R.P.: Convergence Estimates in Approximation Theory. Springer, Berlin (2014)
6. Gupta, V., Ahmad, A.: Simultaneous approximation by modified Beta operators. Istanbul Universitesi Fen Fakultesi Mat Dergsi **54**, 11–22 (1995)
7. Gupta, V., Gupta, P.: Direct theorem in simultaneous approximation for Szasz-Mirakyan Baskakov type operators. Kyungpook Math. J. **41**(2), 243–249 (2001)
8. Gupta, V., Gupta, M.K., Vasishtha, V.: Simultaneous approximation by summation-integral type operators. Nonlinear Funct. Anal. Appl. **8**, 399–412 (2003)
9. Gupta, V., Ispir, N.: On simultaneous approximation for some modified Bernstein-type operators. Int. J. Math. Math. Sci. **71**, 3951–3958 (2004)
10. Gupta, V., Kim, T., Singh, V.K., Lee, B.: Rate of simultaneous approximation for Baskakov-Szász operators. Proc. Jangjeon Math. Soc. **14**(3), 267–275 (2011)
11. Gupta, V., Noor, M.A.: Convergence of derivatives for certain mixed Szasz Beta operators. J. Math. Anal. Appl. **321**(1), 1–9 (2006)
12. Gupta, V., Sinha, J.: Simultaneous approximation for generalized Baskakov-Durrmeyer type operators. Mediterr. J. Math. **4**(4), 483–495 (2007)
13. Gupta, V., Srivastava, G.S.: Simultaneous approximation by Baskakov-Szasz type operators. Bull. Math. Soc. Sci. Roum. (N.S.) **37**(85)(3–4), 73–85 (1993)
14. Gupta, V., Srivastava, G.S.: Convergence of derivatives by summation-integral type operators. Revista Colombiana de Mate. **29**(1), 1–11 (1995)

15. Gupta, V., Yadav, R.: Direct estimates in simultaneous approximation for BBS operators. Appl. Math. Comput. **218**(22), 11290–11296 (2012)
16. Mihesan, V.: Uniform approximation with positive linear operators generated by generalized Baskakov method. Automat. Comput. Appl. Math. **7**(1), 34–37 (1998)
17. Sahai, A., Prasad, G.: On simultaneous approximation by modified Lupas operators. J. Approx. Theory **45**(2), 122–128 (1985)
18. Sinha, R.P., Agrawal, P.N., Gupta, V.: On simultaneous approximation by modified Baskakov operators. Bull. Soc. Math. Belg. Ser. B **43**(2), 217–231 (1991)

Mechanochemical Corrosion: Modeling and Analytical Benchmarks for Initial Boundary Value Problems with Unknown Boundaries

Yulia Pronina

Abstract In this paper various corrosion models are considered. Difficulties of the modeling of stress corrosion of constructional elements and the need for developing closed-form solutions are highlighted. A new analytical solution is presented for the plane problem of the mechanochemical corrosion of an elastic plate with an elliptical hole under uniform remote tension. The rate of corrosion is supposed to be linear with the maximum principal stress at a corresponding point on the hole surface. The solution obtained can serve for the study of the mechanochemical effect on the corrosion damage propagation. It is proved that the stress concentration factor at a noncircular hole can either increase or decrease, or stay invariant during the corrosion process, depending on the relationship between the corrosion kinetics constants and applied stress.

Keywords Mechanochemical corrosion · General corrosion · Corrosion kinetics · Pitting · Lifetime · Analytical solution

1 Mechanochemical Corrosion Models

"The problems of corrosion are universal, but the control measures are not," N. Sethurathinam, Executive Director, (Refineries Division), Indian Oil Corporation, said [14]. For example in India, the annual loss due to corrosion has been estimated at about 4 per cent of the country's Gross Domestic Product [14]. Corrosion is a natural phenomenon defined as the deterioration of a material or its properties due to an interaction with its environment. Corrosion can cause not only expensive but also extremely dangerous damage of constructions from underground pipelines to aircraft fuselages.

Y. Pronina (✉)
St. Petersburg State University, 7/9 Universitetskaya Nab.,
199034 St. Petersburg, Russia
e-mail: y.pronina@spbu.ru

V.K. Singh et al. (eds.), *Modern Mathematical Methods and High Performance Computing in Science and Technology*, Springer Proceedings in Mathematics & Statistics 171, DOI 10.1007/978-981-10-1454-3_25

Most structures are exploited being subjected to both mechanical loads and operating environments. The combined action of mechanical loads and chemically active media has been the subject of study for more than 100 years. It was observed that such conditions may activate the process of so-called stress corrosion, which is more severe than the simple superposition of damages induced by stresses and electrochemical corrosion acting separately [17, 23, 27]. With regard to general corrosion facilitated by stress, the term "mechanochemical corrosion" was introduced [8, 9]. According to E.M. Gutman, corrosion may often be considered as uniform in the case of elastic deformation. In plastic region, significant electrochemical heterogeneity of the surface may be developed; therefore, the term "mechanochemical corrosion" is not always applicable. General wear can occur both under the formation of a closed protective coating, and in the absence of oxide or biofilms. The formation of a passive film, shift in solution pH, and the change in concentration of reactants can show inhibiting effects, when the corrosion rate can be supposed to follow an exponential decay with time [18, 26].

There exists a number of different approaches to the problems of stress corrosion, based on physical and chemical mechanics of materials, thermodynamics, continuum mechanics, and fracture mechanics. E.M. Gutman proposed an exponential dependence of the rate of the anodic dissolution of deformed metal on the stress value. On the basis of his theory several elegant mathematical models of the corrosion of pipe elements were developed [1, 9, 10]. The theory of the mechanochemical effect of dissolution in terms of the chemical affinity was formulated by A.I. Rusanov. According to his work [23], dissolution rate is a quadratic function of strain-components. The case of dissolution/evaporation of a bent plate was examined in details theoretically and experimentally; the effect of the strain sign observed in the experiments was explained by the existence of surface tension [23]. Interesting discussion of the mentioned results is available in literature. Based on the concept of chemical affinity tensor, the authors of [5–7] studied the kinetics of the stress-assisted chemical reaction sustained by the diffusion of gas through an elastic solid. Some spherically symmetric problems were solved there. The effect of the sign and value of the reaction front curvature were also examined.

Note that due to the highly complex structure of metals and alloys [26], development of a thermodynamic model that takes into account all the details of their structure and competing processes, seems to be very difficult to realize. Thus, we have to rely on experimental results. A lot of experimental data demonstrated a linear dependence of the metal corrosion rate on the effective stress [18]. Beginning with the pioneering work by V.M. Dolinskii [3], this dependence is often used for engineering calculations [10, 11, 16, 19].

When corrosion rates depend on stresses, and stresses, in turn, depend on changing (due to corrosion) geometry of an element, one has to solve an initial boundary value problem with unknown boundaries. Such problems are mostly studied by numerical methods. However, several analytical solutions have been found for the uniform mechanochemical dissolution of structural elements, e.g., by the authors of [1, 3, 4, 9, 11, 16, 25].

For example, consider the system of equations for the problem of the double-sided mechanochemical corrosion of an elastic thick-walled spherical vessel under internal p_r and external p_R pressure [25]. The inner r and outer R radii of the sphere change with time t because of corrosion. The corrosion velocities on the inside and outside, denoted by υ_r and υ_R, respectively, can be approximated by the expressions [18]:

$$\upsilon_r = \frac{dr}{dt} = \upsilon_r^0 \exp(-bt) \quad \text{at} \quad |\sigma_1(r)| \le \left|\sigma_r^{th}\right|, \tag{1}$$

$$\upsilon_R = -\frac{dR}{dt} = \upsilon_R^0 \exp(-bt) \quad \text{at} \quad |\sigma_1(R)| \le \left|\sigma_R^{th}\right|, \tag{2}$$

and

$$\upsilon_r = \frac{dr}{dt} = [a_r + m_r\sigma_1(r)] \exp(-bt) \quad \text{at} \quad |\sigma_1(r)| \ge \left|\sigma_r^{th}\right|, \tag{3}$$

$$\upsilon_R = -\frac{dR}{dt} = [a_R + m_R\sigma_1(R)] \exp(-bt) \quad \text{at} \quad |\sigma_1(R)| \ge \left|\sigma_R^{th}\right|. \tag{4}$$

Here, $b,\ \upsilon_r^0,\ \upsilon_R^0,\ m_r,\ m_R,\ \sigma_r^{th}$, and σ_R^{th} are experimentally determined constants; $a_r = \upsilon_r^0 - m_r\sigma_r^{th}$; $a_R = \upsilon_R^0 - m_R\sigma_R^{th}$; σ_r^{th} and σ_R^{th} are threshold stresses; σ_1 is the maximum principal stress on the relevant surface:

$$\sigma_1(r) = \frac{p_r r^3 - p_R R^3}{R^3 - r^3} + \frac{(p_r - p_R)R^3}{2\left(R^3 - r^3\right)}, \tag{5}$$

$$\sigma_1(R) = \frac{p_r r^3 - p_R R^3}{R^3 - r^3} + \frac{(p_r - p_R)r^3}{2\left(R^3 - r^3\right)}. \tag{6}$$

As one can see, these stress components increase (in absolute value) with time due to the change in the radii r and R and accelerate corrosion process more and more. Thus, we have to solve simultaneous equations (1)–(6). Analytical solution to this problem is presented in [25].

Numerical investigation of the problems with unknown changing boundaries requires high qualification. Unfortunately, using finite element software even for static problems not always leads to appropriate results. In such situations, analytical solutions can serve as benchmarks for numerical analysis and can help to identify the role of mechanochemical effect in damage propagation observed in specimens under study.

In practice, structural elements are often designed to have supplementary thickness as a corrosion allowance that can increase to a considerable amount of additional metal. However, these calculations are not wholly adequate and can lead to a substantial cost increase [2]. Using the models with reduced thickness for the strength calculation of solids with nonuniform damages can also lead to significant errors [24].

2 Problem of the Mechanochemical Corrosion of a Plate with an Elliptical Hole

Previously obtained solutions (e.g., [20–22]) for the mechanochemical corrosion of elastic or elastic–plastic thick-walled cylinders and spheres can be applied to the problems of corrosion of a large enough solid with a small cylindrical or spherical cavity under uniform tension or compression. However, those solutions do not allow to observe the change in the shape of the cavity. In the framework of the theory involved, the hole remains circular during the corrosion process. Nevertheless, the results presented below demonstrate that even a nearly circular hole can grow nonuniformly under uniform remote tension.

2.1 *Problem Formulation*

Consider the first fundamental problem for a linearly elastic, isotropic infinite plane S bounded by an elliptic contour L with the semi-axes A and B ($A \geq B$). The plane is supposed to be subjected to remote uniform tension p. The cavity surface is stress free and exposed to mechanochemical corrosion defined as material dissolution. In this case the hole, associated with the contour L, grows with time t. Let A_0 and B_0 be the semi-axes of the ellipse L at the initial moment $t = 0$. According to [18], the rate of corrosion, v, is linear with the maximum principal stress at corresponding points on the surface:

$$v(s) = \frac{d\delta(s)}{dt} = a + m\,\sigma(s), \quad s \in L(t), \tag{7}$$

where a and m are empirically determined constants of corrosion kinetics; $d\delta$ is an increment (due to material dissolution) of the hole size in the direction of the normal to its contour L.

It is required to track the change of the hole geometry with time.

2.2 *Problem Solution*

Stress distribution on the elliptic contour L in the plane S under remote tension have been found in [15] by the use of the transformation of the region S on to the infinite plane with a circular hole, $|\zeta| > 1$. The relevant transformation is

$$z = R\left(\zeta + \frac{M}{\zeta}\right), \quad R > 0, \quad 0 \leq M < 1, \tag{8}$$

where $z = x + iy$ and $\zeta = \rho e^{i\theta}$. The ellipse L (with the center at the origin of the coordinate system $0xy$) is then mapped on to the circle $|\zeta| = 1$, so that $A = R(1 + M)$ and $B = R(1 - M)$. Corresponding stress components on the contour $|\zeta| = \rho = 1$ are

$$\sigma_{\theta\theta}(\theta) = 2p\,\frac{1 - M^2}{1 - 2M\cos 2\theta + M^2}, \quad \sigma_{\rho\theta}(\theta) = \sigma_{\rho\rho}(\theta) = 0. \tag{9}$$

According to some experimental data, we can assume that the hole remains elliptical during the corrosion process. Then, Eqs. (8) and (9) should hold true at any t for A and B (and consequently, R and M) growing with time.

Therefore, we have to solve simultaneous equations (7)–(9) at $\theta = 0$ and $\theta = \pi/2$, where the values of $\sigma(0)$, $\sigma(\pi/2)$, A, and B change synergetically. Solution of this problem can be expressed in an implicit form through a new variable $\eta = A/B$:

$$t = -\frac{B_0}{a - 2pm}\left(\frac{(\eta_0 - 1)^{a+2pm}}{\eta_0^{2pm}}\right)^{1/(a-2pm)} \int_{\eta_0}^{\eta} \left(\frac{\eta^{2pm}}{(\eta - 1)^{2a}}\right)^{1/(a-2pm)} d\eta, \tag{10}$$

where $\eta_0 = A_0/B_0$.

Equation (10) gives a point-to-point correspondence between t and η. For every η we can then find

$$B = B_0\left(\frac{\eta^{2pm}}{\eta_0^{2pm}}\,\frac{(\eta_0 - 1)^{a+2pm}}{(\eta - 1)^{a+2pm}}\right)^{1/(a-2pm)} \tag{11}$$

and

$$A = \eta B. \tag{12}$$

Thus, we obtain a one-to-one relationship between t, A, and B.

If $A_0 = B_0 = R_0$, then the shape of the hole remains circular for the corrosion process and its radius R grows with the constant rate

$$\frac{dR}{dt} = a + 2pm \tag{13}$$

for any values R_0, a, m, and p. Therefore,

$$R = R_0 + (a + 2pm)\,t. \tag{14}$$

2.3 Calculation Results

The evolution of the hole under corrosion condition can be quite different depending on the relationship between the corrosion kinetics constants a and m, the traction value p, and the initial axes ratio A_0/B_0.

When the mechanochemical effect is weak enough as compared to the constant rate component a, the hole grows almost uniformly. Limiting case of the constant rate corrosion—when $m = 0$ and $a = 0.2(l_c/t_c)$—is demonstrated in Fig. 1 for the holes with the initial semi-axes $A_0 = 1.25(l_c)$, $B_0 = 1(l_c)$ (dashed lines) and $A_0 = 3(l_c)$, $B_0 = 1(l_c)$ (solid lines). Gradually increasing contours of both the holes correspond to the times $t = 0;\ 0.56;\ 1.25;\ 2.14;\ 3.33$; and $5(t_c)$, respectively.

Here and below, l_c, t_c, and p_c are appropriate units of length, time, and stress, respectively.

Another limiting case of the pure mechanochemical corrosion—when $m = 0.008(l_c/[t_c p_c])$ and $a = 0$—is shown in Fig. 2 for the holes with the same as above initial semi-axes $A_0 = 1.25(l_c)$, $B_0 = 1(l_c)$ (dashed lines) and $A_0 = 3(l_c)$, $B_0 = 1(l_c)$ (solid lines). Gradually increasing contours of the first hole (dashed lines) correspond to the times $t = 0;\ 3.29;\ 6.27;\ 8.99;\ 11.49$; and $13.81(t_c)$. Growing contours of the second hole (solid lines) correspond to $t = 0;\ 0.99;\ 1.88;\ 2.70;\ 3.45$; and $4.14(t_c)$. The graph is built for $p = 10(p_c)$.

It was proved that when the mechanochemical effect is weak ($a > 2mp$), the ratio $\eta = A/B$ decreases with time, tending to unity. For example, for the cases demonstrated in Fig. 1, the ratio η is equal to 1.25; 1.225; 1.2; 1.175; 1.15; and 1.125 (for the gradually increasing dashed contours) and 3; 2.8; 2.6; 2.4; 2.2; and 2 (for the gradually increasing solid contours), respectively. Therefore, the stress concentration factor near the hole decreases as well and approaches 2. In this case the durability of the plate is not reduced.

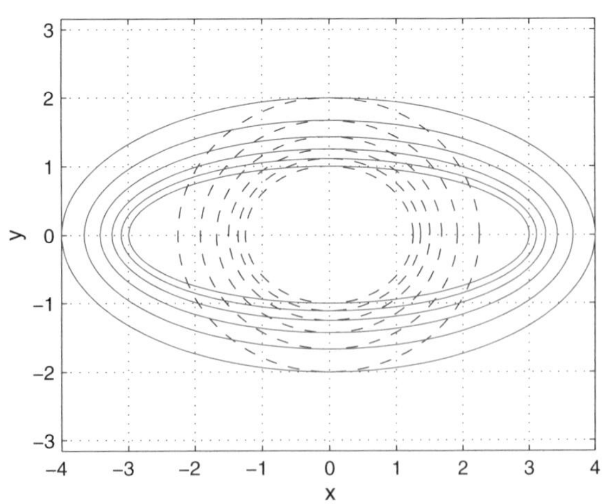

Fig. 1 Gradually growing contours of the holes with $A_0 = 1.25$, $B_0 = 1$ (*dashed lines*) and $A_0 = 3$, $B_0 = 1$ (*solid lines*) corresponding to the times $t = 0;\ 0.56;\ 1.25;\ 2.14;\ 3.33$; and 5. The case of constant rate corrosion

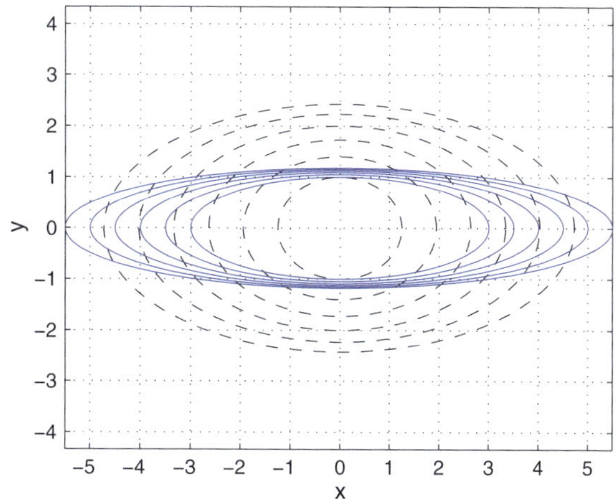

Fig. 2 Gradually increasing contours of the hole with $A_0 = 1.25$, $B_0 = 1$ (*dashed lines*) corresponding to the times $t =$ 0; 3.29; 6.27; 8.99; 11.49; and 13.81 and of the hole with $A_0 = 3$, $B_0 = 1$ (*solid lines*) corresponding to the times $t =$ 0; 0.99; 1.88; 2.70; 3.45; and 4.14. The case of pure mechanochemical corrosion

When the mechanochemical effect takes place at $a < 2mp$, the ratio η grows. This fact must be borne in mind when using Eq. (10) to plot the dependencies $t(\eta)$. For the cases demonstrated in Fig. 2, the ratio η is equal to 1.25; 1.39; 1.53; 1.67; 1.81; and 1.94 (for the gradually increasing dashed contours) and 3; 3.(3); 3.(6); 4; 4.(3); and 4.(6) (for the gradually increasing solid contours), respectively. It is seen that the greater the initial aspect ratio η_0 is, the faster η grows. Moreover, the corrosion in the direction of A-axis is accelerated with time. Therefore, the stress concentration factor increases and the durability of the plane decreases. In this case the lifetime of the plane can be determined by formula (10) with a certain critical value $\eta*$ (corresponding to a strength limit) for η.

If $A_0 = B_0 = R_0$, then the stress concentration factor is equal to 2 at any t and for any values R_0, a, m, and p and there is no need to use Eqs. (10)–(12) for lifetime assessment. Despite the mechanochemical effect, the rate of corrosion remains constant for the corrosion process (see Eqs. (13)–(14)).

3 Conclusion

All the discussed analytical solutions can serve as benchmarks for numerical analysis implemented by the use of an appropriate corrosion rate model. Moreover, they can help to identify the role of mechanochemical effect in damage propagation observed in experiments.

The analytical results proposed here show that the stress concentration factor at a noncircular hole can either increase or decrease, or stay invariant during the corrosion process, depending on the relationship between the corrosion kinetics constants and applied stress.

I would like to note that in addition to the developing numerical methods and computational techniques it is reasonable to create a single integrated data bank of closed-form solutions for various initial/boundary value problems. That would be a powerful aid for solving applied problems worldwide.

Acknowledgments I am very grateful to the organizing committee of the International conference on "Modern Mathematical Methods and High Performance Computing in Science & Technology (M3HPCST-2015)" for the invitation and support for my participation in this conference.
I would like to thank S.M. Khryashchev (specialist in differential calculus [12, 13]) for his helpful advises. This work is partially supported by the Russian Foundation for Basic Research (project No. 16-08-00890).

References

1. Bergman, R.M., Levitsky, S.P., Haddad, J., Gutman, E.M.: Stability loss of thin-walled cylindrical tubes, subjected to longitudinal compressive forces and external corrosion. Thin-Walled Struct. **44**(7), 726–729 (2006)
2. Bhaskar, S., Iyer, N.R., Rajasankar, J.: Cumulative damage function model for prediction of uniform corrosion rate of metals in atmospheric corrosive environment. Corros. Eng. Sci. Tech. **39**(4), 313–320 (2004)
3. Dolinskii, V.M.: Calculations on loaded tubes exposed to corrosion. Chem. Pet. Eng. **3**(2), 96–97 (1967)
4. Elishakoff, I., Ghyselinck, G., Miglis, Y.: Durability of an elastic bar under tension with linear or nonlinear relationship between corrosion rate and stress. J. Appl. Mech. Trans. ASME **79**(2), 021013 (2012)
5. Freidin, A., Morozov, N., Petrenko, S., Vilchevskaya, E.: Chemical reactions in spherically symmetric problems of mechanochemistry. Acta Mech. (2015). doi:10.1007/s00707-015-1423-2
6. Freidin, A.B.: Chemical affinity tensor and stress-assist chemical reactions front propagation in solids. In: ASME International Mechanical Engineering Congress and Exposition, Proceedings (IMECE), vol. 9 (2013)
7. Freidin, A.B., Vilchevskaya, E.N., Korolev, I.K.: Stress-assist chemical reactions front propagation in deformable solids. Int. J. Eng. Sci. **83**, 57–75 (2014)
8. Gutman, E.M.: Mechanochemistry of Solid Surfaces. World Scientific, Singapore (1994)
9. Gutman, E.M., Zainullin, R.S., Shatalov, A.T., Zaripov, R.A.: Strength of Gas Industry Pipes under Corrosive Wear Conditions. Nedra, Moscow (1984). (in Russian)
10. Gutman, E.M., Haddad, J., Bergman, R.: Stability of thin-walled high-pressure vessels subjected to uniform corrosion. Thin-Walled Struct. **38**, 43–52 (2000)
11. Karpunin, V.G., Kleshchev, S.I., Kornishin, M.S.: Calculation of plates and shells taking general corrosion into account. In: Proceedings, 10th All-Union Conference of the Theory of Shells and Plates, vol. 1, pp. 166–174 (1975)
12. Khryashchev, S.M.: Controllability and number-theoretic properties of dynamical polysystems. Nonlinear Phenom. Complex Syst. **16**(4), 388–396 (2013)
13. Khryashchev, S.M.: On control of continuous dynamical polysystems in discrete times. AIP Conference Proceedings, vol. 1648 (2015). doi:10.1063/1.4912664
14. Loss due to corrosion can be 4 per cent of GDP. The Hindu, http://www.thehindu.com/todays-paper/tp-national/tp-tamilnadu/loss-due-to-corrosion-can-be-4-per-cent-of-gdp/article6340613.ece
15. Muskhelishvili, N.I.: Some Basic Problems of the Mathematical Theory of Elasticity. Nordhoff, Groningen (1954)

16. Ovchinnikov, I.G., Pochtman, YuM: Calculation and rational design of structures subjected to corrosive wear (review). Mater. Sci. **27**(2), 105–116 (1992)
17. Pavlov, P.A., Kadyrbekov, B.A., Borisevich, V.V.: Uniform stress corrosion and corrosion cracking of structural steels. Sov. Mater. Sci. **21**(3), 248–251 (1985)
18. Pavlov, P.A., Kadyrbekov, B.A., Kolesnikov, V.A.: Strength of Steels in Corrosive Environments. Nauka, Alma-Ata (1987). (in Russian)
19. Pronina, Y.G.: Estimation of the life of an elastic tube under the action of a longitudinal force and pressure under uniform surface corrosion conditions. Rus. Metall. (Metally) **2010**(4), 361–364 (2010)
20. Pronina, Y.G.: Thermoelastic stress analysis for a tube under general mechanochemical corrosion conditions. In: Proceedings of the 4th International Conference on Computational Methods for Coupled Problems in Science and Engineering, COUPLED PROBLEMS 2011, pp. 1408–1415 (2011)
21. Pronina, Y.G.: Analytical solution for the general mechanochemical corrosion of an ideal elastic-plastic thick-walled tube under pressure. Int. J. Solids Struct. **50**, 3626–3633 (2013)
22. Pronina, Y.G.: Analytical solution for decelerated mechanochemical corrosion of pressurized elastic-perfectly plastic thick-walled spheres. Corros. Sci. **90**, 161–167 (2015)
23. Rusanov, A.I.: Mechanochemistry of dissolution: Kinetic aspect. Russ. J. Gen. Chem. **77**(4), 491–502 (2007)
24. Sedova, O.S., Khaknazarova, L.A., Pronina, Y.G.: Stress concentration near the corrosion pit on the outer surface of a thick spherical member. In: IEEE 10th International Vacuum Electron Sources Conference, IVESC 2014, pp. 245–246 (2014)
25. Sedova, O., Pronina, Y.: Generalization of the Lame problem for three-stage decelerated corrosion process of an elastic hollow sphere. Mech. Res. Commun. **65**, 30–34 (2015)
26. Srivatsan, T.S., Sudrashan, T.S.: Soni, Kamal: General corrosion characteristics of a quaternary aluminum-lithium alloy in aqueous environments. Mater. Trans. JIM **31**(6), 478–486 (1990)
27. Zheng, Y., Li, Y., Chen, J., Zou, Z.: Effects of tensile and compressive deformation on corrosion behaviour of a MgZn alloy. Corros. Sci. **90**, 445–450 (2015)

Retraction Note to: A Collocation Method for Integral Equations in Terms of Generalized Bernstein Polynomials

Vinai K. Singh and A.K. Singh

Retraction Note: Chapter "A Collocation Method for Integral Equations in Terms of Generalized Bernstein Polynomials" in: V.K. Singh et al. (eds.), *Modern Mathematical Methods and High Performance Computing in Science and Technology*, Springer Proceedings in Mathematics & Statistics 171, DOI 10.1007/978-981-10-1454-3_23

The chapter published in the book 'Modern Mathematical Methods and High Performance Computing in Science and Technology', pages 271–285, DOI 10.1007/978-981-10-1454-3_23 has been retracted because it contains significant parts plagiarizing another publication: 'A Collocation Method for Linear Integral Equations in Terms of the Generalized Bernstein Polynomials', New Trends in Mathematical Sciences, Volume: 4, Issue: 1 (Jan 2016), pp: 203–213, DOI: 10.20852/ntmsci.2016115855.

The updated original online version for this chapter can be found at
10.1007/978-981-10-1454-3_23

V.K. Singh (✉)
Department of Applied Mathematics, Raj Kumar Goel Institute of Technology,
NH-58, Delhi-Meerut Road, Ghaziabad 201003, India
e-mail: drvinaiksingh@rkgit.edu.in

A.K. Singh
Department of Science and Technology,
Government of India, Technology Bhavan, New Mehrauli Road, New Delhi 110016, India
e-mail: ashokk.singh@nic.in

V.K. Singh et al. (eds.), *Modern Mathematical Methods and High Performance Computing in Science and Technology*, Springer Proceedings in Mathematics & Statistics 171, DOI 10.1007/978-981-10-1454-3_26

MIX
Papier aus verantwortungsvollen Quellen
Paper from responsible sources
FSC® C105338
FSC
www.fsc.org

If you have any concerns about our products,
you can contact us on
ProductSafety@springernature.com

In case Publisher is established outside the EU,
the EU authorized representative is:
Springer Nature Customer Service Center GmbH
Europaplatz 3, 69115 Heidelberg, Germany

Printed by Libri Plureos GmbH
in Hamburg, Germany